Navigation: Science and Technology

The series presents new developments and advances in various aspects of navigation—from land navigation, marine navigation, aeronautic navigation, to space navigation; and from basic theories, mechanisms, to modern techniques. It publishes monographs, edit volumes, lecture notes and professional books on topics relevant to navigation—quickly, up to date and with a high quality. A special focus of the series is the technologies of the Global Navigation Satellite Systems (GNSSs), as well as the latest progress made in the existing systems (GPS, BDS, Galileo, GLONASS, etc.). To help Readers keep abreast of the latest advances in the field, the key topics in NST include but are not limited to:

- Satellite Navigation Signal Systems
- GNSS Navigation Applications
- Position Determination
- Navigation instrument
- Atomic Clock Technique and Time-Frequency System
- X-ray pulsar-based navigation and timing
- Test and Evaluation
- User Terminal Technology
- Navigation in Space
- New theories and technologies of navigation
- Policies and Standards

More information about this series at http://www.springer.com/series/15704

Renbiao Wu · Wenyi Wang
Dan Lu · Lu Wang · Qiongqiong Jia

Adaptive Interference Mitigation in GNSS

Science Press
Beijing

Springer

Renbiao Wu
Civil Aviation University of China
Tianjin
China

Lu Wang
Civil Aviation University of China
Tianjin
China

Wenyi Wang
Civil Aviation University of China
Tianjin
China

Qiongqiong Jia
Civil Aviation University of China
Tianjin
China

Dan Lu
Civil Aviation University of China
Tianjin
China

ISSN 2522-0454 ISSN 2522-0462 (electronic)
Navigation: Science and Technology
ISBN 978-981-10-5570-6 ISBN 978-981-10-5571-3 (eBook)
https://doi.org/10.1007/978-981-10-5571-3

Jointly published with Science Press, Beijing, China
ISBN: 978-7-03-054231-1 Science Press, Beijing

The print edition is not for sale in China Mainland. Customers from China Mainland please order the print book from: Science Press.

Library of Congress Control Number: 2017950278

Original Chinese edition published by Science Press, China, 2015.

Printed on acid-free paper

This Springer imprint is published by Springer Nature
The registered company is Springer Nature Singapore Pte Ltd.
The registered company address is: 152 Beach Road, #21-01/04 Gateway East, Singapore 189721, Singapore

Preface

I

Global Navigation Satellite System (GNSS) generally refers to various global satellite navigation systems, and their augmentation systems. It can provide accurate position, velocity and time (PVT) for any person or object at any place and at any time. GNSS, as an important space information infrastructure, reflects a country's comprehensive national strength. Major countries and organizations all around the world have been vigorously developing satellite navigation systems with various characteristics, such as the Global Positioning System (GPS) of the USA, the GLObal NAvigation Satellite System (GLONASS) of Russia, the Galileo System of Europe (Galileo), the BeiDou Satellite System of China (BeiDou), the Indian Regional Navigational Satellite System (IRNSS) of India, and the Quasi-Zenith Satellite System (QZSS) of Japan. Since 1996, the United States has initiated the "GPS modernization" project. In recent years, we have seen the most extensive GPS satellite launching schedule since 1993, and the first GPS III is scheduled to be launched in 2018. China proposed a "three-step" development plan for the Beidou system. In 2012, the Asia-Pacific Regional Navigation Satellite System was constructed and put into use (making China the third country in the world to put a GNSS system into operation); and since then, China has been making steady progress toward establishing a global system by launching the first global GNSS satellite in July 2017. In the future, all these systems will be integrated into a Global Navigation Satellite System of Systems, to provide more reliable and accurate services for global users.

GNSS applications are almost ubiquitous, and can be applied in almost any imaginable situation, including air, sea, and ground transportation and management, smart grid, telecommunications systems, mobile phone positioning, smart carrier tools, exploration and mapping, criminal tracking, emergency rescue, disease control, fishing operations, oil exploration, precision agriculture, as well as military applications such as precise weapon guidance and targeting. GNSS are embedded in all the above applications as a stealthy technology running in the background. It can also provide support for many critical infrastructures tightly connected to the operations of a nation and the livelihood of its people, e.g., smart grids (timing service), bank operations (timing service), transportation systems (location and timing services), and communication systems (location and timing services).

GNSS is vulnerable because it is so valuable! Since GNSS satellites are usually positioned 20,000–30,000 kilometers above the earth, their signals are very weak (it can be compared to observing a 50W light bulb from approximately 20,000 kilometers away), and they are usually more than 20 dB below the noise level, meaning that they are very vulnerable to a variety of malicious and unintentional interferences. Unintentional interferences include ionospheric scintillation, solar radio pulse interference, multipath interference, Radio Frequency Interference (RFI), and pulse interference generated by DME (Distance Measuring Equipment) and TACAN (Tactical Air Navigation System) equipment working in the aeronautical radio protection band. Malicious interferences include jamming and spoofing. Jamming can make the receiver out of lock, but the monitoring and suppression of jamming are easier and more established methods are available, as many related techniques can be borrowed from the radar field. Spoofing takes advantage of the open transparency and predictability of GNSS civil signals, and it can create interference signals very similar to authentic GNSS signals. Consequently, spoofing has superior concealment and can result in greater repercussions. Spoofing can generate timing and positioning errors, or even take control of the target receiver, without the user even being aware of its presence. Due to the existence of these interference sources, GNSS cannot provide safe and reliable services.

The robustness and safety of GNSS have led to great concerns around the world for the past dozen years. Below are comments made by some renowned political characters, GNSS experts, and related organizations in the US:

(1) Former U.S. President Barack Obama (February 2013)

    "Critical infrastructure must be secure and able to withstand and rapidly recover from all hazards."
    "The nation's critical infrastructure provides the essential services that underpin American society. Proactive and coordinated efforts are necessary to strengthen and maintain secure, functioning, and resilient critical infrastructure —including assets, networks, and systems…."

(2) Former US President George W. Bush (December 2004)

    "…the Global Positioning System has grown into a global utility…integral to the U.S. national security, economic growth, transportation safety, and homeland security, and are an essential element of the worldwide economic infrastructure…."
    "The Secretary of Transportation shall…develop, operate, and maintain backup position, navigation, and timing capabilities that can support critical transportation, homeland security, and other critical civil and commercial infrastructure …in the event of a disruption of the Global Positioning System or the other space-based positioning, navigation, and timing services."

(3) US National Positioning, Navigation, and Timing (PNT) Advisory Committee (November 2010)

"The United States is now critically dependent on GPS…cell phone towers, power grid synchronization, new aircraft landing systems, and the future FAA Air Traffic Control System cannot function without it…increasing incidents of deliberate or inadvertent interference that render GPS inoperable for critical infrastructure operations."
"We strongly recommend that the previously announced decision (to deploy eLoran as the primary Alternate PNT) should be reconfirmed and quickly implemented."

(4) The "Father of GPS," Prof. Brad Parkinson at Stanford University (2013)

"Reliance on satellite navigation and timing systems has become a single point of failure for much of America and is our largest, unaddressed critical infrastructure problem."
"Positioning, Navigation and Time (PNT) service has become a worldwide utility-thanks to GPS. In fact, PNT service is now worth billions of dollars a year, yet taken for granted. On the other hand this service is potentially threatened by jamming and related threats. I subscribe to PTA-Protect, Toughen and Augment this valuable asset. Particularly appealing is the use of eLoran to Augment or as a standalone service. In this role, eLoran would be a powerful deterrent to malicious interference."

(5) Renowned GNSS interference mitigation expert, Prof. Todd Humphreys at University of Texas at Austin (2013)

"The next few decades will see pervasive autonomous control systems become critical to the world economy-from autonomous cars and aircraft to smart homes, smart cities, and vast energy, communication, and financial networks controlled at multiple scales. Protecting these systems from malicious navigation and timing attacks is a matter of urgent societal interest."
"The greatest upcoming challenge in PNT security will be providing proof of location or time to a skeptical second party."

In the present day, there are many types of jamming equipment in the market, including the Personal Privacy Device (PPD), which can be purchased cheaply online. In November 2011, in an event that shocked the whole world, an Unmanned Aerial Vehicle (UAV) owned by the US Central Intelligence Agency (CIA) was captured by Iran with only minor damage to the landing gear, which was estimated to be caused by the landing. Until today, the exact reason has not been released. Engineers from the Iranian side claimed that they had interfered with the communication link between US UAV operators and the UAV, causing the UAV to switch to an autopilot mode. Then the UAV was guided to a base in Afghanistan relying solely on the GPS equipment on the UAV. When the UAV was in this working mode, Iranian engineers claimed that they had tricked the UAV to fly to Iran using spoofing. Some US experts questioned the claims made by Iranian engineers because the CIA UAV usually uses a military code, so it should have been very difficult to use spoofing against it. Instead, the experts envisaged the following scenario: The Iranians used jamming and spoofing simultaneously to interfere with GPS military and civil signals respectively, and the UAV was set in a mode in which the civil signals were used when the military signals were jammed. Regardless of which explanation is correct, the evidences imply that the UAV was experiencing malicious interferences of jamming and spoofing simultaneously.

In 2001, the US Department of Transportation evaluated the impacts on transportation facilities by vulnerabilities of GPS, and expressed concerns on the threats posed by spoofing. In 2008, researchers created low-cost spoofing devices that could be made based on software radio technology and components from general stores, and showed that spoofing signals can be successfully generated using this low-cost approach. In June 2012, Prof. Todd E. Humphreys, the head of radio navigation lab of the University of Texas at Austin, was invited by the US Department of Homeland Security to perform two public test studies in order to evaluate the impacts of spoofing and jamming on civil UAVs (which will be merged into the national airspace system) and smart grid. The results showed that these systems were easily affected by spoofing. In July 2013, in an actual field demo, the same lab successfully tricked a ship into deviating from its original route using spoofing.

The above events show that the robustness and safety of GNSS face very big challenges. The Science and Technology Directorate (S&T) is a branch within the United States Department of Homeland Security. Part of the S&T's mission is to evaluate the impacts of jamming and spoofing, study interference mitigation measures, and provide better practical means to protect critical infrastructures. In June 2014, former US Defense secretary Ashton Carter expressed concerns about the lack of robust PNT. One of the senior US officers even proclaimed that "GPS is much too vulnerable, we must replace it with new inertials and chip scale atomic clocks." It is reported that the US has been evaluating various possibilities on alternative PNT schemes that can complement GPS. The goal is to improve GPS robustness and safety through integration.

Interference monitoring and suppression for GNSS have drawn great attentions around the world. In 2011, the International Committee on GNSS promoted the establishment of a special workshop on Interference Detection and Mitigation (IDM). In 2012, the first workshop session was held at the Vienna International Convention Center in Austria with the main topics being compatible interoperability among different GNSS systems, and interference monitoring and suppression. The workshop has been held for five consecutive sessions. In addition, at the GNSS academic annual meeting hosted by ION (Institute of Navigation), IDM has always been a hot topic and focus for discussion.

II

I studied in the Northwestern Polytechnic University's first accelerated Bachelor's/Master's pilot program from 1985 to 1991. When I was studying for my master's degree, I participated in research works on super-resolution array signal processing. From 1991 to 1994, I worked on my Ph.D. degree in the field of radar signal processing at Xidian University. During this period, I participated in the research on Airborne Early Warning Radar Space-Time Adaptive Processing (STAP) led by Prof. Zheng Bao, who is the academician of the Chinese Academy of Sciences. This was the earliest group to start systematic STAP research in China. Since then, this research group has achieved advanced research results on par with the findings of other leading international scholars in the field. For example, Dr. Richard Klemm, the author of the first STAP monograph, stated the following in the foreword section of his book "There are activities all over the world, especially in the USA and China." Between 1997 and 2002, as a postdoctoral fellow and a visiting professor, I worked in the Spectrum Analysis Lab at the University of Florida, USA. During that period, I collaborated with Prof. Jian Li, who is an IEEE fellow and a winner of the American Presidential Young Investigator (PYI) award, in studying spectral estimation based on decoupled parameter estimation theory and robust Capon beamforming. In 2004, I participated in research on anti-jamming technology for the Galileo system in the Imperial College of London as part of the China's first group of distinguished research scholars. In summary, my research interests have always been focused on sensor array signal processing (super-resolution direction of arrival estimation, adaptive beamforming, and space-time adaptive processing), modern spectrum analyses and their applications.

In the summer 2002, I was invited to be a visiting professor at the Spectrum Analysis Lab at the University of Florida. My main research focus there was to study robust Capon beamforming and its applications. During that tenure, I read the introduction on GNSS robust beamforming written by Professir Sayed's research group at Stanford University. Another paper brought to my attention was titled "Wideband cancellation of interference in a GPS receiver array," written by Dr. Ronald L. Fante from the MITRE corporation and published in IEEE Transactions on Aerospace and Electronic Systems, in which the author expanded the STAP technology used for airborne early warning radar to GNSS applications, and used the technology to suppress wideband interference and dispersive multipath interference. Similarly to me, Dr. Fante was originally engaged in airborne radar signal processing works. These two papers have cultivated my strong interest in GNSS interference mitigation technology.

III

This book is the first most comprehensive monograph on the subject of GNSS adaptive interference mitigation. The book covers the topics of jamming (including high dynamic jamming), spoofing, multipath interference, and pulse interference suppressions. We have mainly studied the above methods based on array signal processing, including spatial domain, and spatial–temporal domain adaptive filtering. These methods take full advantage of the characteristics of GNSS signals, e.g., small power, periodic repetition, and known spread spectrum code. Only autonomous systems are considered, i.e., no other sensors are used (e.g., inertia navigation) to provide supplementary information. The focus of this book is on the robust adaptive filtering method that is not sensitive to array calibration errors.

Direction of Arrival (DOA) estimation and multipath time delay estimation can both take advantage of the methods based on decoupled parameter estimation theory. By using this approach, an estimation problem of multiple overlapping signal parameters can be converted into a series of estimation problems on a single signal parameter using a cyclic optimization algorithm with a special structure (the key is to assure the accuracy and computation efficiency of these single signal parameter estimations). This approach has the benefits of simple and effective computation, good convergence, and robustness to model errors.

IV

The research works and achievements listed in the book have been sponsored by the Chinese National Outstanding Youth Fund (60325102) and National Natural Science Foundation of China (61179064, 61172112, 61471363, 61271404). And thanks to the sustained support of the National Natural Science Foundation of China, these research works have lasted 12 years.

First of all my grateful thanks go to my Ph.D. advisor, Academician Zheng Bao at Xidian University. I have benefited throughout my whole life from my research experiences on STAP under his guidance. His attitudes and methods toward academic research have made profound impacts on my life and career. My thanks also go to Prof. Jian Li at the University of Florida USA. During my three working tenures in her lab, we have co-authored and published more than 20 articles in IEEE and IET/IEE journals, which have greatly improved my academic prowess. These papers were mainly focused on using the decoupled parameter estimation theory and method in various applications. Many of the achievements of this book would not have been possible without the guidance from my two mentors. My appreciations are also extended to my lab mates in the former STAP group at Xidian University (Dr. Yuhong Zhang, Dr. Guisheng Liao, Dr. Yongliang Wang, Dr. Lingrang Zhang, and Dr. Xiaochu Chen). Together we have enjoyed memorable school years. I am also grateful to my other co-authors at the Key Lab for Advanced Signal Processing at China Civil Aviation University, as well as Associate Professor Tieqiao Hu, Dr. Lunlong Zhong, and many other Ph.D. and master students involved in related research works.

Parts of the formal texts of this book were tentatively finished on September 11, 2015. From September 14 to September 18, at the ION GNSS+ 2015 meeting held in Tampa, Florida, USA, I led two co-authors of this book in efforts to further improve our draft by communicating face to face with our colleagues from all around the world. I fondly remember how I became interested in GNSS interference mitigation 12 years ago in Florida. Now, after attending this meeting in the beautiful city of Tampa, I feel extremely grateful and delighted to have come back with this book to where this journey started. This foreword was composed all at once on my return flight from Tampa to Beijing in a single session.

On the return flight from Tampa, Florida
September 2016

Renbiao Wu

The original version of the book was revised:
For detailed information please see Erratum.
The erratum to the book is available at
https://doi.org/10.1007/978-981-10-5571-3_7

Contents

Author's Profile

Renbiao Wu is the Tianjin Professor and the Director of the Tianjin Key Lab for Advanced Signal Processing at the Civil Aviation University of China. He received his B.Sc. and M.Sc. in Electrical Engineering from the Northwest Polytechnic University in 1988 and 1991 respectively, and his Ph.D. in Electrical Engineering from Xidian University in 1994. He has worked at the Imperial College of London, the University of Florida, and Virginia Tech as a Distinguished Research Scholar, Visiting Professor, and Postdoctoral Fellow for 5 years. His research interests include adaptive array signal processing and spectral estimation, especially in regards to their applications in GNSS and radar. He has published over 300 peer-reviewed papers, and ten plus books and book chapters. He is the recipient of the Chinese National Outstanding Young Investigator Award in 2003.

About the Authors

Wenyi Wang is a professor of the Tianjin Key Lab for Advanced Signal Processing at the Civil Aviation University of China. He received his M.Sc. in Electrical Engineering from the Huazhong University of Science and University in 2005, and his Ph.D. in Electrical Engineering from the Institute of Acoustics, Chinese Academy of Sciences, in 2008. His research interests include adaptive array signal processing and spectral estimation regarding their applications to GNSS and ADS-B. He has published over 40 papers, and co-authored one book and three book chapters.

Dan Lu is an associate professor of the Tianjin Key Lab for Advanced Signal Processing at the Civil Aviation University of China. She received her B.Sc. from the Heifei University of Technology in 1999, M.Sc. from the Civil Aviation University of China in 2004 and Ph.D. from Xidian University in 2013, all in electrical engineering. Her research interests include adaptive array signal processing and spectral estimation regarding their applications to GNSS. She has published over 30 papers and co-authored two books.

Lu Wang is a lecturer of the Tianjin Key Lab for Advanced Signal Processing at the Civil Aviation University of China. She received her B.Sc. and M.Sc. from the Civil Aviation University of China in 2007 and 2010, and her specialized master degree in space communication system from ISAE in France in 2011 under the support of the Thales scholarship. Her research interests include adaptive array signal processing and spectral estimation regarding their applications to GNSS. She has published 15 papers and co-authored one book.

Qiongqiong Jia is a lecturer of the Tianjin Key Lab for Advanced Signal Processing at the Civil Aviation University of China. She received her B.Sc. and M.Sc. from the Civil Aviation University of China in 2008 and 2011, and her specialized master degree in navigation engineering from ENAC in France in 2015. Her research interests include adaptive array signal processing and spectral estimation regarding their applications to GNSS. She has published 20 papers, and co-authored three books and book chapters.

Chapter 1
Principles of Satellite Navigation System

1.1 Introduction

Satellite navigation is the technology that uses navigation satellites to transmit positioning signals, in order to provide real-time positioning for users in the air, on the ground, at sea, and in space. Since it can provide high-precision information such as three-dimensional position, velocity and time (PVT) for any location and on any people and objects, it has unparalleled advantages over other navigational technologies. Thus, it can be widely applied in civil fields such as transportation, surveying, mapping, telecommunications, water conservancy, fishery, forest fire prevention, disaster reduction, and disaster relief. It can also be used in military fields such as aerospace and weapon guidance. Consequently, the satellite navigation system has become a keystone for a country's space information infrastructure, and an important indicator to reflect its status as a modern country, a great power, and the country's comprehensive national strength. Major countries and organizations all around the world have been vigorously developing satellite navigation systems with various characteristics.

Global Navigation Satellite System (GNSS) is an integrated constellation system, and generally refers to all existing global satellite navigation systems, and regional and enhanced navigational systems, such as Global Positioning System (GPS) of USA, GLObal NAvigation Satellite System (GLONASS) of Russia, Galileo System of Europe, and BeiDou Satellite System of China. The related augmentation systems include the Wide Area Augmentation System (WAAS) of USA, the European Geostationary Navigation Overlay Service (EGNOS) of Europe, the Multi-functional Satellite Augmentation System (MSAS) of Japan and the GPS Aided Geo Augmented Navigation (GAGAN) of India, etc.

The original version of this chapter was revised: For detailed information please see Erratum. The erratum to this chapter is available at https://doi.org/10.1007/978-981-10-5571-3_7

R. Wu et al., *Adaptive Interference Mitigation in GNSS*, Navigation: Science and Technology, https://doi.org/10.1007/978-981-10-5571-3_1

However, since the distance from a GNSS satellite to a receiver can range from 20,000 km to more than 30,000 km, and a typical satellite transmitter's power is generally only a few tens of Watts, the GNSS signal becomes very weak when it reaches the ground. For example, the signal power of a GPS L1 signal is approximately −160 dBw when it reaches the earth's surface, which is lower than a receiver's thermal noise level by about 20 dB [1], therefore GNSS can be easily impacted by various types of external interference. To enhance GNSS performance under complex electromagnetic environments, effective interference mitigation algorithms must be studied.

First of all, this chapter briefly introduces developments of various GNSS. Then, to help readers to understand the contents of subsequent chapters, the C/A (Coarse/Acquisition) code is used as an example, to briefly introduce the composition of a satellite navigation system and the principles of acquisition, tracking and positioning for a receiver. Lastly, the impacts of interference on GNSS receivers are briefly introduced.

1.2 Development and Current Status of GNSS

As stated above, the four existing major GNSS constellation systems are GPS of USA, GLONASS of Russia, Galileo of Europe, and BeiDou of China. Table 1.1 compares satellite constellations and current statuses on satellite constellation networks for the four major GNSS. The cutoff date for the statuses on satellite constellation networks is July 30th, 2015.

Next, we give brief introductions on the four major GNSS.

Table 1.1 Constellations of different GNSS

GNSS constellation systems		GPS	GLONASS	Galileo	Beidou
Designed satellite constellations	Number of satellites	24	24	30	35
	Number of orbital planes	6	3	3	3
	Orbit inclination angle (°)	55	64.8	56	55
	Operation cycle	11 h 58 min	11 h 15 min	13 h	12 h 55 min
	Satellite orbit (km)	MEO 20220	MEO 19130	MEO 23222	GEO 36000 MEO 21500 IGSO 36000
Status	Number of satellites	31	28	8	19

1.2.1 GPS

The US department of Defense started to develop GPS in the 1970s, and it was the third large-scale space program created by US following the Apollo Moon program and the Space Shuttle program. On February 22nd, 1978, the first GPS satellite during the development stage was launched successfully. But the GPS system did not begin its formal operation until 1995. GPS signals use CDMA (Code Division Multiple Access) modulation to allow multiple satellites to share the same time and frequency in a given band. The initial GPS signal design has two spreading codes: coarse/acquisition code (C/A code) and precision code (P code). The C/A code is a pseudo-random code open to the public, with a code length of 1024 and a code period of 1 ms. The P code's usage has been restricted, and this special pseudo-random code has been reserved for US military applications, with a longer code length and code cycle. Initial GPS carrier signals include L1 signal (center frequency of 1575.42 MHz) and L2 signal (center frequency of 1227.6 MHz). The C/A code is broadcasted on the L1 carrier only, while the P code is broadcasted on both the L1 and L2 carriers [1, 2]. The United States, in the mid-1990s, for national security reasons, included SA (Selective Availability) on the GPS signal, which intentionally increased positioning errors on receivers, so the precision was only around 100 m. In May 2000, the use of SA was discontinued, which has greatly promoted the applications of GPS.

To conduct a comprehensive upgrade of the GPS system, and to replace failed satellites to ensure normal system operations, the United States has initiated the "GPS modernization" project since 1996. With the advances of the GPS modernization process, L2C signal, M code signal, L5 signal and L1C signal, as modernized GPS signals, have expanded the initial GPS signal family [3]. The L2C signal is a newly added civil signal on the L2 frequency that was formerly used for military application only. Due to the more flexible signal structure it has, navigational precision has been improved. M-code is a newly added military code to improve the original military code's interference mitigation capability and safety. Unlike the P code, M code uses an autonomous positioning method in the design: the receiver can directly solve the position solely based on the M code. After overcoming many technological challenges and constant launch time delays, the GPS IIR-20(M) satellite, carrying the newly added third civil signal L5, was launched on March 24th, 2009, and then it started to successfully transmit the L5 signal on April 10th, 2009. One of the objectives for launching L5 signal is to provide service for Safety-of-Life (SoL) applications. L5's carrier frequency is 1176.45 MHz, which is located on the internationally protected ARNS (Aeronautical Radio Navigation Service) frequency band, making it convenient for aviation authorities to effectively manage radio interference encountered in this frequency band. But, this also results in some electromagnetic compatibility problems between the L5 signal and existing navigation equipments working on the ARNS frequency band. Compared with the C/A code on the L1 frequency, the L5 code has a higher code rate (10.23 Mcps), greater transmission power, and better interference mitigation performance. The L1C signal is the fourth civil signal that is designed for facilitating interoperability between the GPS and other GNSS. The GPS III satellite carrying the L1C signal has been planned to launch in 2017. The constellation and network status for the GPS are shown in Table 1.1.

1.2.2 GLONASS

GLONASS was developed by the former Soviet Union in 1976, and has been managed by Russian government since the collapse of the Soviet Union [4]. On October 12th, 1982, the first GLONASS satellite was launched successfully, indicating the birth of the second GNSS after the GPS. Unlike GPS, initial GLONASS signals used Frequency Division Multiple Access (FDMA) technology, so that all satellites within the GLONASS system used the same spreading code, but different satellites used different signal carrier frequencies. This arrangement is very different from the CDMA technology used by the GPS system to distinguish its satellites. All GNSS signals broadcasted by any one of the GLONASS satellites were modulated on two types of carrier signals, i.e. L1 and L2 frequencies, where $L1 = 1602 + 0.5625 \times k$ (MHz), and $L2 = 1246 + 0.4375 \times k$ (MHz). The k in the formula represents the serial number of a satellite.

Although GLONASS constellation had already completed a full layout composed of 24 satellites and one spare satellite in 1995, Russia could not afford to launch substitutes for failed satellites due to lack of funds after the collapse of the former Soviet Union. After an economic upturn, on August 20th 2001, Russian government approved the initialization of 2002–2011 GLONASS development plan. The plan aimed to recover and modernize the GLONASS. Huge amount of funds was raised to promote the construction of GLONASS. From 2003, the GLONASS system had entered a stage of comprehensive upgrading, during which time new GLONASS-M and GLONASS-K1 were launched in succession. In 2008, GLONASS declared that it would add CDMA civil signals, thereby making it more compatible and interoperable with other GNSS systems. At the same time, Russia declared that future GLONASS satellites would broadcast CDMA signals while maintaining the original FDMA signals. On October 2nd, 2011, the launch of the 24th satellite marked the first comprehensive restoration of GLONASS since 1996. The constellation and network status for the GLONASS are shown in Table 1.1.

1.2.3 Galileo

The Galileo satellite navigation system is developed and established by European Union (EU). The plan was published by the European Commission in February 1999, and has been a joint responsibility of the European Commission and the European Space Agency [5]. The first Galileo satellite was launched on December

28th 2005. Galileo, like GPS, uses CDMA signal encoding. Galileo signals have four carrier frequency bands: L1 frequency band (center frequency of 1575.42 MHz), E5A frequency band (center frequency of 1176.45 MHz, the same as that of GPS L5), E5B frequency band (center frequency of 1207.1 MHz), and E6 frequency band (center frequency of 1278.75 MHz). Every satellite can broadcast on all four frequency bands. After the establishment of Galileo system, five types of service modes can be provided: Open Service (OS), Commercial Service (CS), Safety-of-Life (SoL), Public Regulated Service (PRS), and Support to Search and Rescue Service (SAR).

On March 27th 2015, the 7th and 8th Galileo satellites were launched. The constellation and network status for the Galileo are shown in Table 1.1. Even though the original Galileo plan aimed to complete the whole system construction in 2007, and started commercial operations in 2008, the Galileo plan has been postponed repeatedly due to economic and other reasons. The new plan is to complete the construction of Galileo system before 2020 and provide full services.

1.2.4 Beidou

Beidou is a satellite navigation and positioning system under construction in China with independent intellectual property rights. Compared with other systems, one of the highlighted features of the Beidou system is that it is capable of short message communication. The feature has important applications in applied areas such as transportation (especially offshore sailing) and emergency rescue [6]. China developed a "three-step" development plan for the Beidou so that the step-by-step implementations have followed a general principle of "Regional first, global next; active first, passive next", as illustrated in Table 1.2. The experimental system (the 1st generation system) had begun development since 1994, and achieved regional active positioning capability in 2000. The second step, the construction of the Beidou (the second generation system) started in 2004, and the regional passive positioning capability was achieved in 2012. Lastly, for the third step, in 2020, the Beidou will have the global passive positioning capability.

On October 31st, 2000, the first Beidou experimental satellite was launched. The construction of the Beidou experimental system indicated that China had become

Table 1.2 Three-step development plan for Beidou

Milestones	2000	2012	2020
Goals	Regional active positioning	Regional passive positioning	Global passive positioning
Current status	Completed	Completed	Under construction

the third country in the world to have its own GNSS. On April 14th 2007, the first Beidou satellite was launched. On December 27th, 2012, Beidou started to formally provide services in the Asia-Pacific region. On December 27th, 2013, the system documentation (ICD documentation) of "*Beidou Navigation System Signal in Space Interface Control Document (Version 2.0)*" was officially released, reflecting the willingness to have more open, in-depth, and extensive cooperation with other countries in the world. The ICD file defines the related contents such as signal structure, basic parameters and ranging codes, for public service signal B1I and B2I between the Beidou space constellation and user terminals. Among them, the B1I signal's carrier frequency is 1561.098 MHz, and the B2I signal's carrier frequency is 1027.14 MHz, and the chip rates for B1I's and B2I's signal ranging codes are both 2.046 Mcps, with a length of 2046 bits.

Presently, the construction for step 3 of the Beidou has been making steady progress based on the plan. On March 30th, 2015, the 17th Beidou satellite (the first experimental global satellite) was launched, with more than 95% parts localization rate, and the LOONGSON CPU made in China was used for the first time. On July 25th, 2015, the 18th and 19th Beidou satellites (both are experimental global satellites) were launched, with more than 95% parts localization rate, and the new generation high-precision rubidium clocks independently developed by China were installed on both satellites respectively. The constellation and network status for the Beidou are shown in Table 1.1.

1.3 Composition of GNSS

For convenience, we briefly introduce the composition of a GNSS system using the GPS as the example [1, 2]. As shown in Fig. 1.1, the GPS mainly comprises three segments: the space constellation segment, the ground monitoring and control

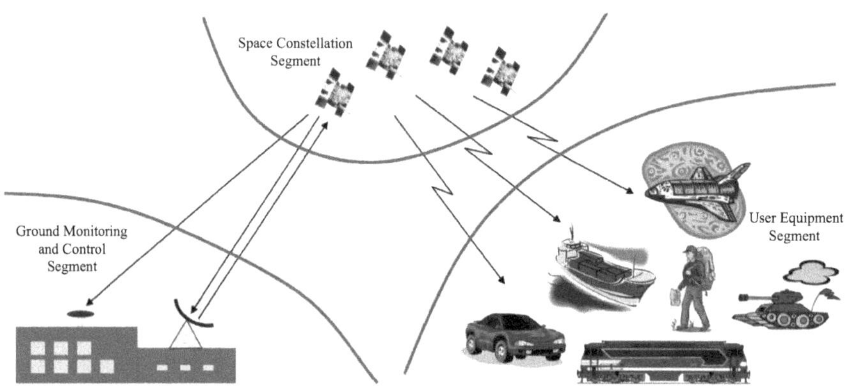

Fig. 1.1 GPS segments

segment, and the user equipment segment. Among the three segments, the first two segments are controlled by the United States, while the user equipment segment can be freely designed and selected by users.

The space constellation is composed of orbiting GPS satellites. The satellites are evenly distributed on six orbital planes maintaining 55° inclination angles with the earth's equatorial plane. The respective orbital planes are separated by 60° from each other, i.e. the right ascensions of the ascending nodes (RAAN) for the respective orbital planes are separated by 60°. The arguments of latitude for the four satellites in every unique orbital plane are 90° apart from each other, while the satellite located within every orbital plane leads the corresponding satellite orbiting the adjacent orbital plane on its west by 30°, to satisfy the requirement of providing even global coverage by GPS satellite signals. The average altitude of the satellite orbits is 20,200 km, and the orbital period is 11 h, 57 min and 57.26 s, i.e. 12 sidereal hours. In short, the satellite constellation orbital parameters and the distribution are determined after a comprehensive optimization, to ensure that four or more GPS satellites with more than 5° elevation angle can be observed simultaneously at any time and place on the earth. Generally 9–12 satellites can be observed simultaneously. The diagram of the GPS constellation is shown in Fig. 1.2.

The ground monitoring and control segment is composed of a master control station, three injection stations, and five monitor stations. The ground monitoring and control stations are mainly located in the United States or in overseas US

Fig. 1.2 GPS constellation

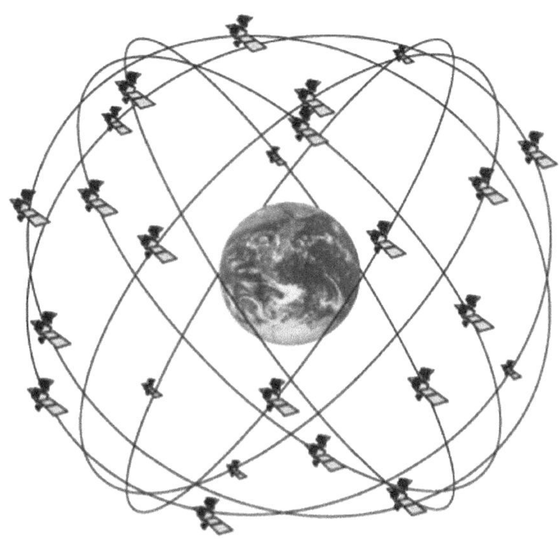

military bases. The master control station is located at Falcon Air Force Base in Colorado, and is equipped with large-scale computers, to calculate satellite ephemeris and satellite clock correction parameters based on observed GPS data given by various monitoring stations. In addition, the master control station performs satellite control as well: when a working satellite experiences failure, the spare satellite can be dispatched. The injection stations are located in Ascension Island, Diego Garcia and Kwajalein, and their main function is to inject the satellite ephemeris and satellite clock correction parameters calculated by the master station to the satellites. The monitor stations are equipped with GPS receivers, atomic frequency standards, and computer processors. The master station also has the functions of a monitor station as well. The other four monitor stations are located at Hawaii, Kwajalein, Diego Garcia, and Ascension Island, which are mainly used to monitor satellites' status. In short, the ground control segment can monitor the statuses of GPS signals, to ensure normal operation of satellite navigation, and to provide real-time information updates.

The main functions of a user receiver are GPS signal acquisition, tracking and demodulation, and positioning calculation based on the intersection principle as illustrated in Fig. 1.3. If three-dimensional positioning information is needed, signals from more than 4 satellites must be received (the impacts of receiver clock bias need to be taken into consideration, more details can be found in Sect. 1.5.2). A complete set of GPS user equipments includes receiver hardware, software, and post-processing software package for GPS data. Presently, various types of receiver have been becoming smaller in volume, lighter in weight, and more convenient for users.

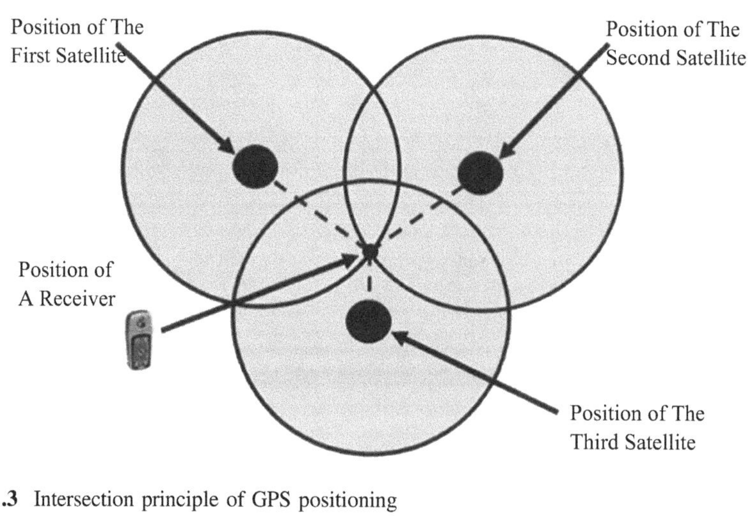

Fig. 1.3 Intersection principle of GPS positioning

1.4 GNSS Signals

1.4.1 GNSS Signal Structure

GPS/GNSS signals consist of three parts: carrier, spreading code and navigation message (D code). Traditional GPS navigation signals refer to the navigation signals broadcasted by Block IIR series satellites (or other satellite series before it), e.g. navigation signals broadcasted by L1 frequency and L2 frequency. For modern GPS satellites, new military M code has been added in addition to the L1 and L2 frequencies, and three civil signals L2C, L5 and L1C have been added as well [3]. All carrier frequencies are modulated by dedicated Pseudo Random Noise (PRN) code, navigation message are used for each satellite, and the CDMA multiplexed mode has been used. Various parameters of present GPS signals are shown in Table 1.3 [3].

The initial GPS system has two ranging codes: Course-Acquisition code (C/A code) and Precision code (P code). The GPS navigation message sets data at a 50 Hz rate, and through the corresponding modulation method, it is modulated onto the 1.023 MHz C/A code and the 10.23 MHz P code.

1.4.2 Transmission Carrier

As described above, modern GPS signals are composed of three carrier frequencies: L1, L2, and L5. The reference clock frequency for the signals is $f_0 = 1.023\,\text{MHz}$, and various frequencies needed are generated using a frequency synthesizer. The center frequency of L1 is 1575.43 MHz, and its wavelength is 19 cm. The center frequency of L2 is 1227.6 MHz, and its wavelength is 24 cm. The center frequency of L5 is 1176.45 MHz, and its wavelength is 26 cm.

Table 1.3 Related parameters of GPS signals

Codes	Carrier frequency (MHz)	Modulation type	Chip rate (Mcps)	Chip length
L1C/A	1575.42	BPSK	1.023	1023
L1P		BPSK	10.23	6.187104e12
L1M		BOC	5.115	Undisclosed
L1C		MBOC	1.023	10230
L2P	1227.6	BPSK	10.23	6.187104e12
L2CM		BPSK	0.5115	10230
L2CL				767250
L2M		BOC	5.115	Undisclosed
L5	1176.45	QPSK	10.23	10230

We highlight the L1 and the L5 frequency signals here. The L1 frequency provides two ranging code signals: the C/A code and the P code. The L5 frequency provides ranging code and synchronization code. On the L1 frequency, the C/A code and the P code are orthogonal to each other, and a single satellite's transmitted signal can be expressed as

$$S_{\mathrm{L1}}(t) = A_{\mathrm{P}}P(t)D(t)\cos(2\pi f_1 t + \phi) + A_{\mathrm{C}}c(t)D(t)\sin(2\pi f_1 t + \phi) \tag{1.1}$$

where $S_{\mathrm{L1}}(t)$ is the L1 frequency signal; A_{p} is the P code amplitude; $P(t) = \pm 1$ denotes the P code; $D(t) = \pm 1$ denotes the data code; f_1 is the L1 carrier frequency; ϕ is the initial phase; A_{C} is the C/A code amplitude; $c(t) = \pm 1$ denotes the C/A code.

The L5 signal is composed of two components with equal magnitudes and in a phase quadrature relationship. Between the two components, one component carries navigation information, which is also known as the data channel. And the other channel that does not carry navigation information is known as the pilot channel or dataless channel [3]. A single satellite's transmitted L5 signal can be defined as

$$S_{\mathrm{L5}}(t) = d(t)NH_{10}(t)c_{\mathrm{I}}(t)\cos(2\pi f_5 t + \phi) + NH_{20}(t)c_{\mathrm{Q}}(t)\sin(2\pi f_5 t + \phi) \tag{1.2}$$

where $S_{\mathrm{L5}}(t)$ is the L5 frequency signal; $d(t)$ is the navigation data stream after encoding; $NH_{10}(t)$ and $NH_{20}(t)$ are the NH codes with 10-symbol and 20-symbol periods, and they can also be called the synchronization code; $c_{\mathrm{I}}(t)$ is the PRN code of the data channel; $c_{\mathrm{Q}}(t)$ is the PRN code of the pilot channel. Each satellite has a pair of unique $c_{\mathrm{I}}(t)$ and $c_{\mathrm{Q}}(t)$ codes, but $c_{\mathrm{I}}(t)$ and $c_{\mathrm{Q}}(t)$ are not correlated to each other, and the PRN codes of different satellites are not correlated either; f_5 is the L5 frequency; ϕ is the initial phase.

Due to the use of spread spectrum, the signals are very weak, and far below the noise level. At the same time, due to relative movement between satellites and receivers, the Doppler effect ensues.

1.4.3 Ranging Code

The C/A code is a short code, with 1.023 MHz chip rate, 1023 chip length, and 1 ms chip period. The main reason for using a short code length is to facilitate a fast signal acquisition. Different satellites use different C/A codes.

The C/A code generator is composed of two sets of 10-stage shift registers: G_1 and G_2, as shown in Fig. 1.4. The characteristic polynomials for the two 10-stage linear feedback shift registers G_1, G_2 can be expressed as

$$G_1 = 1 + x^3 + x^{10} \tag{1.3}$$

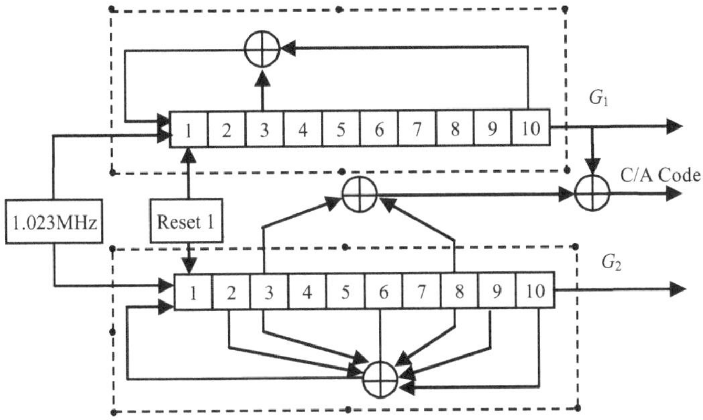

Fig. 1.4 C/A code generator

$$G_2 = 1 + x^2 + x^3 + x^6 + x^8 + x^9 + x^{10} \tag{1.4}$$

where G_1 and G_2 can both generate 1023-bit long maximal linear shift register sequence (m sequence), and the initial conditions are both 1, which can usually be depicted using a polynomial form $1 + \sum x^i$, where x^i refers to the i th stage's output by a modulo-2 addition (XOR) input shift register, and 1 refers to feed the output of the adder to the first stage. The C/A code of every satellite is obtained by XORing G_2's and G_1's direct output sequences. G_2's delay effect is obtained by selecting the positions of the two taps and performing an XOR operation, so the result is still a Pseudo-random code sequence with a changed phase. The C/A codes for different satellites are produced by using different tapping method, and the tapping method for every satellite can be seen found in reference [1, 2].

Different GNSS signals can be detected and distinguished by utilizing the characteristics that the C/A codes have good autocorrelation and cross-correlation and each GNSS signal's C/A code is different. This is the basic working principle of GNSS receivers.

ICD-GPS-200 stipulates that the C/A code is generated by forming a time delay through two taps of the shift register G_2, therefore there are only 45 Gold code combinations. Among them, 32 Gold codes associated with the best characteristics are used for space GNSS signals. Since the C/A code has the characteristics of circulation correlation, when we study correlation functions for the C/A code, we essentially only consider its characteristics inside one period. When the delay time is within a C/A code period, the autocorrelation function of the C/A can be approximated as

$$R(\tau) = \begin{cases} 1 - \frac{|\tau|}{T_C} & |\tau| \leqslant T_C \\ 0 & \text{others} \end{cases} \tag{1.5}$$

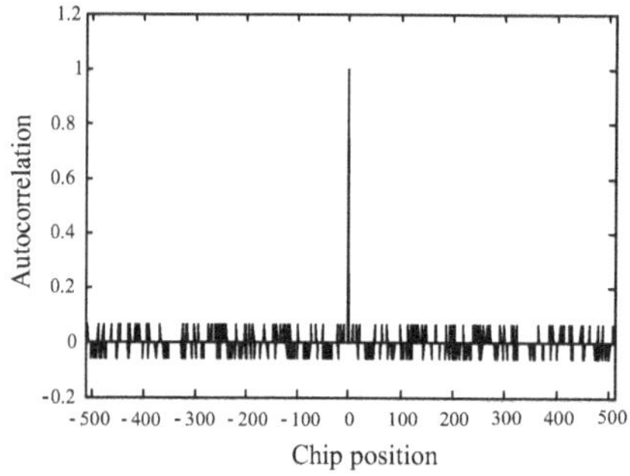

(a) Autocorrelation function for the C/A codes sent by the same satellite

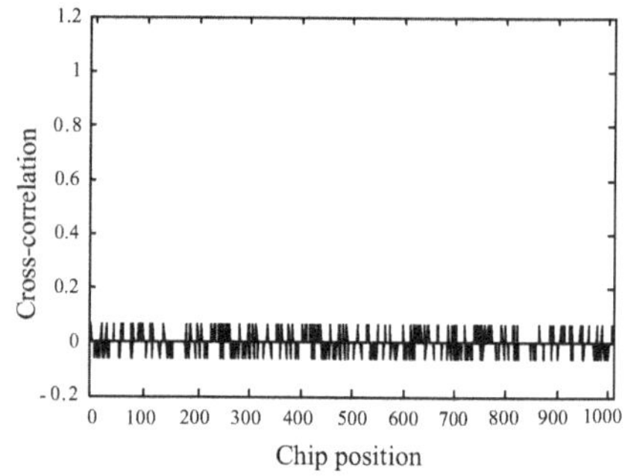

(b) Cross-correlation function for the C/A codes sent by different satellites

Fig. 1.5 Correlation characteristics of C/A codes

where T_C is the chip length of the C/A code and τ represents the time delay in the autocorrelation function.

Different GNSS satellites are allocated with different C/A codes. In the process of de-spreading, a receiver generates all satellites' C/A codes locally, and then makes phase shifts in order to perform correlation calculation with received GNSS signals. Since autocorrelation is strong with the C/A code sent by the corresponding satellite, a peak value is formed when the phase delays are the same. While for different satellites, the cross-correlation coefficients are very small. Consequently, by using the C/A codes' autocorrelation and cross-correlation characteristics, different GNSS signals can be distinguished. Figure 1.5 compares autocorrelation and cross-correlation functions for GNSS signal's C/A codes.

The power of a GPS/GNSS signal's C/A code is approximately −160 dBw, therefore its carrier-to-noise ratio (CNR) is 45.2 dB-Hz. The navigation message, obtained by the receiver terminal after de-spreading and demodulation, has a 50 Hz chip rate, while the chip rate for the C/A code is 1.023 MHz. Consequently the C/A code has approximately 43 dB of spread spectrum processing gain. After de-spreading, the signal-to-noise ratio (SNR) is approximately 25.2 dB, which is sufficient to ensure that the receiver can perform normally when receiving weak signals.

The chip rate of the P code is 10.23 MHz. The spectral width of the main lobe is 20.46 MHz and it is generated by two PRN code sequences with the same chip rates. The first pseudo noise sequence has a length of 15 345 000 chips, and the length for the other one is 15 345 037 chips. The difference between both sequences' lengths is 37 chips, and the lengths for both sequences are mutually prime to each other. Consequently, the synthesis code for the two codes has a chip length of 23 017 555.5 s, which is a little longer than 38 weeks. Since the code is reset once a week, the actual P code chip length is 1 week. The 38-week long code is divided into 37 different P codes, so that each satellite can use different part of the same long code.

1.4.4 Navigation Message

Navigation message is the navigation positioning data sent to users in the form of binary codes. It is mainly composed of satellite ephemeris, clock bias correction parameters, signal propagation delay parameters, satellite status information, time synchronization information, and an almanac for all satellites. The navigation message is transmitted to users through pseudo random code after spread spectrum modulation. A receiver performs demodulation on the received signals, so the related navigation message can be obtained, and the position information for the receiver can be calculated.

One basic unit of the navigation message is the mainframe. Each mainframe has a length of 1500 bits, which can be further divided into five sub-frames. Each sub-frame contains 10 words, and every word contains 30-bit navigation data. The format of a navigation message is shown in Fig. 1.6. Since the navigation message's transmission rate is 50 bits/s, 6 s is needed to transmit a sub-frame, and 30 s is needed to transmit a main frame. The first two words in each sub-frame are a telemetry word (TLM) and a handover word (HOW). The first 8 bits of the TLM is the synchronization header. The first 17 bits of the HOW is the time of the week (TOW) count, using 6 s as the unit, and the 20th–22nd bits are used for sub-frame ID.

The reminder of the message constitutes the first and second data blocks, which include satellite clock calibration parameters, clock reference time, atmospheric correction parameters, and satellite ephemeris or orbit parameters. The sub-frame 4 and 5 constitute the third data block, which carries the almanac information for all GPS satellites. Consequently, to accurately calculate the satellite ephemeris data, the complete data for at least the first three sub-frames is needed.

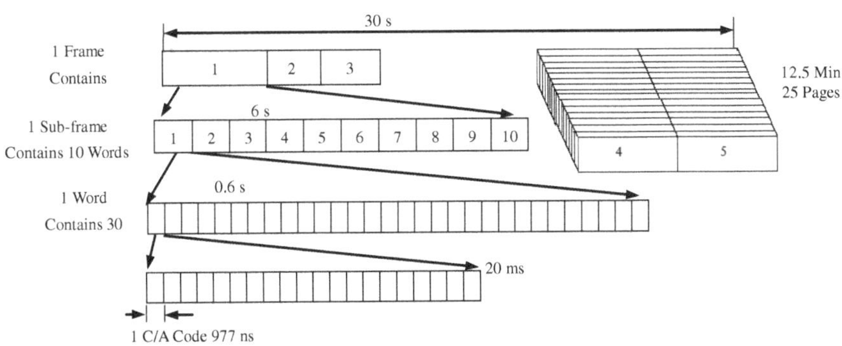

Fig. 1.6 Navigation message structure

1.5 GNSS Receiver

1.5.1 Types of GNSS Receivers

A GPS signal receiver is the terminal device used to implement GPS satellite navigation positioning. It is a type of wireless receiver equipment used for receiving, tracking, converting and measuring GPS positioning signals. It has all the characteristics of commonly used wireless receiver equipments, and is capable of acquiring, tracking and processing weak GPS signals.

GPS receivers can be categorized into the following types:

1. Categorization based on working principles

For GPS satellite measurement techniques based on the passive ranging principle, the key is how to measure the distance between a GPS signal receiver antenna and the GPS satellites. Based on different ranging signals used for measuring distance, GPS receivers can be classified as below: ① Code phase receivers use the ranging code to determine the time delay for a satellite signal to reach the receiver. Its real-time positioning precision is relatively low (generally around 10 m). But this type of receiver is very cheap, so it has been widely applied. ② Carrier phase receivers compares the reference carrier signal generated by the receiver and the received carrier signal with Doppler shift. This type of receiver has better positioning precision than the code phase type, and it is mainly used for precise geodetic and engineering survey applications. This type of receiver has a complex structure and is fairly expensive.

2. Categorization based on applications

Since GPS signal receivers can measure its moving carrier's parameters under all weather conditions, at anytime, anywhere with high-precision, they can be classified based on the applications as below: ① geodetic receivers; ② navigation receivers; ③ timing receivers.

3. Categorization based on the number of used carrier frequencies

Based on the number of used carrier frequencies, the receivers can be classified as: ① A single-frequency receiver uses only a single carrier frequency signal to perform navigation positioning measurement. ② A multi-frequency receiver simultaneously uses multiple frequencies to provide navigation positioning measurement.

4. Categorization based on the number of used GNSS

Based on how many GNSS are used, receivers can be classified as: ① A single-mode receiver only uses a single GNSS system to provide navigation positioning measurements. ② A multi-mode receiver simultaneously uses multiple GNSS systems to provide navigation positioning measurements, e.g. a GPS/Beidou double-mode receiver.

1.5.2 *Working Principle of GNSS Receivers*

To use the working principle of a code phase receiver as an example, a satellite navigation signal, after being transmitted by an antenna, goes through the ionosphere and the troposphere to reach the frontend of a RF antenna. After down-conversion and A/D conversion, the received RF signal is amplified, and converted to a digital intermediate frequency (IF) signal. A software receiver acquires the received signal and obtains the satellite's assigned serial number, initial code phase shift, and the corresponding signal frequency containing Doppler shift components. Then the above parameters are sent to the tracking module, to estimate each satellite's precise code phase shift and precise signal frequency containing Doppler shift components. The data is then demodulated. At the same time, the synchronization head of the demodulated data can be found using the navigation message's preamble, and then the navigation message can be decoded based on the synchronized data. And these can be combined with the code phase shift to calculate the Pseudo-range value. Finally, the Pseudorange value is used to calculate the position of the user receiver.

1. Acquisition

The objective of acquisition is to obtain the satellite's assigned serial number, the initial code phase shift and a rough estimation on the signal frequency containing Doppler shift components, in order to provide initial values for the tracking loop.

Signal acquisition inside the GPS receiver can be regarded as a three dimensional search: the first dimension is PRN code; the second dimension is spread code phase; and the third dimension is Doppler frequency shift.

For the PRN code, if there is no supplemental information available when a receiver turns on, and there is no knowledge on the present satellite constellation distribution, then the number of all PRN codes is 32. Under the circumstance, possible PRN codes must be tried one by one exhaustively. This is called a Cold-Start since no prior information can be leveraged. If supplemental information can reduce the search volume, then the search time can be reduced significantly. For example, the Warm-Start or Hot-Start, using existing historic almanac or ephemeris data and the local time, and under the condition of knowing the approximate position, can approximately derive the GPS constellation distribution under the current zenith. Thus it greatly reduces the search space of PRN codes, so the time required for the search process can be shortened significantly.

For searching the pseudorandom spreading code phase, firstly local spreading codes need to be generated. By setting different local spreading code phases, the correlations between the local spreading codes and the input signal are calculated. Due to the strong autocorrelation of the spreading code, strong correlation can only be produced when the phase of a local code and the input signal's spreading code are aligned. Once the corresponding correlation value for a spreading code phase surpasses a preset threshold value, the correct local spreading code phase can be regarded as being identified. Since the position of the corresponding peak is very sensitive to the phases, we would lose the peak quickly if the phase difference is more than one chip. Consequently, once a related higher peak appears, the difference between the spreading code phase of the input signal and the local spreading code can be regarded as within one chip.

For Doppler frequency search, the local carrier is generated and adjusted first, and then it is multiplied with the input signal. If a local carrier and the carrier of the input signal are very similar to each other, the high frequency carrier components in the input signal can be removed. Since possible carrier frequencies for the input signal are not known in advance, different local carrier frequency values need to be tried. In addition, since Doppler frequency shift can be generated due to relative movements between receivers and satellites, all possible Doppler frequencies need to be covered as well, and an usual range of ±10 kHz [1] is used. The selection of the frequency step is a compromise between search sensitivity and search efficiency: for a fixed frequency search interval, a smaller frequency step increases the number of searches needed, thus directly increases the workload needed to cover the whole frequency search range, leading to a prolonged search time. But it has the benefit of increased search precision. Figure 1.7 illustrates an acquisition search.

Presently, commonly used signal acquisition methods include traditional time-domain sliding correlation acquisition, parallel time domain sliding correlation acquisition, and FFT parallel acquisition. Among them, the time domain sliding correlation acquisition method is most straightforward, but the acquisition time is longer. The other two methods can reduce acquisition time using parallelization.

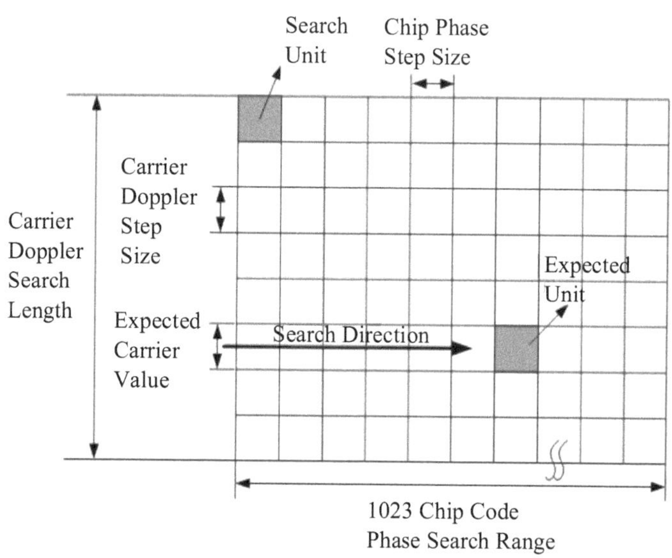

Fig. 1.7 GPS acquisition search

(1) Traditional Time Domain Sliding Correlation Acquisition

The time domain sliding correlation acquisition method is a method to obtain correlation by performing a sliding operation on the C/A code in the time domain, and then acquiring the GPS signal. The working principle is to acquire a satellite signal, and then the receiver inputs the satellite signal into the acquisition function module after mixing and sampling. Simultaneously, within the frequency search range, the local carrier frequency is determined, and then the satellite's C/A code can be reproduced locally. After setting the local carrier frequency and the C/A code, the received satellite signal can be multiplied with the local carrier, so the carrier can be separated from the signal. Then the correlation calculation is performed with the local C/A code to finally detect the peak value of the correlation calculation, and record the corresponding C/A code phase of the peak value. According to the autocorrelation characteristics of the C/A code, when the locally generated C/A code phase is the same as the input C/A ode phase, the correlation is at its maximum. Otherwise, the correlation value decreases. To compare the correlation value to a threshold, if it is larger, then the acquisition is successful. Otherwise, the C/A code phase needs to be changed. If all possible code phases for a frequency have been tried out, but no peak value has been found to surpass the threshold, then the local carrier frequency is not accurate enough to demodulate the carrier frequency. In that case, the local carrier frequency can increase or decrease by a given step size, and the above procedure can be repeated. In this fashion, the search is conducted inside the frequency range, until the correlation peak value can be acquired and the current local carrier frequency can be recorded. The frequency shift and the initial position of the C/A code caused by the Doppler effect can be reduced by repeated computations.

(2) Parallel Time Domain Sliding Correlation Acquisition

The time consumed for acquisition is determined by the time consumed for C/A code correlation calculation. Under the premise that there is no means to reduce the number of correlation calculations, to adopt a parallel computational structure helps to reduce the computation time. There are three types of parallel acquisition: C/A code parallel acquisition, carrier parallel acquisition, and simultaneous parallel acquisitions of the C/A code and the carrier.

The C/A code parallel acquisition method adds N code correlators on the basis of the traditional time-domain sliding correlation method. That is to add $N - 1$ computation channels in addition to the original correlation computation channel. The C/A code correlation computations for N shifts can be performed in parallel. If N results surpass a threshold, then the C/A code's initial position information can be obtained from the acquisition process. If the results do not surpass the threshold, then all N channels would update the C/A code after shifting. Consequently the acquisition speed of the method is N times faster than that of the traditional method.

The approach of carrier parallel acquisition is similar. The only difference is that the parallel computations are performed on the carrier instead of on the C/A code. In the process of satellite signal demodulation, multiple carrier signals are generated and the demodulated signals are fed into multiple correlators, to perform correlation computations with the same C/A code signal.

By performing simultaneous parallel acquisitions on the C/A code and the carrier, the results can be combined. By adopting multiple local NCOs (Numerically Controlled Oscillator) and code correlators, shifts of the carrier and the C/A code can be performed in parallel. Compared with the traditional acquisition method, the time needed for the method can be reduced significantly, but its hardware structure tends to be complex, so more resources are consumed.

(3) FFT Parallel Acquisition

For the time domain sliding correlation acquisition method, the acquisition process is a process to reproduce different C/A code phases and local carriers, and then perform correlation computations with the GPS after the sampling to obtain the maximum value. The algorithm is very easy to implement, but the required amount of computations is larger, and the computation time is longer. To achieve rapid acquisition of GPS signals, the parallel acquisition method based on FFT can be used to significantly reduce acquisition time [7].

2. Tracking

After acquiring the satellite signal, the initial estimates on the frequency and the code phase for the input signal carrying Doppler components are obtained and fed into the receiver's tracking module. The role of the tracking loop is to track the C/A code's initial phase and the signal frequency carrying Doppler components, and to perform the accurate synchronization between the receiver's local signal and the input signal. After the tracking loop finishes the satellite signal tracking, accurate navigation messages can be extracted. Thus the accurate positions of the satellite and the receiver can be calculated. The satellite signal tracking loop is composed of two loops: the carrier tracking loop and the code tracking loop. The carrier tracking loop performs the synchronization on the carrier frequency and carrier phase, while the code tracking loop aims to obtain the C/A code phase in the satellite signal.

Figure 1.8 illustrates the working principle of the tracking loop. Firstly, the input signal multiplies with the local carrier's quadrature and in-phase components to demodulate the carrier, and two channels of mixed signals can be obtained. Then, by multiplying the local C/A code with the signal after the mixing, the satellite signal is de-spreaded, and low-pass filters are used to filter out the high-frequency components of the mixed signals. The discriminator gives the phase difference between the modulating carrier in the signal and the local carrier. The main role of the loop filter is to reduce noise, and to ensure that an accurate estimate on the original signal's carrier frequency can be obtained. The filter's order and noise bandwidth also determine the dynamic range of the tracking loop. The error signal can be generated by subtracting the loop filter's output signal and the original signal, and the error signal is fed back to the filter's input to form a closed loop. After the phase difference message passes the loop filter, the output

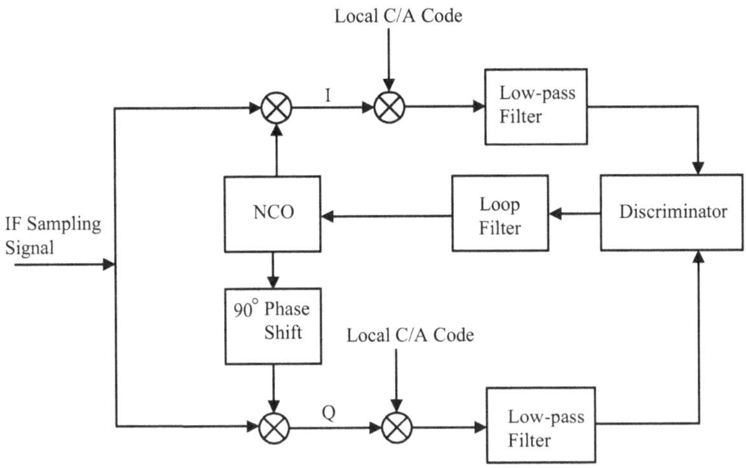

Fig. 1.8 Carrier tracking loop

information is fed back to a numerical digital controlled oscillator (NCO). This is used to correct the local carrier, to achieve the objective of carrier tracking.

Figure 1.9 shows the working principle of the code tracking loop. *I* and *Q* represent the quadrature and in-phase components respectively. *E*, *P* and *L* represent the local C/A code as the early code, the prompt code and the late code. The accumulators in the loop, i.e. pre-detection integrator, code discriminator and code loop filter, determine the characteristics of the code tracking loop.

After multiplying the input signal of the tracking loop and the straightly aligned carrier, the input is demodulated onto the baseband. Then it multiplies with the

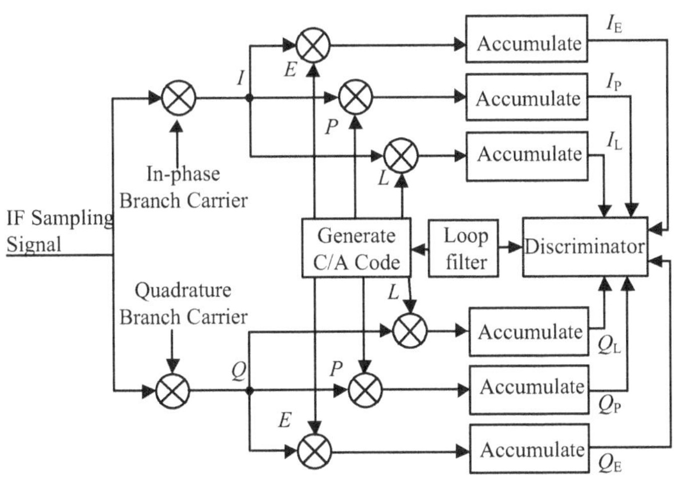

Fig. 1.9 Code tracking loop

three locally generated C/A code variations (the early code, the late code and the prompt code). The integral accumulation is then performed, i.e. calculating the correlations between the demodulated signal and the three variations of the C/A codes. The locally generated C/A code's chip rate can be adjusted based on the correlation values of the three C/A code variations, and the code sequence's positions can be changed to maintain the same correlation values for the early and late codes, thus the correlation value for the prompt code is kept at the maximum. Figure 1.10 shows the chip positions where the correlation values for the late code and the prompt code are the maximum.

In Fig. 1.10a, the late code's correlation value reaches the maximum, verifying that at the present time, the code phase's correlation value has already advanced, so the code phase needs to be adjusted. The results in Fig. 1.10b are obtained after the code phase adjustment. At the time the correlation peak values for the early and late codes are the same, and the prompt code's correlation peak value reaches the maximum, therefore the loop tracking performs properly.

The carrier tracking loop provides the local carrier information for the code loop, and the code loop also provides the code phase delay information for the carrier tracking loop. Essentially, the implementation mechanism for the carrier tracking loop and the code tracking are the same. Figure 1.11 shows the coupling structure of the two tracking loops in GNSS.

3. Positioning Solution Algorithm

There are many methods to perform positioning using GPS. They can be categorized as two types, i.e. absolute positioning and relative positioning, based on different reference points. The absolute positioning methods determine the position of the station relative to the Earth's center of mass, in the protocol earth coordinate system (i.e. WGS-84 coordinate system). In this case, it can be regarded that the reference point coincides with the geocenter. The relative positioning methods

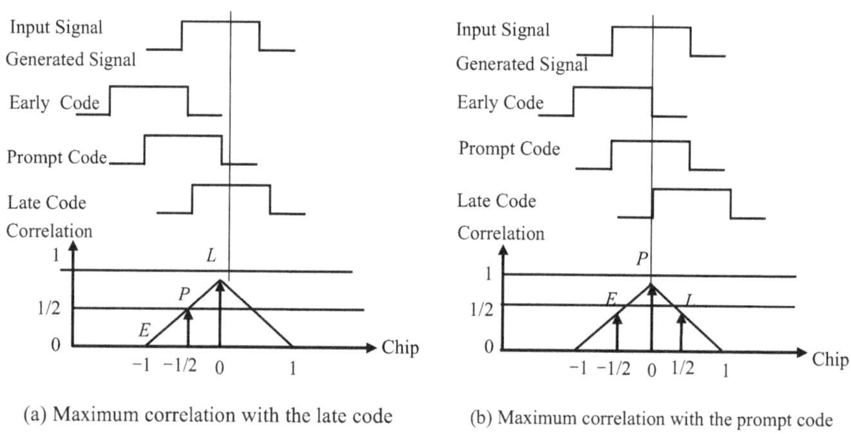

Fig. 1.10 Illustration of code correlation

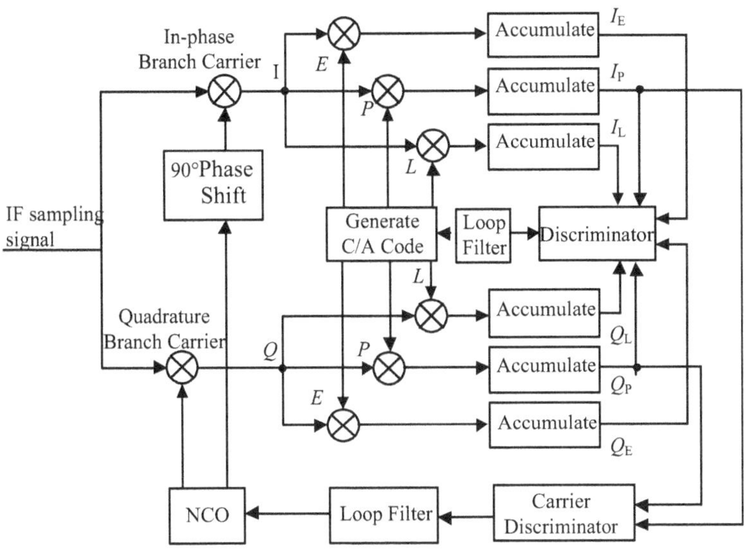

Fig. 1.11 Carrier and code phase joint tracking loops

determine the relative position between the station and a ground reference point, in the WGS-84 coordinate system.

The main roles of the positioning module in the GNSS receiver include:

(1) Find the synchronization header of the navigation message after the tracking demodulation.
(2) Perform bit and frame synchronizations.
(3) Calculate the initial pseudorange from the satellite to the receiver using the precise code phase shift and synchronization header position.
(4) Use the error correct parameters to eliminate errors caused by factors affecting positioning accuracy such as satellite clock difference, Ionospheric and tropospheric errors, and perform pseudorange adjustment to calculate the accurate satellite position at a given time.
(5) Based on the satellite position, nonlinear least squares criterion is used to calculate the position of the user receiver. For user position calculation, four unknown parameters (i.e. three dimensional coordinates and receiver clock difference) need to be solved. Consequently, to determine a user's position, generally signals from more than four satellites are needed.

After the pseudorange adjustment, the corresponding pseudoranges for the four GNSS signals can be calculated as $(\rho_{A1}, \rho_{A2}, \rho_{A3}, \rho_{A4})$, the equivalent distance for the receiver clock bias is b_A, and then the positioning equations become:

$$
\begin{aligned}
\rho_{A1} &= \sqrt{(x_A - x_1)^2 + (y_A - y_1)^2 + (z_A - z_1)^2} + b_A \\
\rho_{A2} &= \sqrt{(x_A - x_2)^2 + (y_A - y_2)^2 + (z_A - z_2)^2} + b_A \\
\rho_{A3} &= \sqrt{(x_A - x_3)^2 + (y_A - y_3)^2 + (z_A - z_3)^2} + b_A \\
\rho_{A4} &= \sqrt{(x_A - x_4)^2 + (y_A - y_4)^2 + (z_A - z_4)^2} + b_A
\end{aligned} \tag{1.6}
$$

where (x_A, y_A, z_A) is the receiver's coordinates under the WGS-84 coordinate system; and (x_i, y_i, z_i) is the ith satellite's coordinates under the WGS-84 coordinate system.

By solving the above equations, parameters such as the receiver's position can be obtained. But the equations are nonlinear. To use the pseudorange positioning algorithm, the nonlinear equations need to be linearized. Processing methods include the iterative method and the direct solution algorithm. The iterative positioning algorithm needs the initial user position, and is highly dependent on the initial value: if the initial value has a larger deviation, then the number of iterations and the amount of computations would increase. The direct solution method does not need any iteration and the initial user position, and its computation is simple and the real-time performance is very good. But pseudoranges from at least five satellites are required.

For the iterative positioning algorithm, an initial position (x_{A0}, y_{A0}, z_{A0}) of a receiver is identified to start the linearization, and usually the center of the earth $(0, 0, 0)$ is selected. By defining increment formula as:

$$
\begin{cases}
x_{A1} = x_{A0} + \Delta x_A \\
y_{A1} = y_{A0} + \Delta y_A \\
z_{A1} = z_{A0} + \Delta z_A
\end{cases} \tag{1.7}
$$

By performing a first-order Taylor expansion on the nonlinear parts in the Eq. (1.6) at (x_{A0}, y_{A0}, z_{A0}), the following can be derived:

$$
\begin{cases}
\rho_{A1} = \rho_{A1}^0 + \dfrac{x_{A0} - x_1}{\rho_{A1}^0} \Delta x_A + \dfrac{y_{A0} - y_1}{\rho_{A1}^0} \Delta y_A + \dfrac{z_{A0} - z_1}{\rho_{A1}^0} \Delta z_A + b_A \\
\rho_{A2} = \rho_{A2}^0 + \dfrac{x_{A0} - x_2}{\rho_{A2}^0} \Delta x_A + \dfrac{y_{A0} - y_2}{\rho_{A2}^0} \Delta y_A + \dfrac{z_{A0} - z_2}{\rho_{A2}^0} \Delta z_A + b_A \\
\rho_{A3} = \rho_{A3}^0 + \dfrac{x_{A0} - x_3}{\rho_{A3}^0} \Delta x_A + \dfrac{y_{A0} - y_3}{\rho_{A3}^0} \Delta y_A + \dfrac{z_{A0} - z_3}{\rho_{A3}^0} \Delta z_A + b_A \\
\rho_{A4} = \rho_{A4}^0 + \dfrac{x_{A0} - x_4}{\rho_{A4}^0} \Delta x_A + \dfrac{y_{A0} - y_4}{\rho_{A4}^0} \Delta y_A + \dfrac{z_{A0} - z_4}{\rho_{A4}^0} \Delta z_A + b_A
\end{cases} \tag{1.8}
$$

where

$$\begin{cases} \rho_{A1}^{0} = \sqrt{(x_{A0} - x_1)^2 + (y_{A0} - y_1)^2 + (z_{A0} - z_1)^2} \\ \rho_{A2}^{0} = \sqrt{(x_{A0} - x_2)^2 + (y_{A0} - y_2)^2 + (z_{A0} - z_2)^2} \\ \rho_{A3}^{0} = \sqrt{(x_{A0} - x_3)^2 + (y_{A0} - y_3)^2 + (z_{A0} - z_3)^2} \\ \rho_{A4}^{0} = \sqrt{(x_{A0} - x_4)^2 + (y_{A0} - y_4)^2 + (z_{A0} - z_4)^2} \end{cases} \tag{1.9}$$

The non-linear equations are converted into a set of linear equations, and then they can be solved using an iterative algorithm.

After receiving and obtaining the pseudo-ranges from at least 5 satellites, the direct positioning algorithm can be used. The direct positioning method performs transposition square on the Pseudo-range equations first to obtain:

$$\begin{cases} (\rho_{A1} - b_A)^2 = (x_A - x_1)^2 + (y_A - y_1)^2 + (z_A - z_1)^2 \\ (\rho_{A2} - b_A)^2 = (x_A - x_2)^2 + (y_A - y_2)^2 + (z_A - z_2)^2 \\ (\rho_{A3} - b_A)^2 = (x_A - x_3)^2 + (y_A - y_3)^2 + (z_A - z_3)^2 \\ (\rho_{A4} - b_A)^2 = (x_A - x_4)^2 + (y_A - y_4)^2 + (z_A - z_4)^2 \\ (\rho_{A5} - b_A)^2 = (x_A - x_5)^2 + (y_A - y_5)^2 + (z_A - z_5)^2 \end{cases} \tag{1.10}$$

where $\rho_{A5} = \sqrt{(x_A - x_5)^2 + (y_A - y_5)^2 + (z_A - z_5)^2} + b_A$.

By subtracting the first equation from all the other equations, the pseudo-range equations can be linearized. Given that $r_i^2 = x_i^2 + y_i^2 + z_i^2$, the equations have the forms below after the subtraction:

$$\begin{cases} 2(x_2 - x_1)x_A + 2(y_2 - y_1)y_A + 2(z_2 - z_1)z_A - 2(\rho_{A2} - \rho_{A1})b_A = (r_2^2 - r_1^2) - (\rho_{A2}^2 - \rho_{A1}^2) \\ 2(x_3 - x_1)x_A + 2(y_3 - y_1)y_A + 2(z_3 - z_1)z_A - 2(\rho_{A3} - \rho_{A1})b_A = (r_3^2 - r_1^2) - (\rho_{A3}^2 - \rho_{A1}^2) \\ 2(x_4 - x_1)x_A + 2(y_4 - y_1)y_A + 2(z_4 - z_1)z_A - 2(\rho_{A4} - \rho_{A1})b_A = (r_4^2 - r_1^2) - (\rho_{A4}^2 - \rho_{A1}^2) \\ 2(x_5 - x_1)x_A + 2(y_5 - y_1)y_A + 2(z_5 - z_1)z_A - 2(\rho_{A5} - \rho_{A1})b_A = (r_5^2 - r_1^2) - (\rho_{A5}^2 - \rho_{A1}^2) \end{cases} \tag{1.11}$$

Then, by solving the above linear equations, the parameters such as receiver position and clock bias can be obtained. When the number of satellite pseudo-ranges obtained by the receiver is larger than the minimum requirement, the least squares algorithm is commonly used to take advantage of pseudo-ranges to further improve ranging precision [1].

1.6 The Impacts of Interference on GNSS Receivers

Since GNSS usually uses the CDMA multi-access technology, a periodic pseudorandom spreading code is used for spread spectrum modulation on the transmitter, and the same pseudorandom spreading code is used for de-spreading based on correlation on the receiver. After the spreading spectrum modulation, the information energy is evenly distributed on a wider bandwidth, and the power spectral density decreases. After the spread spectrum signal de-spreading, the wideband signal is reconstituted into a narrow band signal, and the power spectral density increases. Consequently the spread spectrum technology itself carries certain interference mitigation capability. Nevertheless, the distance between the satellite and the receiver is very far and the satellite's transmitting power is rather low. One analogy to illustrate this phenomenon is to watch a 50 W light bulb from approximately 20,000 km away, as shown in Fig. 1.12. Similarly, the GNSS signal received by the receiver is very weak, usually submerged in the thermal noise on the receiver. When the receiver is under the impacts of various intentional and unintentional RF interferences, the GNSS signal cannot be extracted from the noise, finally leading to a decrease of positioning accuracy and locking loss on the tracking loop. Sometimes even the satellite cannot be acquired normally [8–10].

Interference on GNSS receivers can be categorized into two types: intentional and unintentional interference, as shown in Fig. 1.13. Intentional interference includes jamming and spoofing; while unintentional interference includes Radio Frequency Interference (RFI), multipath interference, and in-band pulse interference etc.

Jammer mainly transmits high-power interference signal to jam the GNSS signal at a receiver's front end, so the receiver cannot accurately acquire and track the GNSS signal [8–11]. A jammer only needs to overlay its high-power random waveform signal with the GNSS signal on the frequency spectrum, so the technical difficulty of implementation is low. Jamming can be further categorized as narrowband interference (aiming) and wideband interference (blocking).

Based on the motion state of a receiver relative to the interference, the interference can be classified as low or high dynamic interference [12]. In a high dynamic environment, a GNSS receiver experiences significant changes on its motion velocity and rate of velocity change, which impact the receiver's acquisition, tracking, and positioning, leading to serious degradation on the performance of traditional adaptive interference mitigation techniques. This is because under high dynamic conditions, due to high-speed motion or rotation of a receiver platform, the

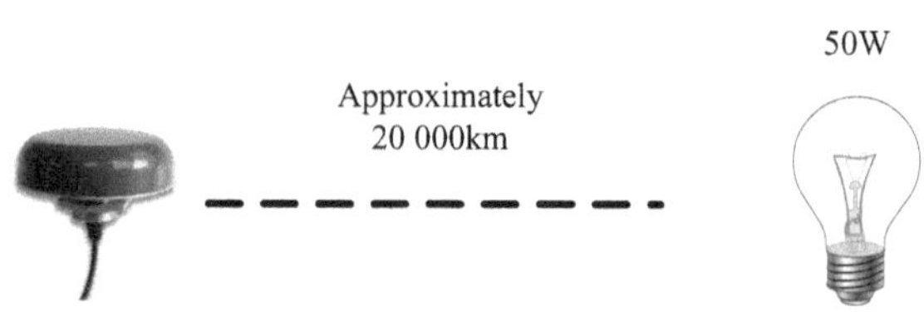

Fig. 1.12 An analogy on GNSS received signal strength

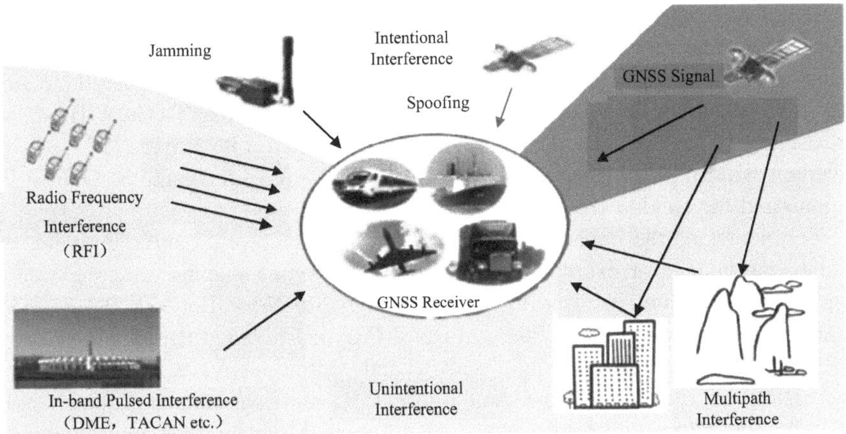

Fig. 1.13 Different types of GNSS interference

incoming direction of the interference related to the receiver experiences rapid changes. Since the actual adaptive process requires stable signals, only few data snapshots can be used to estimate the adaptive weight values. Under the condition, if a regular adaptive beamforming method is used, the null formed around the interference direction is too narrow. The interference could then easily shift out of the null zone and avoid suppression, leading to the failure of regular adaptive method.

Spoofing transmits fake signals similar to the GNSS signal to disrupt the receiver, making it to generate faulty positioning information, or even take control of the targeted receiver [13, 14]. Based on different working principles, spoofing techniques can be further classified as "generating" or "meaconing" types. To use the "generating" interference, the GNSS signal structure must be grasped, so the implementation difficulty is high. The "meaconing" interference, after intercepting the receiver's GNSS signal, delays and amplifies the signal and then forwards to the target GNSS receiver, causing the target receiver to miscalculate the signal propagation time from the satellite to the receiver, and leading to a faulty positioning. Compared with the "generating" interference, the implementation difficulty of the "meaconing" type is much lower.

Multipath interference refers to the phenomenon that when a signal transmitted by a satellite, after being reflected by the objects around the receiver antenna, arrives at the receiver at the same time as the signal following a direct path. This changes the phase discrimination characteristics at the receiving loop, leading to positioning errors [15–17]. The existing study results have shown that the pseudorange errors produced by multipath interference can range from several meters to even hundreds of meters. In addition, for geodetic receivers with higher precision requirements, the impacts of multipath interference are greater. Consequently, multipath interference mitigation is one of the hot topics in the GNSS receiver design field.

Different from the continuous interference on the time domain, the pulse interference appears from time to time. When the power of pulse interference is large, it leads to decoding difficulties for GNSS receivers as well, or even make them stop normal operation. A typical pulse interference is the DME/TACAN (Distance Measuring Equipment/ Tactical Air Navigation System) interference [18, 19] that overlaps with the working frequencies of the GNSS GPS L5 signal, the Galileo E5 signal, and the Beidou B2 signal.

To understand the impacts of interference on the receivers more intuitively, we use the jamming as an example for illustration. In a single antenna structure system, 8 simulated satellite signals with a −20 dB Signal-to-Noise Ratio (SNR) are generated, and they are PRN1, PRN2, PRN3, PRN6, PRN14, PRN20, PRN22 and PRN25 respectively. By adding a narrowband continuous wave interference, and increasing the Interference-to-Signal Ratio (ISR) until we cannot acquire GNSS signals. The acquisition results based on different ISR conditions are obtained as shown in Fig. 1.1.

It can be seen from Fig. 1.14 that, as the SIR increases, the number of acquired satellites reduces significantly. When the SIR is 28 dB, only some satellites can be acquired and when the SIR reaches 32 dB, no satellites can be acquired normally.

This shows that various types of interference pose serious threats on GNSS systems. In November 2013, during the PNT (Positioning, Navigation, and Timing) workshop hosted by Stanford University, a report written by FAA in US clearly pointed out that "GNSS is vulnerable because it is so valuable!". In the Fourth Chinese National Conference on GNSS in May 2013, the conference report written by Yang Yuanxi, a Beidou satellite navigation expert and a fellow of the Chinese Academy of Sciences, also pointed out that interference mitigation is one of the few main problems faced by GNSS. In recent years, supported by multiple national-level projects, the Civil Aviation University of China where the authors have been working at, has been conducting studies on GNSS adaptive interference mitigation, and has made a series of achievements. The national-level research projects include one national outstanding youth fund project, an oversea outstanding young scholars funds collaborative research project, a national "863" high-tech project, and four National Natural Science Foundation projects etc. A series of research achievements on topics such as low dynamic jamming suppression [20, 21], high dynamic jamming suppression [22–26], spoofing countermeasures [27–29], multipath interference suppression [30–35] and pulse interference suppression [36–39] have been achieved. In later parts of the book, we will elaborate on related interference mitigation algorithms for different types of interferences and different application scenarios.

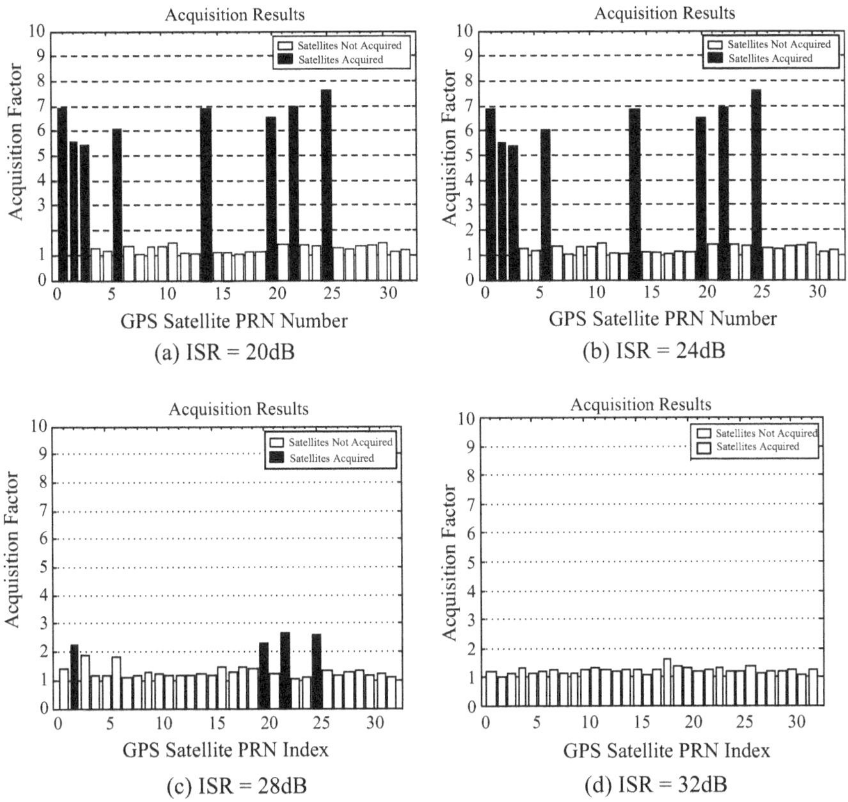

Fig. 1.14 Acquisition results based on different ISRs

1.7 Summary

Firstly, this chapter briefly introduces the GNSS and its development status. Then, using the GPS as an example, we briefly present the basic principles of GNSS from various perspectives such as the composition, the signal structure, and the characteristics of GNSS, and the basic principles on GNSS receivers. Finally, we present the impacts of interference on GNSS receivers, and elaborate the necessity of GNSS interference mitigation. The simple introductions on the principles in this chapter lay a foundation for incoming chapters in which we will present various satellite interference mitigation techniques.

References

1. Misra P, Enge P. Global positioning system: signals, measurements, and performance. Lincoln, MA: Ganga-Jamuna Press; 2004.
2. Kaplan ED, Hegarty CJ. Understanding GPS: principles and applications. Artech House; 2005.
3. http://www.gps.gov/systems/gps/modernization/.
4. http://www.nis-glonass.ru/en/glonass/.
5. http://www.esa.int/Our_Activities/Navigation.
6. http://www.beidou.gov.cn/xtjs.html.
7. Van Nee DJR, Coenen AJRM. New fast GPS code-acquisition technique using FFT. Electron Lett. 1991;27(2):158–60.
8. Sklar JR. Interference mitigation approaches for the global positioning system. Lincoln Lab J. 2003;14(2):489–509.
9. O'Brien AJ. Adaptive antenna arrays for precision GNSS receivers. PhD dissertation. The Ohio State University; 2009.
10. Amin MG, Sun W. A novel interference suppression scheme for global navigation satellite systems using antenna array. IEEE J Sel Areas Commun. 2005;23(5):999–1012.
11. Borio D. GNSS acquisition in the presence of continuous wave interference. IEEE Trans Aerosp Electron Syst. 2010;46(1):47–60.
12. Hinedi S, Statman JI. High-dynamic GPS trackings final report. JPL Publication. 1988;88-35:9–11.
13. Humphrey TE, Ledvina BM, Psiaki ML, et al. Assessing the spoofing threat: development of a portable GPS civilian spoofer. In: The 21th international technical meeting of the satellite division of the Institute of Navigation (ION GNSS), 16–19 Sep. 2008, Savannah Georgia, USA. 2008. p. 2314–25.
14. Daneshmand S, Jafarnia-Jahromi A, Broumandan A, et al. A low complexity GNSS spoofing mitigation technique using a double antenna array. GPS World Mag. 2011;22(12):44–6.
15. Fante RL, Vaccaro JJ. Cancellation of jammers and jammer multipath in a GPS receiver. IEEE Aerosp Electron Syst Mag. 1998;13(11):25–8.
16. Kalyanarman SK, Braasch MS, Kelly JM. Code tracking architecture influence on GPS carrier multipath. IEEE Trans Aerosp Electron Syst. 2006;42(2):548–61.
17. Daneshmand S, Broumandan A, Asl N, et al. GNSS multipath mitigation with a moving antenna array. IEEE Trans Aerosp Electron Syst. 2013;49(1):693–8.
18. Hegarty C, Kim T, Ericson S. Methodology for determining compatibility of GPS L5 with existing systems and preliminary results. In: Proceedings of the Institute of Navigation annual meeting. 1999. no 1, p. 1–10.
19. Gao GX, Denks H, Steingassnd A, et al. DME interference mitigation based on flight test data over European hot spot. GPS Solut. 2013;17(4):561–73.
20. Lu D, Wu R, Liu H. Global positioning system anti-jamming algorithm based on period repetitive CLEAN. IET Radar Sonar Navig. 2013;7(2):164–9.
21. Wang W, Du Q, Wu R, et al. Interference suppression with flat gain constraint for satellite navigation systems. IET Radar Sonar Navig. 2015;9(7):852–6.
22. Wu R, Li C, Lu D. Power minimization with derivative constraints for high dynamic GPS interference suppression. Sci China: Inf Sci. 2012;55(4):857–66.
23. Lu D, Wu R, Wang W. Robust widenull anti-jamming algorithm for high dynamic GPS. In: International conference on signal processing proceedings. 2012. p. 378–81.
24. Ma Y, Lu D, Wang W, et al. A high-dynamic null-widen GPS anti-jamming algorithm based on statistical model of the changing interference DOA. China Satell Navig Conf. 2014;(I):695–702.
25. Wu R, Ge L, Lu D, et al. A high-dynamic wideband interference suppression in GNSS via reduced-rank STAP. In: Proceedings of the 28th international technical meeting of the satellite division of the Institute of Navigation (ION GNSS + 2015). 2015.

26. Lu D, Li S, Wu R. Simultaneous null and main lobe widening for jamming suppression in attitude high-dynamic GNSS receiver. In: Proceedings of the 28th international technical meeting of the satellite division of the Institute of Navigation (ION GNSS + 2015). 2015.
27. Zhang Y, Wang L, Wang W, et al. Spoofing jamming suppression techniques for GPS based on DOA estimating. China Satell Navig Conf. 2014;(I):683–93.
28. Zhang Y, Wang L, Wang W, et al. Spoofing interference suppression for GNSS based on estimating steering vectors. China Satell Navig Conf. 2015;I:765–71.
29. Wang L, Wu R, Zhang Y, et al. Multi-type interference suppression for GNSS based on despread-respread method. In: Proceedings of the 28th international technical meeting of the satellite division of the Institute of Navigation (ION GNSS + 2015). 2015.
30. Wu R, Jia Q, Wang W. Efficient FFT-based algorithms for multipath interference mitigation in GNSS. In: Fourier transform signal processing and physical sciences. INTECH; 2015.
31. Li J, Wang W, Lu D, et al. A correlate-based GPS multipath time delay estimation algorithm. China Satell Navig Conf. 2014;I:53–62.
32. L J, Wu R, Lu D. GPS multipath interference suppression algorithm based on decoupled parameter estimation theory. Signal Process. 2011;27(12):1884–8.
33. Li J, Wu R, Wang W, et al. A novel GPS signal acquisition algorithm. Adv Inf Sci Serv Sci. 2012;4(17):597–604.
34. Li J, Wu R, Wang W, et al. GPS fine time delay estimation based on decoupled parameter estimation theory. In: International conference on signal processing proceedings. 2012. p. 236–40.
35. Jia Q, Wu R, Wang W, et al. Multipath interference mitigation in GNSS via WRELAX. GPS Solut. 2017;21(2):487–98.
36. Li L, Wang W, Lu D, et al. Wavelet packet transformation based technique in mitigation DME pulsed interference for GNSS. China Satell Navig Conf. 2014;I:715–24.
37. Wu R, Wang W, Li L, et al. Distance measuring equipment interference suppression based on parametric estimation and wavelet-packet transformation for global navigation satellite systems. IEEE Trans Aerosp Electron Syst. 2016;52(4):1607–17.
38. Hu T, Wang L, Wu R. Study on wideband and pulse interference mitigation techniques with RAW BD data. ION GNSS + 2015.
39. Fang W, Wu R, Wang W, et al. DME pulse interference suppression based on NLS for GPS. In: International conference on signal processing proceedings. 2012. p. 174–8.

Chapter 2
Jamming Suppression

2.1 Introduction

The vulnerabilities of GNSS systems expose them to impacts from various intentional and unintentional interferences. Of these various interferences, jamming is the most common type, which mainly includes narrowband jamming and wideband jamming. Even though GNSS adopts spread spectrum technology so that it has certain capability of interference mitigation, this capability is constrained by spreading gain. Usually the power of jamming is much larger than the power of the GNSS signal. Under these conditions, the jamming power, even after de-spreading, is still much larger than the signal power, and consequently a receiver cannot acquire GNSS signal.

Techniques used for GNSS jamming mitigation can be roughly divided into three categories: temporal domain and frequency domain filtering techniques, spatial domain filtering techniques, and space-time domain joint filtering techniques. Temporal domain and frequency domain filtering techniques [1–7] have the advantages of low cost and simplicity. But when there are multiple narrowband jammings or one single wideband jamming (temporal domain wideband signal), these techniques cannot deliver effective jamming suppression performance. Space filtering technique is a commonly used jamming mitigation approach. It can effectively suppress narrowband and wideband jammings (array wideband signals) using adaptive antenna arrays. Existing space filtering techniques include power minimization algorithm (also known as power inversion algorithm) [8–11], Capon algorithm [12, 13], and blind adaptive beamforming method using GNSS signal characteristic (the known spread spectrum code characteristic of periodic repetition) [14, 15]. However, only space filtering technique consumes one degree of freedom to suppress one narrowband jamming. To suppress wideband jamming, the number

The original version of this chapter was revised: For detailed information please see Erratum. The erratum to this chapter is available at https://doi.org/10.1007/978-981-10-5571-3_7

R. Wu et al., *Adaptive Interference Mitigation in GNSS*, Navigation: Science and Technology, https://doi.org/10.1007/978-981-10-5571-3_2

of antenna array elements needs to be increased. This increases the cost of GNSS receivers and also is difficult to implement in the case of restricted antenna aperture (e.g. airborne and missile-borne conformal arrays). Space Time Adaptive Processing (STAP), which was originally designed for Airborne Early Warning (AEW) radar to suppress ground clutter [16, 17], can be used to solve the above problem. This technique, under the premise of not increasing the number of array elements, can increase the degree of freedom of an adaptive filtering system by using multiple taps.

Due to the limitations of temporal domain and frequency domain techniques, this chapter only briefly introduces the temporal domain and frequency domain filtering techniques, and highlights the spatial domain and space-time filtering techniques. First, we introduce the general model for array signal processing, then we introduce the conventional spatial domain adaptive filtering technique. Secondly we study two methods that use array signal processing gain to improve GNSS signal's carrier-to-noise ratio (CNR) [18]. Based on this foundation, we list an implementation scheme for a type of digital multi-beam jamming mitigation receiver and its experiment results. Finally, we study the space-time adaptive filtering algorithm and the corresponding reduced rank algorithm and equalization algorithm.

2.2 Temporal Domain and Frequency Domain Adaptive Filtering

Temporal domain and frequency domain filtering methods were the earliest GNSS jamming mitigation algorithms. The temporal domain filtering method suppresses the jamming signal by designing adaptive temporal domain filtering. Frequency domain filtering method first converts the signal into frequency domain for processing, then a threshold can be set, and all spectrum values larger than the threshold are set to be zero to suppress jamming signals. Compared with the temporal domain filtering algorithm, the frequency domain processing is easier to be implemented, so it is a commonly used jamming suppression technique at the present time.

2.2.1 Temporal Domain Filtering

When Q temporal domain narrowband jamming sources exist, the model of the received signals by a GNSS receiver is

$$x(t) = \sum_{l=1}^{L} s_l(t) + \sum_{q=1}^{Q} j_q(t) + e(t) \tag{2.1}$$

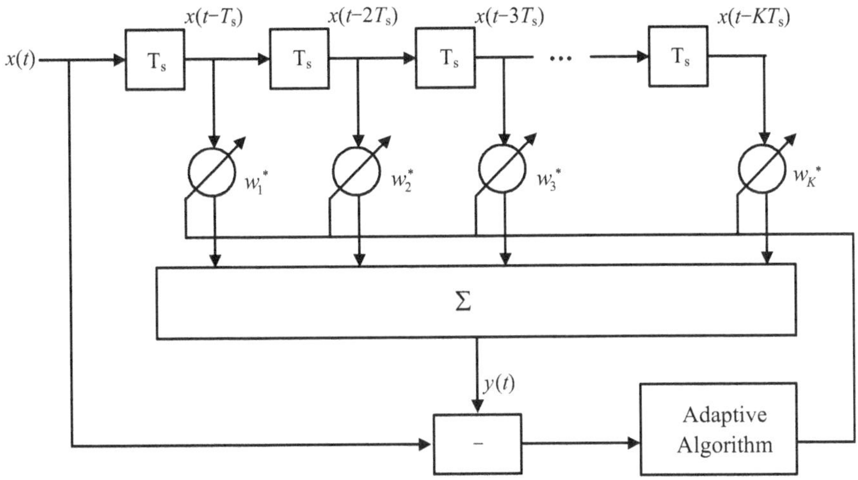

Fig. 2.1 Block diagram of an adaptive FIR filter

where $s_l(t) = A_l D_l(t-\tau_l) c_l(t-\tau_l) e^{j\omega_{dl} t}$ represents the lth GNSS signal $(l = 1, 2, \ldots, L)$; ω_{dl} is the angular frequency of the signal; A_l is the carrier amplitude; $D_l(t)$ is the navigation data bit information; $c_l(t)$ is the C/A code for the lth satellite signal; τ_l is the propagation delay for the lth satellite signal; $j_q(t)$ is the qth jamming signal $(q = 1, 2, \ldots, Q)$; $e(t)$ is the thermal noise of the receiver (usually it is additive Gaussian white noise with zero mean and σ_e^2 variance). We can assume that the GNSS signal, jamming signal, and noise are not correlated with each other.

Among the signals received by the receiver, the GNSS signal and noise are temporal domain wideband signals, and the correlation between their sample values is relatively small. But the jamming signal is a narrowband signal, so its sample values have a very strong correlation. At the same time, the power of GNSS signal is much smaller than the receiver's noise power (by about 20 dB). The sum of both can be approximated as receiver thermal noise. Therefore the jamming signal can be estimated by designing an adaptive linear prediction filter, as a result the jamming signal can be subtracted from the received signal [19]. This type of filter usually adopts the non-recursive transversal structure, also known as adaptive Finite Impulse Response (FIR) filter. Figure 2.1 shows the block diagram of an adaptive FIR filter used for GNSS receivers.

In Fig. 2.1, given the number of FIR taps is K, the input signal vector of the filter can be denoted as

$$\mathbf{x}(t) = [x(t-T_s), \quad x(t-2T_s), \quad \ldots, \quad x(t-KT_s)]^{\mathrm{T}} \tag{2.2}$$

where T_s is the sample time; $(\bullet)^{\mathrm{T}}$ represents the matrix transpose operation. Assuming the filter's weighted vector is

$$\mathbf{w} = \begin{bmatrix} w_1, & w_2, & \ldots, & w_K \end{bmatrix}^{\mathrm{T}} \tag{2.3}$$

The filter's output signal is

$$y(t) = \mathbf{w}^{\mathrm{H}}\mathbf{x}(t) \tag{2.4}$$

where $(\bullet)^{\mathrm{H}}$ represents the conjugate transpose operation. To suppress the jamming signal, we can select a filter's weighted vector to make the filter's output signal $y(t)$ the estimated jamming signal. Since the GNSS signal is very weak, and the jamming signal is a strong narrowband signal, the input signal $x(t)$ can be used as the reference signal $d(t)$. By using the Minimum Mean Square Error (MMSE) for filter design, the weighted vector can be determined. The cost function can be defined as

$$F(\mathbf{w}) = E\left\{\left|x(t) - \mathbf{w}^{\mathrm{H}}\mathbf{x}(t)\right|^2\right\} \tag{2.5}$$

where $E\{\bullet\}$ denotes mathematical expectation. By minimizing $F(\mathbf{w})$, we can derive

$$\mathbf{w}_{opt} = \mathbf{R}_x^{-1}\mathbf{r}_{xd} \tag{2.6}$$

where $\mathbf{R}_x = E\{\mathbf{x}(t)\mathbf{x}^{\mathrm{H}}(t)\}$ is the autocorrelation matrix of the filter input signal; $\mathbf{r}_{xd} = E\{\mathbf{x}(t)d^*(t)\}$ is the cross-correlation between the filter input signal vector and the reference signal; and $(\bullet)^*$ denotes the conjugate operation. At present time, many algorithms can be chosen to solve Eq. (2.6), and the most representative methods include Levinson-Durbin algorithm, Burg algorithm, Least Mean Square (LMS) algorithm, recursive least squares algorithm etc. The details of these algorithms can be found in Ref. [20], and we will not elaborate on them in this chapter.

2.2.2 *Frequency Domain Filter*

A simple type of frequency domain method can be adopted for narrowband jamming suppression, and its block diagram is shown in Fig. 2.2. The algorithm first performs Fast Fourier Transform (FFT) on the input signal. Then it detects the jamming signals and suppresses them by setting the corresponding spectral values in the frequency domain to zero. Finally it converts the signal back into temporal

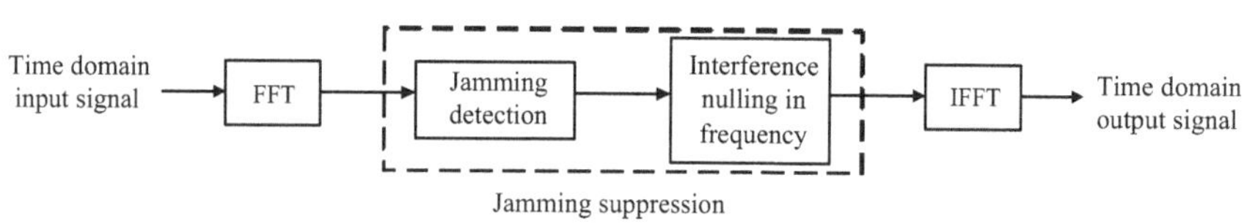

Fig. 2.2 Frequency domain jamming suppression block diagram

domain using Inverse Fast Fourier Transform (IFFT), and sends the converted signal into a common receiver for positioning calculation [21, 22].

In Fig. 2.2, the frequency domain processing performs jamming detection first, and then sets the corresponding spectral values to zero. Using a method similar to the Constant False Alarm Rate (CFAR) method used in radar application (more details can be found in Chap. 6 of this book), a threshold can be pre-determined, and the spectral values after FFT can be compared with the threshold. If a value surpasses the threshold, it can then be regarded as a jamming signal. The jamming signal can then be suppressed by performing zero-setting on the detected spectral values in the frequency domain.

When multiple narrowband jammings exist, even if we can suppress narrowband jamming by using zero-value setting in the frequency domain, the method could cause some loss on the GNSS signal. In particular, if the jamming signal is a co-channel wideband signal related to the GNSS signal, the GNSS signal can be filtered out using this method. Therefore even though the temporal domain and frequency domain methods are easy to implement, their performances can

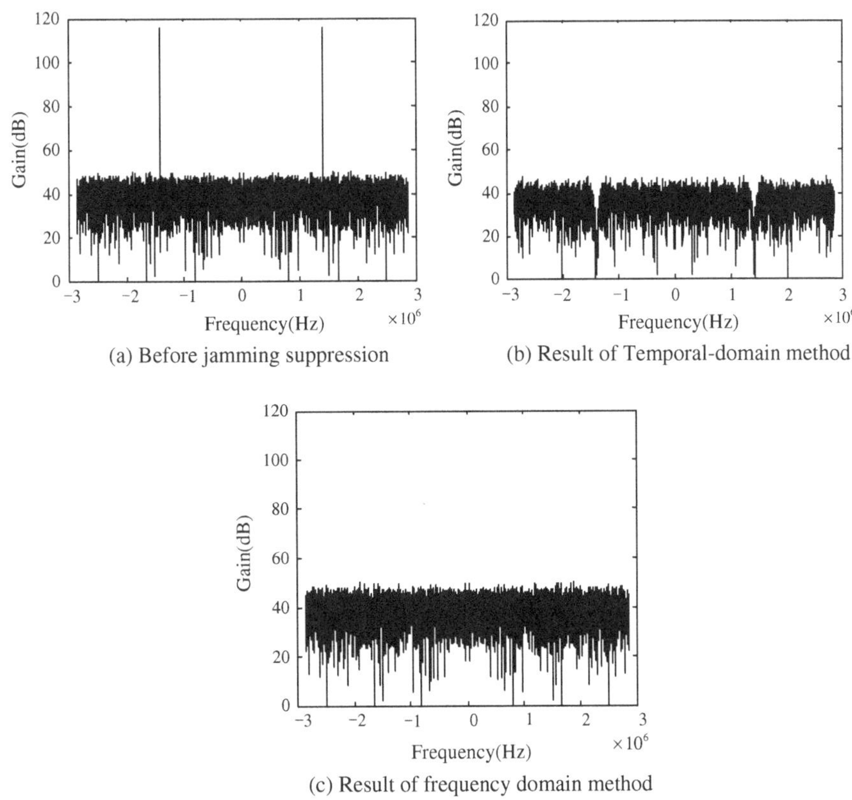

Fig. 2.3 Comparisons of signal spectrums before and after using temporal domain and frequency domain methods (single frequency jamming scenario)

drastically degrade, or even totally lose effectiveness in the cases of multiple narrowband jammings or a wideband jamming.

When a jamming signal's frequency changes over time (e.g. a linear frequency modulated signal), a more effective method is to use time-frequency analysis tool [23] to separate the jamming signal and the GNSS signals, and then perform jamming suppression. References [24–27] studied the time-frequency domain GNSS jamming mitigation method based on Short Time Fourier Transformation, and subspace techniques etc.

2.2.3 *Simulation Results*

In this section we verify the effectiveness of temporal domain and frequency domain methods through simulations. For the first simulation, the received signals include one GPS signal and one single frequency jamming signal with 40 dB

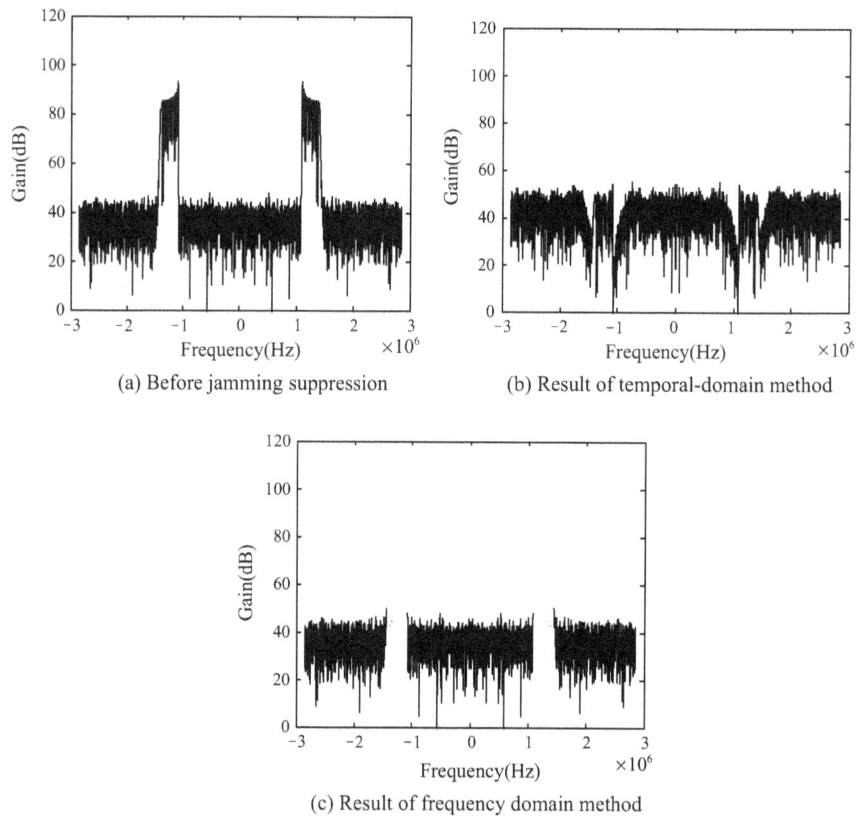

Fig. 2.4 Comparisons of signal spectrums before and after using temporal domain and frequency domain methods (0.3 MHz jamming bandwidth scenario)

jamming-to-noise ratio, −20 dB signal-to-noise ratio and the order of temporal domain filter is 20. The received signal has a digital intermediate frequency of 4.309 MHz, with a sampling rate of 5.714 MHz. Figure 2.3 compares the temporal domain and frequency domain methods. It can be seen in Fig. 2.3 that for single frequency jamming, both methods can effectively suppress the jamming signal.

For the second simulation, the received signals include a jamming signal with an approximately 0.3 MHz bandwidth, and the other simulation conditions are the same as those for the first simulation. Figure 2.4 compares the temporal domain and frequency domain methods. It can be seen in Fig. 2.4 that when the jamming signal has wider bandwidth, the temporal domain processing cannot effectively suppress the jamming signal, and even though the frequency domain processing may suppress the jamming signal, it could cause significant satellite signal loss.

2.3 Conventional Spatial Domain Adaptive Filtering

Temporal domain and frequency domain methods have the advantages of low cost and simplicity. But since they lack the capability to distinguish desired signal and jamming signal in the spatial domain, they cannot cope with a large number of narrowband jammings and wideband jammings. Spatial domain methods can distinguish spatial directions of GNSS signals and jamming signals, so they have become an effective means used for GNSS jamming mitigation. Among them, the most widely used is the power minimization algorithm [9], also known as the power inversion algorithm. The algorithm uses array antennas to suppress jamming. Before we introduce the power minimization algorithm in this section, we first introduce the basic principle and methodology for adaptive array processing.

2.3.1 Adaptive Array Overview

Array processing mainly processes the signal in the spatial domain, including signal filtering, signal detection, and parameter estimation. It has a dual relationship with the temporal domain FIR filter. FIR filter performs discrete sequential sampling on the temporal domain signal, while array processing performs discrete parallel sampling on the spatial domain signal. Usually we can assume the space signal is a far field incident signal. The so-called far field means that the signal source is far enough from the array that the wave front arrives at the array can be approximated as a plane wave. A near field signal, on the other hand, can be regarded as a spherical wave.

For array processing, whether a signal is narrowband or wideband is determined relative to the array itself, which is different from the usual criterion of using the ratio between the signal bandwidth and the center frequency. A coherent array corresponds to a narrowband signal and a non-coherent array corresponds to a wideband signal. If an array performs as a coherent array towards a certain signal, then the maximum time difference for the signal to arrive at various array elements should be small enough so that the signal envelops received by various array elements are consistent. Assuming $\tau_{\max}$ represents the longest time difference, and B_{w} represents the signal bandwidth, then a coherent array should satisfy the following condition:

$$\tau_{\max} \ll \frac{1}{B_{\mathrm{w}}} \tag{2.7}$$

i.e. the time for the signal to go through the array is much shorter than the equivalent time width of the signal. Otherwise, it should be regarded as a non-coherent array.

By considering an arbitrary array composed of M array elements in the far field, there is a narrowband signal incidenting on the array, with an incident polar angle θ, and an azimuth ϕ, as shown in Fig. 2.5. Each array element is assumed to be isotropic, and point O is the reference point for array reception data.

Under the assumed condition of far field narrowband, the received signal by the array can be represented as

$$\mathbf{x}(t) = \mathbf{a}(\theta, \phi)s(t) + \mathbf{e}(t) \tag{2.8}$$

where $\mathbf{x}(t) = \left[x_1(t),\quad x_2(t),\quad \ldots,\quad x_M(t)\right]^{\mathrm{T}}$ is an $M \times 1$ dimension array data vector (the signal sample obtained by performing a single sampling pass on the signal is called the single snapshot data vector); $\mathbf{e}(t)$ is a $M \times 1$ dimension array representing the received noise vector; it can usually be assumed to follow a Gaussian distribution; $s(t)$ is the complex envelop of the signal; $\mathbf{a}(\theta,\ \phi)$ is the signal's steering vector, and

$$\mathbf{a}(\theta, \phi) = \left[\mathrm{e}^{-\mathrm{j}\,\mathbf{u}^{\mathrm{T}}\mathbf{p}_1},\quad \mathrm{e}^{-\mathrm{j}\,\mathbf{u}^{\mathrm{T}}\mathbf{p}_2},\quad \ldots,\quad \mathrm{e}^{-\mathrm{j}\,\mathbf{u}^{\mathrm{T}}\mathbf{p}_M}\right]^{\mathrm{T}} \tag{2.9}$$

Fig. 2.5 Received signal by an arbitrary array

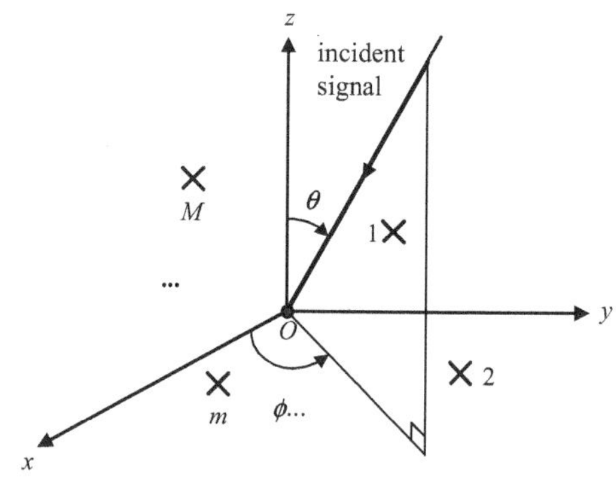

where

$$\mathbf{u} = \frac{2\pi}{\lambda}\begin{bmatrix} \sin\theta\,\cos\phi \\ \sin\theta\,\sin\phi \\ \cos\theta \end{bmatrix} \tag{2.10}$$

is the wave number vector.

$$\mathbf{p}_m = [x_m,\ \ y_m,\ \ z_m]^{\mathrm{T}}, \quad m = 1, 2, \ldots, M \tag{2.11}$$

is the three-dimensional position vector of the mth array element.

For array signal processing, a uniform linear array is a relatively common form of array structure. For convenience, unless otherwise specified, in this chapter we use a uniform linear array as the example to discuss adaptive array processing algorithm. For a uniform linear array, we usually can assume the first array element as a reference point, and then we have

$$\mathbf{a}(\theta) = \left[1,\ \ \mathrm{e}^{\frac{-\mathrm{j}2\pi d\,\sin\theta}{\lambda}},\ \ \ldots,\ \ \mathrm{e}^{\frac{-\mathrm{j}2\pi(M-1)d\,\sin\theta}{\lambda}}\right]^{\mathrm{T}} \tag{2.12}$$

where θ represents the angle between the incident signal and the array's normal line; d represents the distance between two array elements; λ represents the wavelength of the incident signal.

If a desired signal and Q jamming signals incident on a uniform linear array, and their arrival angles are θ_0 and θ_q $(q = 1, 2, \ldots, Q)$ respectively, then the array's received signal model becomes

$$\mathbf{x}(t) = \mathbf{a}(\theta_0)s_0(t) + \sum_{q=1}^{Q}\mathbf{a}(\theta_q)j_q(t) + \mathbf{e}(t) \tag{2.13}$$

Adaptive array processing is also known as spatial domain adaptive filtering, and it is one of the most important research directions of array signal processing (the other direction is super-resolution direction of arrival estimation). Its basic principle is shown in Fig. 2.6.

In Fig. 2.6, $\mathbf{w} = [w_1, w_2, \ldots, w_M]^{\mathrm{T}}$ is an array weight vector. The weight vector performs weighted sum on the received array signals (beamforming), and the output is used as the array output signal. The key of adaptive array processing is to select a proper array weight vector to filter out the jamming signal and preserve the desired signals. Various adaptive array processing methods differ on their algorithms of calculating array weight vector. Usually we can intuitively evaluate the performance of an adaptive array processing method using array pattern. Array pattern is

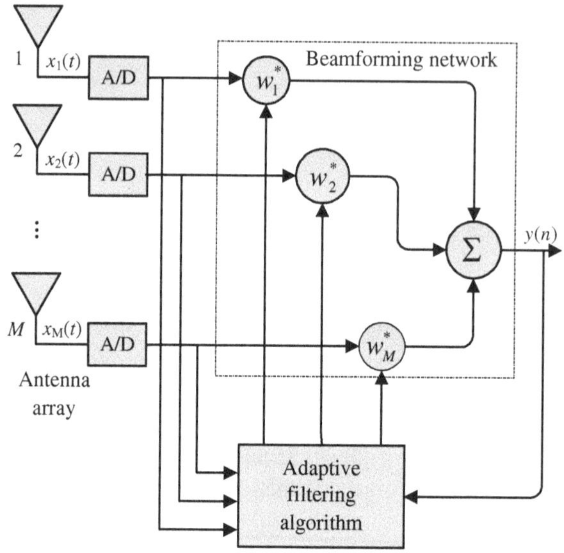

Fig. 2.6 Adaptive array processing

defined as the vector weight vector's responses to signals arrived from different angles, i.e.

$$F(\theta) = \mathbf{w}^{\mathrm{H}}\mathbf{a}(\theta) \tag{2.14}$$

2.3.2 *Statistical Optimal Beamforming*

Statistical optimal beamforming, under the assumption of accurately knowing the received array signal's second order statistics, calculates an array's weight vector to adaptively form null steering towards the direction of jamming. Therefore it can guarantee the reception of desired signals. However, in actual application, an array signal's second order statistics usually cannot be obtained based on finite snapshot data, therefore an optimal weight vector cannot be obtained. The beamforming based on finite snapshot data is usually called adaptive beamforming. Statistical optimal beamforming is an analysis tool for adaptive array signal processing, and it provides a theoretical foundation for implementing adaptive beamforming.

The statistical optimal beamforming problem can be summarized as solving for an optimal weight vector based on certain criteria. Usually these criteria are Maximum Signal to Interference add Noise Ratio (MSINR), Minimum Mean Square Error (MMSE), and Minimum Noise Variance (MNV). It can be proved that these three criteria are equivalent under certain conditions [28].

1. Maximum SINR Criterion

Let **w** represents the array weight array, and $y(t)$ represents array output, then

$$y(t) = \mathbf{w}^{\mathrm{H}}\mathbf{x}(t) \tag{2.15}$$

By substituting (2.13) into (2.15), we can obtain the desired signal and the residue jamming plus noise from the array output.

$$y_s(t) = \mathbf{w}^{\mathrm{H}}\mathbf{a}(\theta_0)s_0(t) \tag{2.16}$$

$$y_{j+e}(t) = \mathbf{w}^{\mathrm{H}}\left(\sum_{q=1}^{Q}\mathbf{a}(\theta_q)j_q(t) + \mathbf{e}(t)\right) \tag{2.17}$$

Then the corresponding output powers are

$$\sigma_{os}^2 = E\left\{y_s(t)y_s^*(t)\right\} = \mathbf{w}^{\mathrm{H}}\sigma_s^2\mathbf{a}(\theta_0)\mathbf{a}^{\mathrm{H}}(\theta_0)\mathbf{w} \triangleq \mathbf{w}^{\mathrm{H}}\mathbf{R}_s\mathbf{w} \tag{2.18}$$

$$\sigma_{o(j+e)}^2 = E\left\{y_{j+e}(t)y_{j+e}^*(t)\right\} = \mathbf{w}^{\mathrm{H}}\left(\sum_{q+1}^{Q}\sigma_q^2\mathbf{a}(\theta_q)\ \mathbf{a}^{\mathrm{H}}(\theta_q) + \sigma_e^2\mathbf{I}\right)\mathbf{w} \triangleq \mathbf{w}^{\mathrm{H}}\mathbf{R}_{j+e}\mathbf{w} \tag{2.19}$$

In (2.18) and (2.19), $\mathbf{R}_s$ and $\mathbf{R}_{j+e}$ are covariance matrices for the desired signal and jamming plus noise; σ_s^2 is the desired signal power; σ_q^2 is the power of the qth jamming signal. The maximum SINR Criterion looks for array weight vector, to maximize the ratio between the array output signal power and the output jamming plus noise power, i.e.

$$\max_{\mathbf{w}}\frac{\mathbf{w}^{\mathrm{H}}\mathbf{R}_s\mathbf{w}}{\mathbf{w}^{\mathrm{H}}\mathbf{R}_{j+e}\mathbf{w}} \tag{2.20}$$

where $\max\limits_{\mathbf{w}}[\bullet]$ represents the selected weight value that maximizes the $[\bullet]$. Using the Lagrange multiplier method, we can derive that the solution for (2.20) is equivalent to the solution for the following generalized eigenvalue problem, i.e.

$$\mathbf{R}_s\mathbf{w} = \lambda\mathbf{R}_{j+e}\mathbf{w} \tag{2.21}$$

Therefore, the optimum weight vector $\mathbf{w}_{opt}$ is the corresponding eigenvector for the maximum eigenvalue in (2.21). It can be proved that, the solution of (2.21) is equivalent to

$$\mathbf{w}_{opt} = \gamma\mathbf{R}_{j+e}^{-1}\mathbf{a}(\theta_0) \tag{2.22}$$

where γ is a constant factor.

2. Minimum Mean Square Error Criterion

Minimum mean square error criterion uses a reference signal to solve for the array weight vector, i.e. to select a weight vector to minimize the mean square error between the reference signal and the array output

$$\min_{\mathbf{w}} E\left\{\left|d(t) - \mathbf{w}^{\mathrm{H}}\mathbf{x}(t)\right|^2\right\} \tag{2.23}$$

where $d(t)$ is the reference signal. We define the covariance matrix of the array received data as $\mathbf{R}_x = E\{\mathbf{x}(t)\mathbf{x}^{\mathrm{H}}(t)\}$, then the solution (2.23) is

$$\mathbf{w}_{opt} = \mathbf{R}_x^{-1}\mathbf{r}_{xd} \tag{2.24}$$

where $\mathbf{r}_{xd} = E\{\mathbf{x}(t)d * (t)\}$ is the cross-correlation between the array received signal and the reference signal. Equation (2.24) is the solution of Wiener-Hoff equation in matrix form and it is the optimal Weiner solution.

If the useful signal $s_0(t)$ in (2.13) is used as the reference signal $d(t)$, since the desired signal, jamming signal and noise are not correlated to each other, then

$$\mathbf{r}_{xd} = \mu\mathbf{a}(\theta_0) \tag{2.25}$$

where $\mu = E\{s_0(t)d^*(t)\}$. By substituting (2.25) into (2.24), we can have

$$\mathbf{w}_{opt} = \mu\mathbf{R}_x^{-1}\mathbf{a}(\theta_0) \tag{2.26}$$

3. Minimum Noise Variance Criterion

Minimum noise variance criterion selects the optimal weight vector by minimizing the array output power. To avoid the case of $\mathbf{w}_{opt} = 0$, some corresponding constraints need to be added. A usual constraint is to guarantee that desired signals pass without distortion, i.e. the array's response on the desired signal is a constant. Therefore, the cost function for the minimum noise variance criterion is

$$\begin{aligned} &\min_{\mathbf{w}} \mathbf{w}^{\mathrm{H}}\mathbf{R}_x\mathbf{w} \\ &s.t.\ \ \mathbf{w}^{\mathrm{H}}\mathbf{a}(\theta_0) = 1 \end{aligned} \tag{2.27}$$

By using the method of Lagrange multiplier, the solution for (2.27) is

$$\mathbf{w}_{opt} = \frac{1}{\mathbf{a}^{\mathrm{H}}(\theta_0)\mathbf{R}_x^{-1}\mathbf{a}(\theta_0)}\mathbf{R}_x^{-1}\mathbf{a}(\theta_0) \tag{2.28}$$

The optimal beamformer expressed by (2.28) is the Minimum Variance Distortless Response (MVDR) beamformer, also known as Capon beamformer.

To compare (2.26) and (2.28), it can be seen that the minimum mean square error criterion and the minimum noise variance criterion differ only by a constant factor. This value does not have impact on an array output's SINR, so that the minimum mean square error criterion and minimum noise variance criterion are equivalent to each other. Based on the matrix inverse lemma it can be proved that, under the conditions of accurately knowing the desired signal steering vector and covariance matrix, the minimum noise variance criterion and the maximum SINR criterion are equivalent, therefore the three statistically optimal beamforming criteria are equivalent to each other.

2.3.3 Power Minimization Algorithm

To compare the three criteria in Sect. 2.3.3, it can be seen that they all need to know the directions of the desired signal or a known reference signal. Therefore, if the two criteria of MSINR and MNV can be used for GNSS jamming mitigation, the directions of the GNSS signals need to be known so that the implementation is relatively complex. If the MMSE criterion is used, then the GNSS signal needs to be known.

Using the GPS system as an example, when the GPS signal and Q jamming signals incident on an M-element uniform linear array, the received signal at time t is

$$\mathbf{x}(t) = \sum_{l=1}^{L} \mathbf{a}(\theta_l)s_l(t) + \sum_{q=1}^{Q} \mathbf{a}(\theta_q)j_q(t) + \mathbf{e}(t) \tag{2.29}$$

where $s_l(t)$ represents the lth GNSS signal $(l = 1, 2, \ldots, L)$; θ_l is the lth GPS signal's direction of arrival; $\mathbf{a}(\theta_l)$ is the corresponding array steering vector; $j_q(t)$ represents the qth jamming signal $(q = 1, 2, \ldots, Q)$; θ_q is the qth jamming signal's direction of arrival; $\mathbf{a}(\theta_q)$ is its corresponding array steering vector; $\mathbf{e}(t)$ represents the thermal noise vector of the receiver, and is usually an additive Gaussian white noise vector with zero mean and σ_e^2 variance, assuming that the GNSS signal, jamming signal and noise do not correlate with each other.

Since the GNSS signal is very weak and buried in the noise (usually around 20 dB lower than the noise), we only need to consider jamming mitigation, i.e. forming a null steering on the direction of jamming without considering where the direction of the beam is pointing. This is the power minimization algorithm that has been widely applied for GNSS jamming mitigation. The algorithm calculates array weight vector by minimizing the output power, but prevents the weight vector from having a zero solution by guaranteeing that the reference antenna's signal in the antenna array has no distorted output, i.e.

$$\begin{aligned} &\min_{\mathbf{w}} \mathbf{w}^{\mathrm{H}} \mathbf{R}_x \mathbf{w} \\ &s.t.\ \mathbf{w}^{\mathrm{H}} \boldsymbol{\delta}_M = 1 \end{aligned} \tag{2.30}$$

where $\boldsymbol{\delta}_M = [1, 0, \ldots, 0]^{\mathrm{T}}$ is a $M \times 1$ dimension vector. The solution for (2.30) can be obtained using the method of Lagrange multiplier

$$\mathbf{w}_{opt} = \frac{1}{\boldsymbol{\delta}_M^{\mathrm{H}} \mathbf{R}_x^{-1} \boldsymbol{\delta}_M} \mathbf{R}_x^{-1} \boldsymbol{\delta}_M \tag{2.31}$$

where $\mathbf{R}_x$ represents the theoretical array covariance matrix, in reality it usually is replaced by the sample covariance matrix defined as

$$\hat{\mathbf{R}}_x = \frac{1}{N} \sum_{n=1}^{N} \mathbf{x}(n) \mathbf{x}^{\mathrm{H}}(n) \tag{2.32}$$

where N represents the number of snapshots.

Power minimization algorithm can adaptively form a null steering towards the jamming direction and there is no need to know the direction of the GNSS signal, and no need to know the array manifold. Therefore this algorithm belongs to the category of blind adaptive beamforming. The implementation is simple and it is compatible with common receivers. But the algorithm can only suppress jamming, and cannot aim the antenna pattern's main lobe towards the GNSS signal. Consequently, it cannot provide signal processing gain brought by the antenna array, and cannot enhance the GNSS signal's carrier-to-noise ratio. The carrier-to-noise ratio is a very important parameter for the receiver, and the receiver's positioning accuracy is highly correlated with this parameter [29]. When the carrier-to-noise ratio is low, the position accuracy is usually worse.

2.4 Spatial Domain Adaptive Filtering Using Periodic Repetitive Characteristic of GNSS Signals

The power minimization algorithm is easy to implement, but cannot improve the carrier-to-noise ratio of a GNSS signal. In this section, on the foundation of the GNSS signal's periodic repetitive characteristic, we discuss the algorithm that uses the signal processing gain brought by antenna arrays to improve GNSS signal carrier-to-noise ratio.

2.4.1 Periodic Repetitive Characteristic of GNSS Signals

The C/A code of GPS signal is a Gold code with 1.023 Mcps chip rate, with a 1 ms chip period, i.e. every period contains 1023 chips. For a civilian GPS signal, the C/A code modulates the D code, and the D code chip rate is 50 Hz. Therefore within a D code period, 20 repetitive cycles of the C/A code exist. Since a D code does not change within one period, the corresponding 20 spread spectrum codes have the same structure, i.e. showing the characteristic of periodic repetition. The periodic repetition characteristic of the C/A code for a GPS signal is a special case of signal cyclostationarity. Below we introduce the signal's cyclostationarity characteristic first.

For a random signal $x(t)$, if for a certain time delay τ, the correlation value between $x(t)$ and the obtained signal after frequency shift $x(t)$ by α is not 0, then we can say that $x(t)$ has cyclostationarity characteristic [30], i.e.

$$r_{xx}^{\alpha}(\tau) \triangleq \frac{\left\langle x(t)\left[x(t-\tau)e^{j2\pi\alpha t}\right]^{*}\right\rangle_{\infty}}{\sqrt{\left\langle |x(t)|^{2}\right\rangle_{\infty}\left\langle |x(t-\tau)e^{j2\pi\alpha t}|^{2}\right\rangle_{\infty}}} = R_{xx}^{\alpha}(\tau)/R_{xx}(0) \neq 0 \tag{2.33}$$

where symbol $\langle\rangle_{\infty}$ represents the average of infinite time observation domain; α is defined as the cycle frequency; $R_{xx}^{\alpha}(\tau)$ is a cycle correlation function of the signal $x(t)$ that can be defined as $R_{xx}^{\alpha}(\tau) = \left\langle x(t)\left[x(t-\tau)e^{j2\pi\alpha t}\right]^{*}\right\rangle_{\infty}$, and $R_{xx}(0) = \left\langle |x(t)|^{2}\right\rangle_{\infty}$ is the signal power.

Similarly, if for a certain time delay τ, the correlation value between $x(t)$ and the signal obtained by frequency shifting its conjugate signal by α is not 0, then $x(t)$ has conjugate cyclostationary characteristic, i.e.

$$r_{xx^*}^{\alpha}(\tau) \triangleq \frac{\left\langle x(t)\left[x^*(t-\tau)e^{j2\pi\alpha t}\right]^{*}\right\rangle_{\infty}}{\sqrt{\left\langle |x(t)|^{2}\right\rangle_{\infty}\left\langle |x^*(t-\tau)e^{j2\pi\alpha t}|^{2}\right\rangle_{\infty}}} = R_{xx^*}^{\alpha}(\tau)/R_{xx}(0) \neq 0 \tag{2.34}$$

where $R_{xx^*}^{\alpha}(\tau)$ is the conjugate cyclic correlation function of the signal $x(t)$. To simplify, we can denote $R_{xx}^{\alpha}(\tau)$ and $R_{xx^*}^{\alpha}(\tau)$ in a unified way:

$$R_{xu} = \begin{cases} R_{xx}^{\alpha}(\tau) & \text{when } u(t) = x(t-\tau)e^{j2\pi\alpha t} \\ R_{xx^*}^{\alpha}(\tau) & \text{when } u(t) = x^*(t-\tau)e^{j2\pi\alpha t} \end{cases} \tag{2.35}$$

Communication signals are usually assumed to be cyclostationary signals. GPS signals adopt a spread spectrum communication system that has the cyclostationary property as well. Based on what was described before, since the C/A code has periodic repetitions, we can regard the cycle frequency as $\alpha = 0$, and τ is an integer

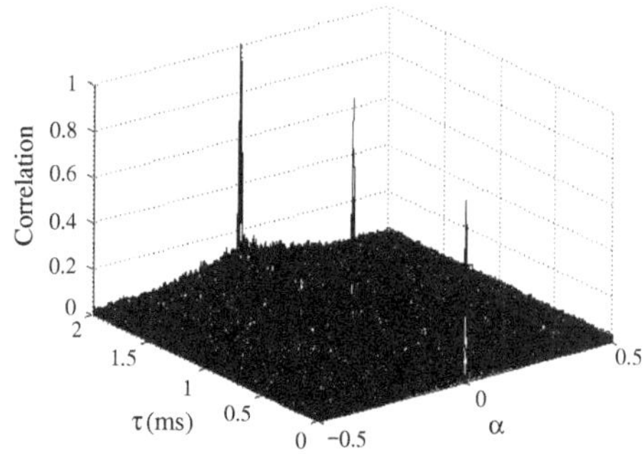

Fig. 2.7 GPS C/A code signal cyclostationary correlation diagram

multiple of the C/A code period (1 ms). This is a special case of cyclostationary signal. Its cycle correlation function is shown in Fig. 2.7.

For array signal processing, in terms of a $M \times 1$ dimension signal vector $\mathbf{x}(t)$, the cyclostationarity correlation matrix and the conjugate periodic stationary correlation matrix are defined as

$$\mathbf{R}_{xu} = \begin{cases} \mathbf{R}_{xx}^{\alpha}(\tau), & \text{when } \mathbf{u}(t) = \mathbf{x}(t-\tau)e^{j2\pi\alpha t} \\ \mathbf{R}_{xx^*}^{\alpha}(\tau), & \text{when } \mathbf{u}(t) = \mathbf{x}^*(t-\tau)e^{j2\pi\alpha t} \end{cases} \tag{2.36}$$

2.4.2 SCORE Algorithm

The R&D team, led by Professor Moeness at Villanova university in the US, proposed the Self-COherence Restoral (SCORE) jamming mitigation algorithm using periodic repetition characteristic of C/A codes [14, 31]. The schematic of the algorithm is shown in Fig. 2.8.

As shown in Fig. 2.8, the SCORE algorithm is composed of two beamformers. The main channel beamformer $\mathbf{w}$ processes the data block signal $\mathbf{x}(t)$, while the auxiliary channel beamformer $\mathbf{f}$ processes the reference block signal $x(t-pT)$. The reference block signal is the data after delaying the data block signal by p $(1 \le p \le 19)$ C/A code periods. Since a navigation data bit includes 20 C/A code periods, we can assume that the data block data and the reference data block are located inside the same navigation data bit. The structure can be shown in Fig. 2.9.

When only one GNSS signal is considered, (2.29) becomes

$$\mathbf{x}(t) = \mathbf{a}(\theta_0)s(t) + \sum_{q=1}^{Q} \mathbf{a}(\theta_q)j_q(t) + \mathbf{e}(t) \tag{2.37}$$

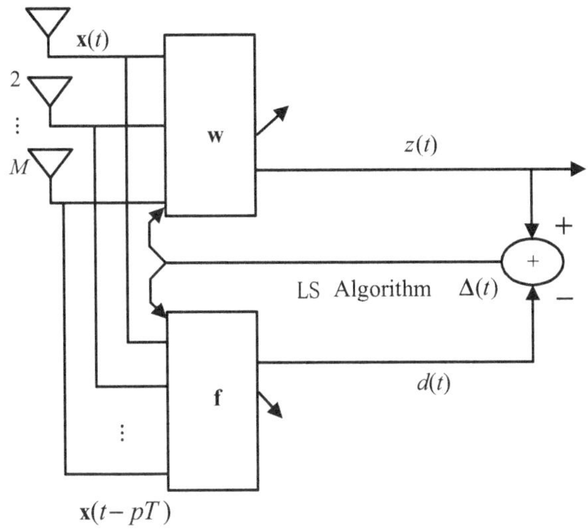

Fig. 2.8 SCORE algorithm

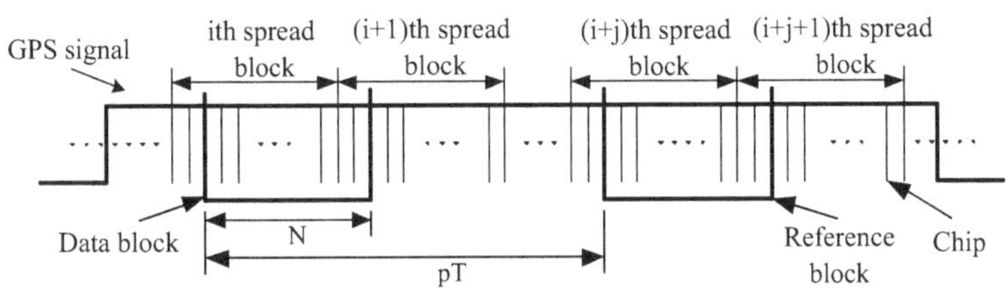

Fig. 2.9 Data block and reference block

SCORE algorithm assumes that the carrier waves have been synchronized completely (i.e. without the impacts from Doppler frequency shift), and the navigation data bit does not change (assuming $D(t) = 1$), then (2.37) becomes

$$\mathbf{x}(t) = A\mathbf{a}(\theta_0)c(t-\tau) + \sum_{q=1}^{Q} \mathbf{a}(\theta_q)j_q(t) + \mathbf{e}(t) \tag{2.38}$$

Let the data after delaying p $(1 \le p \le 19)$ C/A code periods equals $\mathbf{u}(t)$, then

$$\mathbf{u}(t) = \mathbf{x}(t-pT) = A\mathbf{a}(\theta_0)c(t-\tau-pT) + \sum_{q=1}^{Q} \mathbf{a}(\theta_q)j_q(t-pT) + \mathbf{e}(t-pT) \tag{2.39}$$

Within a navigational data bit, the C/A code has the characteristic of periodic repetition, i.e.

$$c(t-\tau-pT)=c(t-\tau) \tag{2.40}$$

By substituting (2.40) into (2.39), we have

$$\mathbf{u}(t)=A\mathbf{a}(\theta_0)c(t-\tau)+\sum_{q=1}^{Q}\mathbf{a}(\theta_q)j_q(t-pT)+\mathbf{e}(t-pT) \tag{2.41}$$

In Fig. 2.8, the respective output signals for the main and auxiliary channel beamformers are:

$$z(t)=\mathbf{w}^{\mathrm{H}}\mathbf{x}(t) \tag{2.42}$$

$$d(t)=\mathbf{f}^{\mathrm{H}}\mathbf{u}(t) \tag{2.43}$$

where $\mathbf{w}$ and $\mathbf{f}$ represent the weight vectors for the main channel and auxiliary channel respectively. We can define

$$R_{zd}=E\{z(t)d^*(t)\}=\mathbf{w}^{\mathrm{H}}E\{\mathbf{x}(t)\mathbf{u}^{\mathrm{H}}(t)\}\mathbf{f} \tag{2.44}$$

$$R_{zz}=E\{z(t)z^*(t)\}=\mathbf{w}^{\mathrm{H}}E\{\mathbf{x}(t)\mathbf{x}^{\mathrm{H}}(t)\}\mathbf{w} \tag{2.45}$$

$$R_{dd}=E\{d(t)d^*(t)\}=\mathbf{f}^{\mathrm{H}}E\{\mathbf{u}(t)\mathbf{u}^{\mathrm{H}}(t)\}\mathbf{f} \tag{2.46}$$

Since the GNSS signal, jamming signal, and noise are independent from each other, then the received data's covariance matrix is

$$\mathbf{R}_x=E\{\mathbf{x}(t)\mathbf{x}^{\mathrm{H}}(t)\}=\mathbf{R}_s+\mathbf{R}_j+\mathbf{R}_e \tag{2.47}$$

where $\mathbf{R}_s$, $\mathbf{R}_j$ and $\mathbf{R}_e$ represent the covariance matrices for GNSS signal, jamming signal, and noise respectively. Similarly, the covariance matrix for the reference signal is

$$\mathbf{R}_u=E\{\mathbf{u}(t)\mathbf{u}^{\mathrm{H}}(t)\}=\mathbf{R}_x \tag{2.48}$$

Since jamming signals and noises have no periodic repetition characteristic, the cross-covariance matrix between the received data and the reference data is

$$\mathbf{R}_{xu}=E\{\mathbf{x}(t)\mathbf{u}^{\mathrm{H}}(t)\}=\mathbf{R}_s=\sigma_s^2\mathbf{a}(\theta_0)\mathbf{a}^{\mathrm{H}}(\theta_0) \tag{2.49}$$

where σ_s^2 represents the GNSS signal's power. Let $\Delta(t)$ represents the difference between the main channel's output signal and the reference signal, i.e.

$$\Delta(t) = z(t) - d(t) \tag{2.50}$$

By fixing $\mathbf{w}$, $E\left\{|\Delta(t)|^2\right\}$ can be minimized in the sense of least squares

$$\mathbf{f}_{LS} = \mathbf{R}_u^{-1}\mathbf{r}_{uz} \tag{2.51}$$

where

$$\mathbf{r}_{uz} = E\{\mathbf{u}(t)z^*(t)\} = \mathbf{R}_{xu}^{\mathrm{H}}\mathbf{w} \tag{2.52}$$

Similarly, by fixing $\mathbf{f}$, we have

$$\mathbf{w}_{LS} = \mathbf{R}_x^{-1}\mathbf{R}_{xu}^{\mathrm{H}}\mathbf{f} \tag{2.53}$$

SCORE algorithm maximizes the cross-correlation between the main channel's output signal and auxiliary channel's output signal in order to select the weight vector $\mathbf{w}$, then we have the following cost function

$$\max J(\mathbf{w},\mathbf{f}) = \frac{|R_{zd}|^2}{R_{zz}R_{dd}} = \frac{\left|\mathbf{w}^{\mathrm{H}}\mathbf{R}_{xu}\mathbf{f}\right|^2}{\left[\mathbf{w}^{\mathrm{H}}\mathbf{R}_x\mathbf{w}\right]\left[\mathbf{f}^{\mathrm{H}}\mathbf{R}_u\mathbf{f}\right]} \tag{2.54}$$

By using $\mathbf{f}_{LS}$, $\mathbf{w}_{LS}$ to substitute $\mathbf{f}$ and $\mathbf{w}$ in (2.54), we can have

$$J(\mathbf{w},\mathbf{f}_{LS}) = \frac{\mathbf{w}^{\mathrm{H}}\mathbf{R}_{xu}\mathbf{R}_x^{-1}\mathbf{R}_{xu}^{\mathrm{H}}\mathbf{w}}{\mathbf{w}^{\mathrm{H}}\mathbf{R}_x\mathbf{w}} \tag{2.55}$$

$$J(\mathbf{f},\mathbf{w}_{LS}) = \frac{\mathbf{f}^{\mathrm{H}}\mathbf{R}_{xu}\mathbf{R}_x^{-1}\mathbf{R}_{xu}^{\mathrm{H}}\mathbf{f}}{\mathbf{f}^{\mathrm{H}}\mathbf{R}_x\mathbf{f}} \tag{2.56}$$

It can be proved that, to maximize (2.55) and (2.56) is equivalent to solve the corresponding eigenvectors for the following maximum generalized eigenvalue problem, i.e.

$$\mathbf{R}_x\mathbf{w} = \lambda\mathbf{R}_{xu}\mathbf{R}_x^{-1}\mathbf{R}_{xu}^{\mathrm{H}}\mathbf{w} \tag{2.57}$$

$$\mathbf{R}_x\mathbf{f} = \kappa\mathbf{R}_{xu}\mathbf{R}_x^{-1}\mathbf{R}_{xu}^{\mathrm{H}}\mathbf{f} \tag{2.58}$$

where λ and κ are eigenvalues.

By substituting (2.49) into (2.57), we can have

$$\mathbf{R}_x\mathbf{w} = \gamma\mathbf{R}_s\mathbf{w} \tag{2.59}$$

where γ is the eigenvalue. It can be known from (2.59) that the solution of (2.57) converges to the maximum signal-to-noise ratio criterion. Therefore the SCORE algorithm can make the array pattern's main beam point towards the direction of GNSS signal. It can be known from the derivation process of the SCORE algorithm that SCORE does not need to know any priori information on the array. Therefore, it is a blind adaptive beamforming technology. Nevertheless, SCORE algorithm only considers the scenario of one GNSS signal. Actually, a receiver can perform positioning calculation by receiving at least 4 or more GNSS signals. When L GNSS signals are received, (2.49) becomes

$$\mathbf{R}_{xu} = \sum_{l=1}^{L} \sigma_{sl}^2 \mathbf{a}(\theta_l)\mathbf{a}^{\mathrm{H}}(\theta_l) \tag{2.60}$$

where σ_{sl}^2 represents the power of the $l\,(l = 1, 2, \ldots, L)$th GNSS signal; $\mathbf{a}(\theta_l)$ represents the steering vector of the lth GNSS signal. Then, (2.57) has larger generalized eigenvalues. Consequently, the eigenvector corresponding to the largest generalized eigenvalue cannot ensure that the mainlobe of the pattern points to every GNSS signal's direction. Therefore it cannot guarantee that the receiver can capture all GNSS signals.

2.4.3 *Single-Channel Single-Delay Crosscorrelation Algorithm*

For the convenience of narration, (2.29) is re-written as below:

$$\mathbf{x}(t) = \sum_{l=1}^{L} \mathbf{a}(\theta_l)s_l(t) + \sum_{q=1}^{Q} \mathbf{a}(\theta_q)j_q(t) + \mathbf{e}(t) \tag{2.61}$$

Since the period of a navigation message is much larger than the period of C/A code, within the weight vector computation time we can assume that navigation message does not change. For the convenience of narration, we set $D_l(t) = 1$. The single channel single delay cross-correlation algorithm [18] takes advantage of the periodic repetitive characteristic of spread spectrum code to estimate the GNSS signal direction, then performs beamforming, thereby increases a GNSS signal's carrier-to-noise ratio.

1. Jamming Mitigation Based on Subspace Projection Technique

When jamming exists, it is hard to estimate a GNSS signal's direction only relying on the periodic repetition characteristic of the spread spectrum code. Therefore the jamming signal must be suppressed first. Since the GNSS signal is very weak, the array's covariance matrix is determined mainly by jamming and noise, i.e.

$$\mathbf{R}_x \approx \mathbf{R}_j + \mathbf{R}_e \tag{2.62}$$

Since jamming and noise are not correlated with each other, then we have

$$\mathbf{R}_x \approx \sum_{q=1}^{Q} \sigma_q^2 \mathbf{a}(\theta_q)\, \mathbf{a}^{\mathrm{H}}(\theta_q) + \sigma_e^2 \mathbf{I}_{M\times M} \tag{2.63}$$

where $\mathbf{I}_{M\times M}$ is an $M \times M$ order unit array. To perform eigenvalue decomposition on $\mathbf{R}_x$, we can have

$$\mathbf{R}_x = \sum_{m=1}^{M} \lambda_m \mathbf{u}_m \mathbf{u}_m^{\mathrm{H}} \tag{2.64}$$

where λ_m and $\mathbf{u}_m (m = 1, 2, \ldots, M)$ are $\mathbf{R}_x$'s eigenvalues and the corresponding eigenvectors, and the λ_m are in descending order. Assuming that all array element noises are independent from each other and have equal power, we can have

$$\lambda_1 \geqslant \lambda_2 \geqslant \cdots \geqslant \lambda_Q \gg \lambda_{Q+1} = \cdots = \lambda_M = \sigma_e^2 \tag{2.65}$$

Then (2.64) turns into

$$\mathbf{R}_x = \sum_{m=1}^{Q} (\lambda_m - \sigma_e^2) \mathbf{u}_m \mathbf{u}_m^{\mathrm{H}} + \sigma_e^2 \mathbf{I}_{M\times M} \tag{2.66}$$

By comparing (2.63) and (2.66), we know that the eigenvectors corresponding to the array covariance matrix's Q largest eigenvalues $\mathbf{u}_m$ $(m = 1, 2, \ \ldots, Q)$ can be expanded as the jamming subspace. So the jamming subspace is

$$\mathbf{U}_{\mathrm{J}} = span\{\mathbf{u}_1, \quad \mathbf{u}_2, \quad \ldots, \quad \mathbf{u}_Q\} \tag{2.67}$$

Then its orthogonal complement space is

$$\mathbf{U}_{\mathrm{J}}^{\perp} = \mathbf{I} - \mathbf{U}_{\mathrm{J}} \mathbf{U}_{\mathrm{J}}^{\mathrm{H}} \tag{2.68}$$

By projecting the data received by the array into this space, the jamming signal can be mitigated. The data after projection becomes

$$\mathbf{y}(t) = \mathbf{U}_{\mathrm{J}}^{\perp} \mathbf{x}(t) = \mathbf{U}_{\mathrm{J}}^{\perp} \sum_{l=1}^{L} \mathbf{a}(\theta_l) s_l(t) + \mathbf{U}_{\mathrm{J}}^{\perp} \mathbf{e}(t) \tag{2.69}$$

where $\mathbf{y}(t) = [y_1(t), y_2(t), \ldots, y_M(t)]^{\mathrm{T}}$ represents the output data on the array channels after projection. Since GPS adopts CDMA spread spectrum, it has certain jamming mitigation capabilities (determined by spread spectrum gain). Therefore when we need to adopt the specialized spatial domain adaptive jamming mitigation algorithm to suppress jamming, we usually assume that the jamming power is much higher than the receiver's thermal noise power. When the noise and the signal are not correlated, the received signal's covariance matrix inversion is proportional to the jamming subspace's orthogonal complement space [32]. The detailed derivation procedure is described as below.

Based on (2.64), we have

$$\mathbf{R}_x^{-1} = \sum_{m=1}^{M} \frac{1}{\lambda_m} \mathbf{u}_m \mathbf{u}_m^{\mathrm{H}} \tag{2.70}$$

Equation (2.70) can be derived from (2.65)

$$\begin{aligned} \mathbf{R}_x^{-1} &= \sum_{m=1}^{Q} \frac{1}{\lambda_m} \mathbf{u}_m \mathbf{u}_m^{\mathrm{H}} + \frac{1}{\sigma_e^2} \sum_{m=Q+1}^{M} \mathbf{u}_m \mathbf{u}_m^{\mathrm{H}} \\ &= \frac{1}{\sigma_e^2} \left(\mathbf{I} - \sum_{m=1}^{Q} \frac{\lambda_m - \sigma_e^2}{\lambda_m} \mathbf{u}_m \mathbf{u}_m^{\mathrm{H}} \right) \end{aligned} \tag{2.71}$$

Then we have

$$\lim_{\sigma_e^2 \to 0} \sigma_e^2 \mathbf{R}_x^{-1} = \mathbf{U}_{\mathrm{J}}^{\perp} \tag{2.72}$$

Consequently, we can use a receiver signal's covariance matrix inversion $\mathbf{R}_x^{-1}$ to replace jamming subspace's orthogonal complement space. Then there is no need to perform eigenvalue decomposition, therefore it can reduce the computation complexity. Equation (2.69) turns into

$$\mathbf{y}(t) = \sum_{l=1}^{L} \bar{\mathbf{a}}(\theta_l) A_l c_l(t - \tau - T)\, \mathrm{e}^{-\mathrm{j}\omega_{dl}(t-T)} + \bar{\mathbf{e}}(t) \tag{2.73}$$

where

$$\bar{\mathbf{a}}(\theta_l) = \mathbf{R}_x^{-1} \mathbf{a}(\theta_l) \tag{2.74}$$

$$\bar{\mathbf{e}}(t) = \mathbf{R}_x^{-1} \mathbf{e}(t) \tag{2.75}$$

2. GNSS Signal Direction Estimation Using Periodic Repetition Characteristic of Spreading Code

It can be known from (2.73) that the GNSS signal is still buried by the noise after projection. If a navigation receiver acquires and tracks the projected data directly, it is equivalent to the power minimization algorithm. Therefore, to improve the carrier-to-noise ratio of a GNSS signal and increase positioning precision, the signal processing gain provided by antenna arrays must be fully leveraged. An effective solution is to produce multiple antenna beams, and make every beam point to a GNSS signal. This forms a null steering towards the jamming directions. The key for such an approach is to first estimate the direction of arrival of every GNSS signal, and then directly use the traditional high resolution DOA (Direction of Arrival) estimation algorithm [33, 34] (e.g. MUltiple SIgnal Classification (MUSIC) and Estimation of Signal Parameters via Rotational Invariance Technique (ESPRIT)). The approach faces two problems: ① The number of satellites could be greater than the number of array elements; ② The GNSS signal is too weak. These two problems result in the difficulty to acquire GNSS signal subspace using the above DOA estimation algorithm. Consequently the characteristic of the GNSS signal must be fully considered to study novel GNSS signal DOA estimation algorithm.

Within the period of a data bit, the C/A code signal repeats itself periodically 20 times, which is also known as the periodic repetition characteristic. Therefore, for the lth GNSS signal, we have

$$c_l(t-\tau_l-pT)=c_l(t-\tau_l)\quad 1\leqslant p\leqslant 19 \tag{2.76}$$

where p is the delayed number of periods, obtained by delaying the projected signal from the first antenna channel (usually serves as the reference antenna) with one C/A code period, as shown in Fig. 2.10.

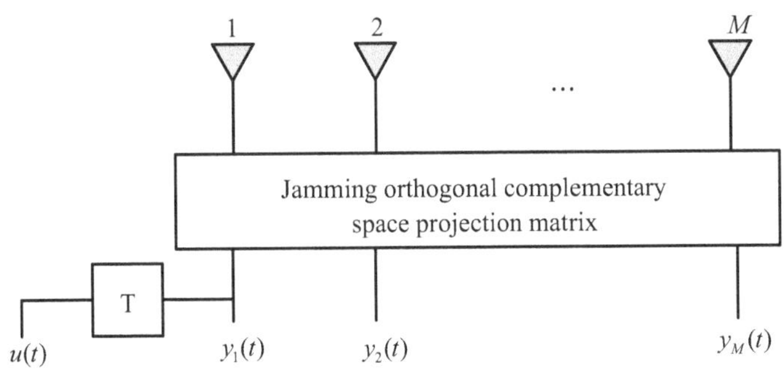

Fig. 2.10 The diagram of reference signal for single channel single delay cross correlation algorithm

Assuming that the delayed signal and the projected signal are located in the same data bit and can be expressed as $u(t)$, then we have

$$u(t) = y_1(t-T) = \sum_{l=1}^{L} \bar{\mathbf{a}}_1(\theta_l) A_l c_l(t-\tau_l-T) \mathrm{e}^{-\mathrm{j}\omega_{dl}(t-T)} + \bar{\mathbf{e}}_1(t-T) \tag{2.77}$$

where $\bar{\mathbf{a}}_1(\theta_l)$ is the first element of $\bar{\mathbf{a}}(\theta_l)$; $\bar{\mathbf{e}}_1(t)$ is the projected noise data of the first array channel. By making the p in (2.76) as 1 and substitute it into (2.77), we have

$$u(t) = \sum_{l=1}^{L} \bar{A}_l c_l(t-\tau_l) e^{-\mathrm{j}\omega_{dl}(t-T)} + \bar{\mathbf{e}}_1(t-T) \tag{2.78}$$

where $\bar{A}_l = \bar{\mathbf{a}}_1(\theta_l) A_l$. Since different C/A codes are independent from each other, the cross correlation between the projected signal $\mathbf{y}(t)$ and the delayed signal $u(t)$ is

$$\mathbf{r} = \sum_{l=1}^{L} \beta_l \bar{\mathbf{a}}(\theta_l) \tag{2.79}$$

where $\beta_l = \sigma_{s_l}^2 \mathrm{e}^{-\mathrm{j}\omega_{dl}T} \bar{\mathbf{a}}_1^*(\theta_l)$ is an unknown complex number. To make

$$\boldsymbol{\beta} = [\beta_1, \quad \beta_2, \quad \ldots, \quad \beta_L]^{\mathrm{T}} \tag{2.80}$$

$$\boldsymbol{\theta} = [\theta_1, \quad \theta_2, \quad \ldots, \quad \theta_L]^{\mathrm{T}} \tag{2.81}$$

$$\mathbf{A} = [\bar{\mathbf{a}}(\theta_1), \quad \bar{\mathbf{a}}(\theta_2), \quad \ldots, \quad \bar{\mathbf{a}}(\theta_L)] \tag{2.82}$$

we could re-write (2.79) in the matrix form below

$$\mathbf{r} = \mathbf{A}\boldsymbol{\beta} \tag{2.83}$$

By using sample cross correlation vector $\hat{\mathbf{r}}$ to replace the $\mathbf{r}$ in (2.83), and minimizing the following cost function g to estimate $\boldsymbol{\beta}$ and $\boldsymbol{\theta}$.

$$g\left(\{\theta_l, \beta_l\}_{l=1}^{L}\right) = \|\hat{\mathbf{r}} - \mathbf{A}\boldsymbol{\beta}\|_2^2 \tag{2.84}$$

We expand (2.84) to have

$$\begin{aligned} g\left(\{\theta_l, \beta_l\}_{l=1}^{L}\right) &= (\hat{\mathbf{r}} - \mathbf{A}\boldsymbol{\beta})^{\mathrm{H}} (\hat{\mathbf{r}} - \mathbf{A}\boldsymbol{\beta}) \\ &= \left[\boldsymbol{\beta} - \left(\mathbf{A}^{\mathrm{H}}\mathbf{A}\right)^{-1}\mathbf{A}^{\mathrm{H}}\hat{\mathbf{r}}\right]^{\mathrm{H}} \left(\mathbf{A}^{\mathrm{H}}\mathbf{A}\right) \left[\boldsymbol{\beta} - \left(\mathbf{A}^{\mathrm{H}}\mathbf{A}\right)^{-1}\mathbf{A}^{\mathrm{H}}\hat{\mathbf{r}}\right] \\ &\quad + \hat{\mathbf{r}}^{\mathrm{H}}\hat{\mathbf{r}} - \hat{\mathbf{r}}^{\mathrm{H}}\mathbf{A}\left(\mathbf{A}^{\mathrm{H}}\mathbf{A}\right)^{-1}\mathbf{A}^{\mathrm{H}}\hat{\mathbf{r}} \end{aligned} \tag{2.85}$$

It can be known from (2.85), to minimize (2.85) is equivalent to make the first term of (2.85) zero and maximizing its last term, i.e.

$$\hat{\boldsymbol{\theta}} = \arg\max_{\boldsymbol{\theta}} \left[\hat{\mathbf{r}}^{\mathrm{H}}\mathbf{A}\left(\mathbf{A}^{\mathrm{H}}\mathbf{A}\right)^{-1}\mathbf{A}^{\mathrm{H}}\hat{\mathbf{r}}\right] \tag{2.86}$$

Correspondingly we have

$$\hat{\boldsymbol{\beta}} = \left(\mathbf{A}^{\mathrm{H}}\mathbf{A}\right)^{-1}\mathbf{A}^{\mathrm{H}}\hat{\mathbf{r}}\Big|_{\boldsymbol{\theta}=\hat{\boldsymbol{\theta}}} \tag{2.87}$$

The cost function in (2.86) has complex multi-peak shape, so it is hard to solve directly. The usual method to solve the problem is to use the loop optimization algorithm, e.g. CLEAN algorithm [35]. The CLEAN algorithm originated from solving astronomical image processing problems in radio astronomy. Since then, it has been widely applied in the fields of microwave imaging, image processing and spectral estimation [36–38]. Based on the characteristic of CLEAN algorithm (computational simplicity), it is used to solve Nonlinear Least Squares (NLS) problems in (2.86). Before the CLEAN algorithm is given, the following preparatory works are performed:

Given $\left\{\hat{\theta}_l, \hat{\beta}_l\right\}_{l=1,\, l\neq k}^{L}$ is known or has been estimated, we define

$$\hat{\mathbf{r}}_k = \hat{\mathbf{r}} - \sum_{\substack{l=1 \\ l\neq k}}^{L} \hat{\beta}_l \hat{\bar{\mathbf{a}}}\left(\hat{\theta}_l\right) \tag{2.88}$$

We also define

$$g_k(\theta_k, \beta_k) = \|\hat{\mathbf{r}}_k - \beta_k \bar{\mathbf{a}}(\theta_k)\|_2^2 \tag{2.89}$$

By minimizing $g_k(\theta_k, \beta_k)$, we can obtain

$$\hat{\theta}_k = \arg\max_{\theta_k} \left[\frac{\left|\bar{\mathbf{a}}^H(\theta_k)\hat{\mathbf{r}}_k\right|^2}{\bar{\mathbf{a}}^H(\theta_k)\bar{\mathbf{a}}(\theta_k)}\right] \tag{2.90}$$

$$\hat{\beta}_k = \frac{\hat{\bar{\mathbf{a}}}^H\left(\hat{\theta}_k\right)\hat{\mathbf{r}}_k}{\hat{\bar{\mathbf{a}}}^H\left(\hat{\theta}_k\right)\hat{\bar{\mathbf{a}}}\left(\hat{\theta}_k\right)} \tag{2.91}$$

Below we list the computation steps of the CLEAN algorithm

(1) Assuming $L = 1$, let $\hat{\mathbf{r}}_1 = \hat{\mathbf{r}}$, (2.90) and (2.91) can be used to estimate $\left\{\hat{\theta}_1, \hat{\beta}_1\right\}$.

(2) Assuming $L = 2$, $\hat{\mathbf{r}}_2$ can be estimated from (2.88) using $\left\{\hat{\theta}_1, \hat{\beta}_1\right\}$, then $\hat{\mathbf{r}}_2$ can be used to estimate $\left\{\hat{\theta}_2, \hat{\beta}_2\right\}$ from (2.90) and (2.91).

(3) Assuming $L = 3$, $\left\{\hat{\theta}_l, \hat{\beta}_l\right\}_{l=1}^{2}$ can be used to estimate $\hat{\mathbf{r}}_3$ from (2.88), then $\hat{\mathbf{r}}_3$ is used to estimate $\left\{\hat{\theta}_3, \hat{\beta}_3\right\}$ from (2.90) and (2.91).

The above steps are repeated, until L equals to the pre-determined number of satellites. Then the signal parameter estimates for all the satellites and $\left\{\hat{\theta}_l, \hat{\beta}_l\right\}_{l=1}^{L}$ can be obtained.

In addition to the CLEAN algorithm, the RELAX algorithm proposed by Li et al. [39–41] is also a type of loop optimization algorithm (also known as a method based on decoupled parameter estimation theory). The RELAX algorithm, compared to the CLEAN algorithm, is more accurate and has higher resolution, but its computation is more complex [42]. If an array antenna receives two GNSS signals with close direction of arrivals, the RELAX algorithm can accurately distinguish the two GNSS signals while the CLEAN algorithm cannot distinguish between the two signals, and it can only treat the two signals as one signal for processing. Nevertheless, this is not a problem for GNSS, because under such a scenario, the pattern's mainlobe obtained using the estimated direction of arrivals of the GNSS signals using the CLEAN algorithm can point to the two GNSS signals simultaneously, thus it still can improve the carrier-to-noise ratio of the two GNSS signals.

3. Beamforming

After estimating directions of arrivals for all GNSS signals, these estimated values can be used to perform beamforming. The array weight vector for the lth GNSS signal is

$$\mathbf{w}_l = \mathbf{R}_x^{-1}\mathbf{a}\left(\hat{\theta}_l\right) \tag{2.92}$$

Since the new algorithm uses the array's single channel single delay data to estimate GNSS signal direction of arrival, the algorithm is named single channel single delay cross correlation algorithm. The output signal of the lth beam is

$$z_l(t) = \mathbf{w}_l^{\mathrm{H}}\mathbf{x}(t) \tag{2.93}$$

The block diagram of the single channel single delay cross correlation algorithm is shown in Fig. 2.11.

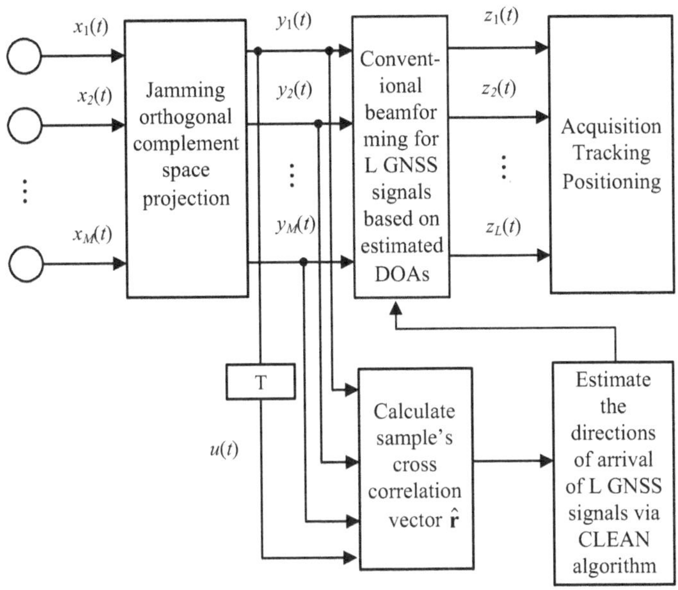

Fig. 2.11 Block diagram of the single channel single delay cross correlation algorithm

4. Cross Correlation Vector Estimation

The single channel single delay cross correlation algorithm needs to calculate the sample cross correlation vector $\hat{\mathbf{r}}$ between the projected data and the data after periodic delay. The respective definitions of the $M \times N$ order sample data matrix after projection and the $1 \times N$ order periodic delay sample data vector are

$$\mathbf{Y}_N = [\mathbf{y}(n), \quad \mathbf{y}(n-1), \quad \ldots, \quad \mathbf{y}(n-(N-1))] \tag{2.94}$$

$$\mathbf{u}_N = [u(n), \quad u(n-1), \quad \ldots, \quad u(n-(N-1))] \tag{2.95}$$

Then we have

$$\hat{\mathbf{r}} = \frac{1}{N}\mathbf{Y}_N\mathbf{u}_N^{\mathrm{H}} \tag{2.96}$$

It can be known from (2.86) that the cross correlation vector estimate $\hat{\mathbf{r}}$ determines the performance of the single channel single delay algorithm. Based on the assumptions in this chapter, $\mathbf{y}(t)$ and $u(t)$ are located in the same navigation message data bit, and consequently $\hat{\mathbf{r}}$ is an effective estimation. Nevertheless, since the data sample is obtained through random sampling, there is no guarantee that $\mathbf{y}(t)$ and $u(t)$ are located in the same data bit. When both of them are located in neighboring data bits, then how should we estimate $\hat{\mathbf{r}}$? Towards this question, Ref. [43] proposed a solution by defining the following events:

H_1: $\mathbf{y}(t)$ and $u(t)$ are located in the same data bit.

H_2: $\mathbf{y}(t)$ and $u(t)$ are located in the neighboring data bits.

H_{21}: $\mathbf{y}(t)$ 和 $u(t)$ are located in the neighboring data bits with the same navigation data bit

H_{22}: $\mathbf{y}(t)$ 和 $u(t)$ are located in the neighboring data bits with the opposite navigation data bit.

The probabilities for the above events to happen are:

$$prob\{H_1\} = 1 - \frac{p}{20} \tag{2.97}$$

$$prob\{H_2\} = \frac{p}{20} \tag{2.98}$$

$$prob\{H_{21}\} = \frac{p}{40} \tag{2.99}$$

$$prob\{H_{22}\} = \frac{p}{40} \tag{2.100}$$

Therefore we have

$$\begin{aligned} \mathbf{r}_{yu} &= E\{y(t)u^*(t)|H_1\}prob\{H_1\} + E\{y(t)u^*(t)|H_2\}prob\{H_2\} \\ &= E\{y(t)u^*(t)|H_1\}prob\{H_1\} + E\{y(t)u^*(t)|H_{21}\}prob\{H_{21}\} \\ &\quad + E\{y(t)u^*(t)|H_{22}\}prob\{H_{22}\} \end{aligned} \tag{2.101}$$

Since

$$E\{y(t)u^*(t)|H_1\} = E\{y(t)u^*(t)|H_{21}\} = \mathbf{r} \tag{2.102}$$

$$E\{y(t)u^*(t)|H_{22}\} = -\mathbf{r} \tag{2.103}$$

where $\mathbf{r} = \mathbf{A}\boldsymbol{\beta}$ represents the cross correlation between the projected GNSS signal and the periodically delayed signal [see (2.83)]. By substituting (2.97)–(2.100), (2.102) and (2.103) into (2.101), we can have

$$\mathbf{r}_{yu} = \left(1 - \frac{p}{20}\right)\mathbf{r} \tag{2.104}$$

For the single channel single delay cross correlation vector method, we have $p = 1$, then

$$\mathbf{r}_{yu} = \frac{19}{20}\mathbf{r} \approx \mathbf{r} \tag{2.105}$$

It can be seen from the above discussion that, when the impact of the navigation data bit is considered, the cross correlation between $\mathbf{y}(t)$ and $u(t)$ is close to $\mathbf{r}$. Thus (2.96) can be used as the estimate of $\mathbf{r}$. In addition, to fully take advantage of the periodic repetition characteristic of the C/A code, the G projected data block, and periodic repetition data block can be selected, and the cross correlation vector's estimate becomes

$$\hat{\mathbf{r}}_G = \frac{1}{G}\sum_{g=1}^{G}\frac{\mathbf{Y}_N(g)\mathbf{u}_N^{\mathrm{H}}(g)}{N} \tag{2.106}$$

It can be proved that [31]

$$E\{\hat{\mathbf{r}}_G\} = \mathbf{r}_{yu} \tag{2.107}$$

i.e. $\hat{\mathbf{r}}_G$ is $\mathbf{r}_{yu}$'s unbiased estimate.

2.4.4 *Multiple Channel Single Delay Cross Correlation Algorithm*

Since the single channel single delay cross correlation algorithm estimates the GNSS signal's direction of arrival using the cross correlation between the projected data and the reference array element's single delay data after subspace projection, it does not fully take advantage of single delay data of the other array elements. When the array element fails, the algorithm might fail as well. Consequently, let $u_i(t)$ ($i = 1, 2, \ldots, M$) represent the signal obtained by delaying the projected data $\mathbf{y}(t)$'s ith antenna channel by one C/A code period, as shown in Fig. 2.12.

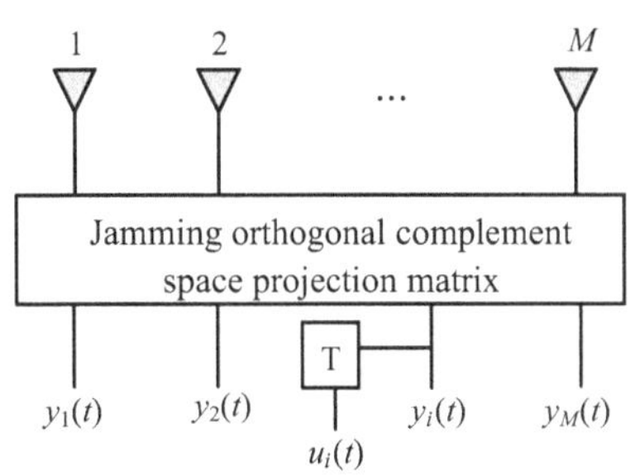

Fig. 2.12 The diagram of reference signal for multiple channels single delay cross correlation algorithm

Then we have

$$u_i(t) = y_i(t-T) = \sum_{l=1}^{L} \bar{\mathbf{a}}_i(\theta_l) A_l c_l(t-\tau-T) \mathrm{e}^{-\mathrm{j}\omega_{dl}(t-T)} + \bar{\mathbf{e}}_i(t-T) \quad (2.108)$$

where $\bar{\mathbf{a}}_i(\theta_l)$ is $\bar{\mathbf{a}}(\theta_l)$'s ith elements; $\bar{\mathbf{e}}_i(t)$ represents the noise data of the ith array channel after projection. By setting the p in (2.76) as 1, and substituting it into (2.108), we have

$$u_i(t) = \sum_{l=1}^{L} \bar{\mathbf{a}}_i(\theta_l) A_l c_l(t-\tau) \mathrm{e}^{-\mathrm{j}\omega_{dl}(t-T)} + \bar{\mathbf{e}}_i(t-T) \quad (2.109)$$

Due to independence among different C/A codes, the cross correlation between the projected signal $\mathbf{y}(t)$ and the delayed signal $u_i(t)$ is

$$\mathbf{r}_i = \sum_{l=1}^{L} \tilde{\beta}_l \bar{\mathbf{a}}_i^*(\theta_l) \bar{\mathbf{a}}(\theta_l) \quad (2.110)$$

where $\tilde{\beta}_l = \sigma_{s_l}^2 \mathrm{e}^{-\mathrm{j}\omega_{dl}T}$. By constructing a new vector $\tilde{\mathbf{r}}$

$$\tilde{\mathbf{r}} = \begin{bmatrix} \mathbf{r}_1^{\mathrm{T}}, & \mathbf{r}_2^{\mathrm{T}}, & \ldots, & \mathbf{r}_M^{\mathrm{T}} \end{bmatrix}^{\mathrm{T}} \quad (2.111)$$

By substituting (2.110) into (2.111) and after simplication, we can have

$$\tilde{\mathbf{r}} = \sum_{l=1}^{L} \tilde{\beta}_l \bar{\mathbf{a}}^*(\theta_l) \otimes \bar{\mathbf{a}}(\theta_l) \quad (2.112)$$

where $\otimes$ denotes Kronecker product. Let

$$\tilde{\mathbf{a}}(\theta_l) = \bar{\mathbf{a}}^*(\theta_l) \otimes \bar{\mathbf{a}}(\theta_l) \quad (2.113)$$

Equation (2.112) becomes

$$\tilde{\mathbf{r}} = \sum_{l=1}^{L} \tilde{\beta}_l \tilde{\mathbf{a}}(\theta_l) \quad (2.114)$$

Let $\tilde{\mathbf{A}} = \begin{bmatrix} \tilde{\mathbf{a}}(\theta_1) & \tilde{\mathbf{a}}(\theta_2) & \cdots & \tilde{\mathbf{a}}(\theta_L) \end{bmatrix}$, (2.113) can be written in matrix form:

$$\tilde{\mathbf{r}} = \tilde{\mathbf{A}}\tilde{\beta} \quad (2.115)$$

By substituting the $\tilde{\mathbf{r}}$ in (2.115) with the sample cross correlation vector $\hat{\tilde{\mathbf{r}}}$, and minimizing the following cost function f to estimate $\tilde{\boldsymbol{\beta}}$ and $\boldsymbol{\theta}$.

$$f\left(\left\{\theta_l, \tilde{\beta}_l\right\}_{l=1}^{L}\right) = \left\|\hat{\tilde{\mathbf{r}}} - \tilde{\mathbf{A}}\tilde{\boldsymbol{\beta}}\right\|_2^2 \tag{2.116}$$

Equations (2.116) and (2.84) have the same structure, so they can be solved using the CLEAN algorithm described in the Sect. 2.4.3 in this chapter. Here we list an estimation formula when there is one single GNSS signal, i.e.

$$\hat{\theta}_1 = \arg\max_{\theta}\left[\frac{\left|\tilde{\mathbf{a}}^{\mathrm{H}}(\theta)\hat{\tilde{\mathbf{r}}}\right|^2}{\tilde{\mathbf{a}}^{\mathrm{H}}(\theta)\tilde{\mathbf{a}}(\theta)}\right] \tag{2.117}$$

$$\hat{\tilde{\beta}}_1 = \frac{\tilde{\mathbf{a}}^{\mathrm{H}}\left(\hat{\theta}_1\right)\hat{\tilde{\mathbf{r}}}}{\tilde{\mathbf{a}}^{\mathrm{H}}\left(\hat{\theta}_1\right)\tilde{\mathbf{a}}\left(\hat{\theta}_1\right)} \tag{2.118}$$

For parameter estimations for other GNSS signals, we can use similar approaches like the CLEAN algorithm, to convert it to a series of single GNSS signal parameter estimation problems. The block diagram of multiple channel single delay cross correlation algorithm is shown in Fig. 2.13.

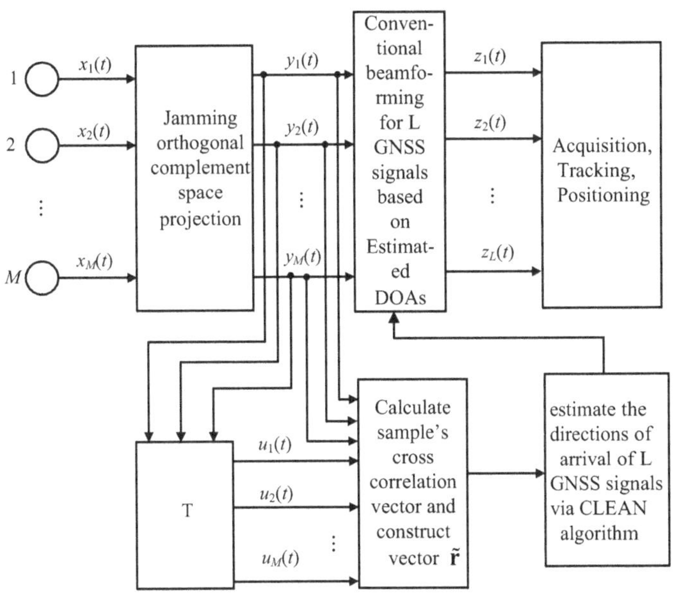

Fig. 2.13 Block diagram of multiple channels single delay cross correlation algorithm

In this section, we discuss two types of adaptive spatial domain filtering algorithm: single channel single delay algorithm and multiple channel single delay algorithm based on GNSS signal's periodic repetition characteristic. It can also be expanded to single channel multiple delays cross correlation algorithm, but two-dimensional search is needed and the computational load is heavy. Compared with the single channel single delay algorithm, multiple channel single delay algorithm takes full advantage of the data from various channels in the array, so it performs better and is more robust. For example, it is not sensitive to a fault in the channel data.

2.4.5 Simulation Results

In this section, we use simulations to verify the effectiveness of the algorithms. For simulations, the antenna array has a 10-element uniform linear array with an interval of a half wavelength. 4 GPS GNSS signals, PRN1,PRN2, PRN3, and PRN20 incident on the array from the directions of 0°, −55°, −20° and 30°, and one jamming signal incidents on the array from the direction of 50°. The jamming-to-noise ratio is 40 dB, and the signal-to-noise ratio is −20 dB. The digital IF frequency for the received signal by the array is 4.309 MHz, and the sampling rate is 5.714 MHz.

1. Adaptive Array Pattern Comparisons

Figure 2.14a–d are the adaptive array patterns for power minimization algorithm, SCORE algorithm, single channel single delay cross correlation algorithm, and multiple channel single delay cross correlation algorithm. The vertical lines in various subfigures represent the incoming GNSS signal directions, and the vertical dashed lines represent the incoming direction of the jamming signal. By comparing Fig. 2.14a–d, we can see that all these methods can suppress jamming. But the array pattern generated by the SCORE algorithm only forms main lobes on the directions of the PRN1 and PRN3 GNSS signals, and for the PRN2 and PRN20 GNSS signals it even has a suppressive effect. The power minimization method cannot provide gains on any satellite. The two new algorithms discussed in this chapter can generate multiple beams, and can ensure that every beam's mainlobe always aims towards the direction of one GNSS signal, so the array pattern with high gain can be obtained.

2. Comparisons of Acquisition Results

Figure 2.15 shows the acquisition results after applying the SCORE algorithm for jamming mitigation. In Fig. 2.15, the x-coordinate represents the GPS satellite's PRN number, and the y-coordinate represents the normalized acquisition factor, which is defined as the ratio between the maximum and the second maximum normalized correlation peaks in the receiver acquisition module. Figure 2.15 shows

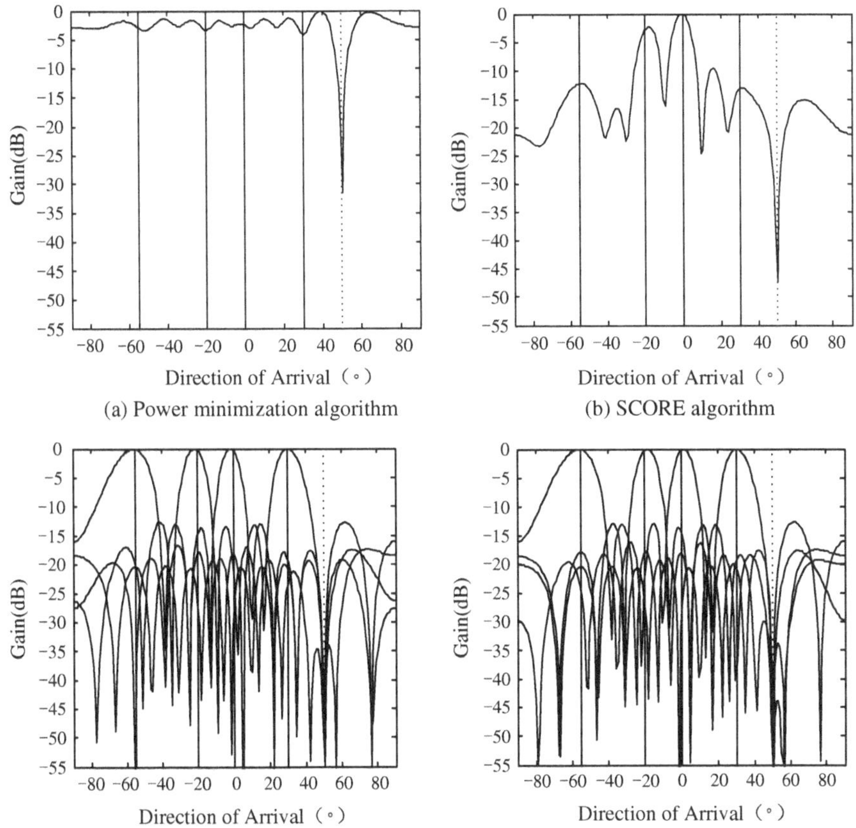

(a) Power minimization algorithm

(b) SCORE algorithm

(c) Single Channel Single Delay Cross Correlation

(d) Multiple Channel Single Delay Cross Correlation.

Fig. 2.14 Comparisons on adaptive array patterns obtained by different methods

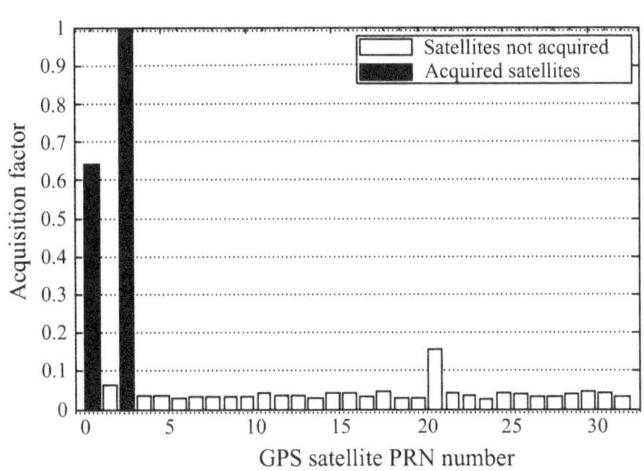

Fig. 2.15 Acquisition results of SCORE algorithm

that the algorithm only acquires two GNSS signals (PRN1 and PRN3). This is because the pattern formed based on this method cannot accurately aim towards every GNSS signal, and it can even degrade some GNSS signals (PRN2 and PRN20). As a result the receiver cannot acquire all GNSS signals.

Figure 2.16 shows the acquisition results using the single channel single delay cross correlation algorithm for jamming mitigation. The multiple channel single delay cross correlation algorithm's simulation results are similar to the single channel single delay cross correlation algorithm, and as a result we cannot draw any redundant graphs for that algorithm. Figure 2.16 and the plots for all the other acquisition have the same coordinate meanings as the Fig. 2.15. Therefore the descriptions are not repeated. It can be seen in Fig. 2.16 that every beam generated by the new algorithm points to a GNSS signal and as a result the receiver can acquire GNSS signals. Since the new algorithm does not use the same weight vector to generate multiple beams, the receiver has to acquire the output of every beam separately, thus we have four plots of acquisition results. It can be seen that the new algorithm is not directly compatible with regular receivers. A receiver adopting the new algorithm needs to perform acquisition, tracking and solve parameters such as

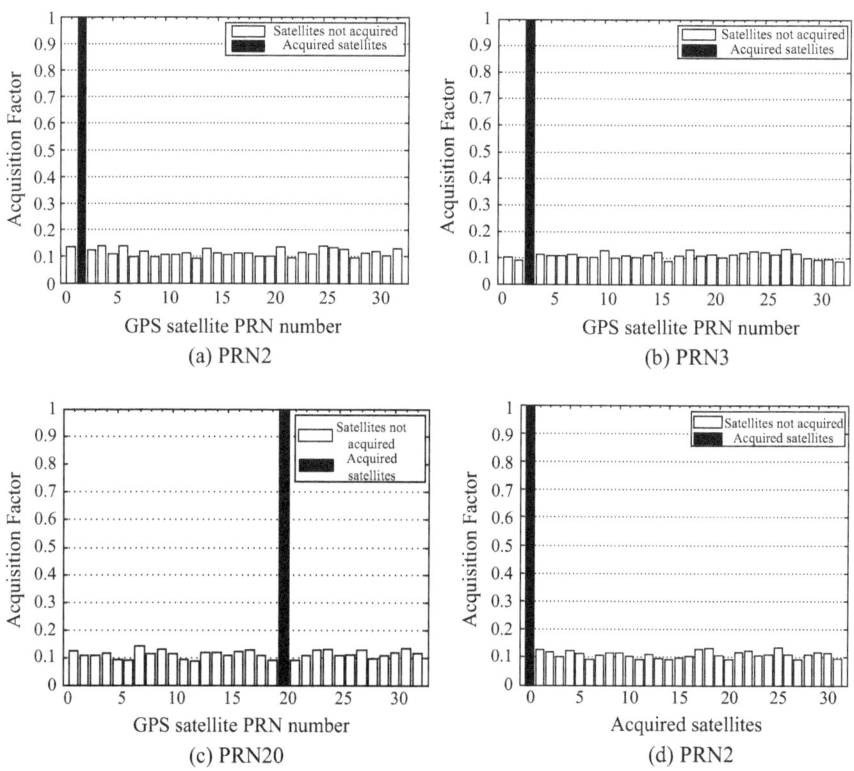

Fig. 2.16 Single channel single delay cross correlation algorithm's acquisition results

pseudo-range on every individual beam, then combine these information to perform positioning calculation. Its procedures to perform acquisition, tracking, pseudo-range calculation, and positioning calculation are the same as regular receivers, and can be implemented using software receivers.

Figure 2.17 compares the acquisition results on the same GNSS signals using the single channel single delay cross correlation algorithm and the power minimization algorithm. Among the subfigures, Fig. 2.17a corresponds to the power minimization algorithm; Fig. 2.17b corresponds to the single channel single delay cross correlation algorithm; Fig. 2.17c corresponds to the multiple channels single delay cross correlation algorithm. It can be seen in Fig. 2.17, after jamming mitigation, even though the power minimization algorithm can acquire the GNSS signal, the sidelobe level of the normalized cross correlation coefficient is rather large. These two algorithms take advantage of the GNSS signal's periodic repetition

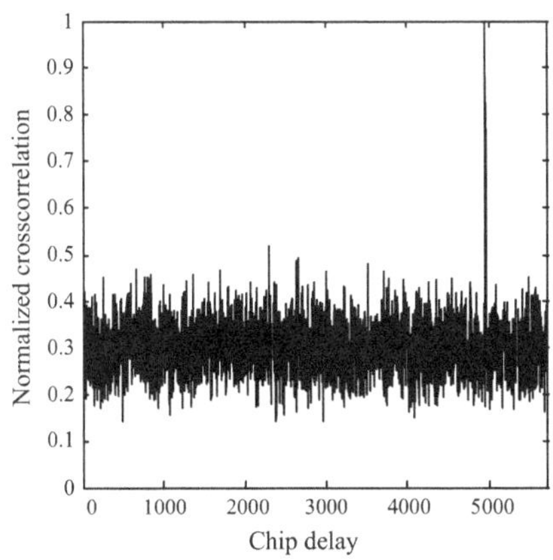

(a) Power minimization algorithm

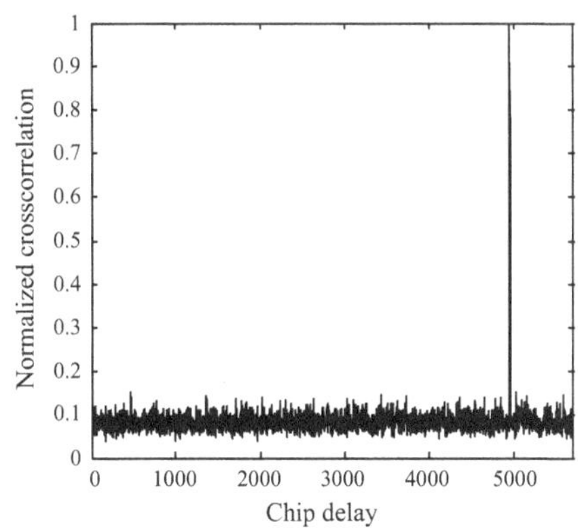

(b) Single channel single delay cross correlation algorithm

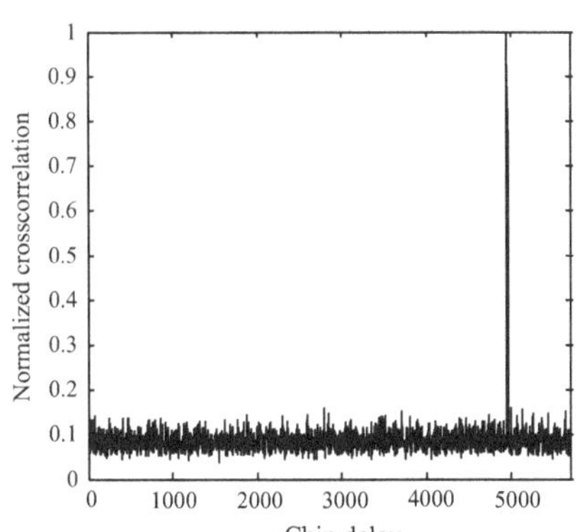

(c) Multiple channel single delay cross correlation algorithm

Fig. 2.17 Comparisons of acquisition results

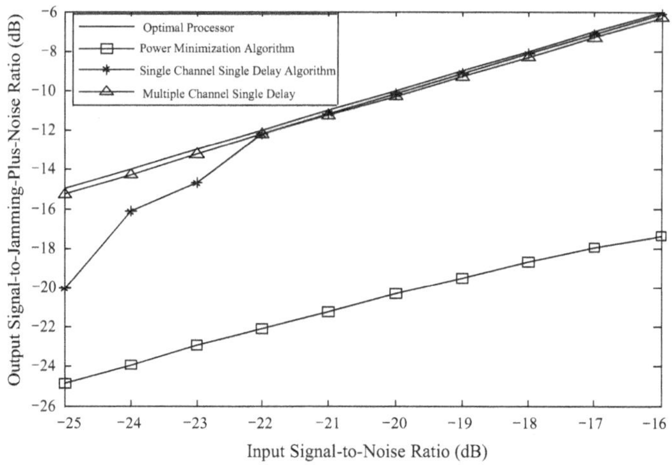

Fig. 2.18 Comparisons of output signal-jamming-plus-noise curves for different algorithms

characteristic to aim the antenna mainlobe towards the direction of the GNSS signal, and consequently its acquisition result is much better than that of the power minimization algorithm.

3. Comparisons of output signal-jamming-plus-noise ratio

Figure 2.18 compares the output signal-jamming-plus-noise ratio curves of the single channel single delay cross correlation algorithm and the power minimization algorithm. Among them, the solid line is the curve for the optimal processor, the "*" line is the curve for the single channel single delay cross correlation algorithm, the "□" line is the curve for the power minimization algorithm, and the "Δ" line is the curve for the multiple channel single delay cross correlation algorithm. It can be seen in Figure 2.18 that since the power minimization algorithm has no beam pointing direction, it cannot improve output signal-jamming-plus-noise ratio, i.e. it cannot improve a GNSS signal's carrier-to-noise ratio. But the single channel single delay cross correlation algorithm can increase a GNSS signal's carrier-to-noise ratio. And as the input signal-to-noise ratio increases, the output signal-jamming-plus-noise ratio can approach the theoretical optimum value.

2.5 Spatial Domain Adaptive Filtering Using Known Spreading Code Information

The adaptive jamming mitigation algorithm using a GNSS signal's periodic repetition characteristic can generate multiple beams aiming towards the GNSS signals. But array manifold information is needed to estimate the directions of the GNSS signals. If the array has errors such as amplitude error, phase error, or array position error, the algorithm performance degrades. In essence, the adaptive jamming mitigation algorithm based on the GNSS signal's periodic repetition characteristic is equivalent to the MMSE criterion described in Sect. 2.3.3. In this section, from the perspective of MMSE criterion, we discuss the algorithm for GNSS signal reconstruction using the de-spread re-spread technique when given known spread spectrum code, and performing beamforming using the reconstructed signal as the reference signal. The algorithm does not need to know the array manifold and the directions of the GNSS signals, so it is very robust. This type of algorithm originated from the blind adaptive jamming mitigation algorithm for the CDMA system, and the most typical variant of the algorithm is the least-squares de-spread re-spread multi-target array [43–46].

2.5.1 *Least-Squares De-spread Re-spread Multi-target Array*

The least-squares de-spread re-spread multi-target array using spread spectrum information [47] is a blind jamming mitigation algorithm aiming towards CDMA systems. This method uses the spread spectrum codes for different users in the CDMA system to obtain the weight vector of a multi-target beamformer, thereby forming multiple beams. To be more specific, the correlations between different users' received signal and spread spectrum code are calculated, to estimate every user's bit information. Then, the spread spectrum code is used again to perform re-spread on the estimated bit information, and the signal after the re-spread is used as the reference signal for the adaptive beamformer to refresh the weight vector. Since GPS signals use CDMA modulation as well, the de-spread and re-spread techniques can be used for GPS jamming mitigation. But compared with other types of CDMA signals, GPS signals are very weak, so the correlator might fail when jamming exists. Consequently, this characteristic has to be fully considered when the de-spread re-spread algorithm is applied. The block diagram for least-squares de-spread re-spread algorithm is shown in Fig. 2.19.

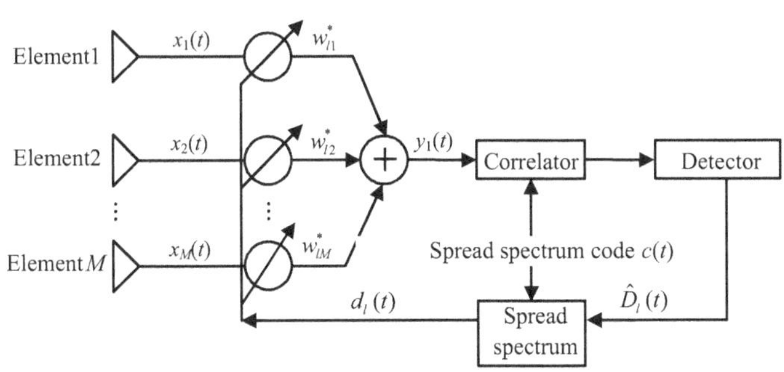

Fig. 2.19 Block diagram for least-squares de-spread re-spread algorithm

In Fig. 2.19, $y_l(t)$ denotes the array output of the lth GNSS signal; $\mathbf{w}_l = [w_{l1}, w_{l2}, \ldots, w_{lM}]^{\mathrm{T}}$ is the array weight vector for the lth GNSS signal; $d_l(t)$ is the re-spread signal of the lth GNSS signal. The Least Squares De-spread Re-spread (LS-DR) determines the optimum weight vector $\mathbf{w}_l$ by minimizing the cost function in (2.119)

$$F(\mathbf{w}_l) = \sum_{n=1}^{N} |y_l(n) - d_l(n)|^2 = \sum_{n=1}^{N} \left|\mathbf{w}_l^{\mathrm{H}}\mathbf{x}(n) - d_l(n)\right|^2 \tag{2.119}$$

where N is the number of samples, and the number of samples within a C/A code period is usually selected. The problem of minimizing (2.119) can be solved using the generalized Gauss method [44]. And the cost function can be expressed as

$$F(\mathbf{w}_l) = \sum_{n=1}^{N} g_n^2(\mathbf{w}_l) \tag{2.120}$$

where $g_n(\mathbf{w}_l) = \left|\mathbf{w}_l^{\mathrm{H}}\mathbf{x}(n) - d_l(n)\right|$, and its gradient is

$$\nabla(g_n(\mathbf{w}_l)) = \mathbf{x}(n)\frac{v_l^*(n)}{|v_l(n)|} \tag{2.121}$$

where $v_l(n) = \mathbf{w}_l^{\mathrm{H}}\mathbf{x}(n) - d_l(n)$. Based on the Gauss method, the weight vector's updating formula is

$$\mathbf{w}_l(k+1) = \mathbf{w}_l(k) - \left[\mathbf{G}(\mathbf{w}_l(k))\mathbf{G}^{\mathrm{H}}(\mathbf{w}_l(k))\right]^{-1}\mathbf{G}(\mathbf{w}_l(k))\mathbf{g}(\mathbf{w}_l(k)) \tag{2.122}$$

where

$$\begin{aligned} \mathbf{g}(\mathbf{w}_l) &= [g_1(\mathbf{w}_l), \quad g_1(\mathbf{w}_l), \quad \ldots, \quad g_N(\mathbf{w}_l)]^{\mathrm{T}} \\ &= [|v_1(n)|, \quad |v_1(n)|, \quad \ldots, \quad |v_N(n)|]^{\mathrm{T}} \end{aligned} \tag{2.123}$$

$$\mathbf{G}(\mathbf{w}_l) = [\nabla(g_1(\mathbf{w}_l)), \quad \nabla(g_2(\mathbf{w}_l)), \quad \ldots, \quad \nabla(g_N(\mathbf{w}_l))] \tag{2.124}$$

By substituting (2.121) into (2.124), we can obtain

$$\mathbf{G}(\mathbf{w}_l) = \left[\mathbf{x}(1)\tfrac{v_l^*(1)}{|v_l(1)|}, \quad \mathbf{x}(2)\tfrac{v_l^*(2)}{|v_l(2)|}, \quad \ldots, \quad \mathbf{x}(N)\tfrac{v_l^*(N)}{|v_l(N)|}\right] = \mathbf{X}\mathbf{V}_l \tag{2.125}$$

where

$$\mathbf{X} = [\mathbf{x}(1), \quad \mathbf{x}(2), \quad \ldots, \quad \mathbf{x}(N)] \tag{2.126}$$

$$\mathbf{V}_l = \mathrm{diag}\left(\left[\frac{v_l^*(1)}{|v_l(1)|}, \quad \frac{v_l^*(2)}{|v_l(2)|}, \quad \ldots, \quad \frac{v_l^*(N)}{|v_l(N)|}\right]\right) \tag{2.127}$$

Based on (2.125)–(2.127), we can obtain

$$\mathbf{G}(\mathbf{w}_l)\mathbf{G}^{\mathrm{H}}(\mathbf{w}_l) = \mathbf{X}\mathbf{V}_l\mathbf{V}_l^{\mathrm{H}}\mathbf{X}^{\mathrm{H}} = \mathbf{X}\mathbf{X}^{\mathrm{H}} \tag{2.128}$$

From (2.123) and (2.125), the following can be obtained

$$\mathbf{G}(\mathbf{w}_l)\mathbf{g}(\mathbf{w}_l) = \mathbf{X}\mathbf{V}_l\begin{bmatrix} |v_l(1)| \\ \vdots \\ |v_l(N)| \end{bmatrix} = \mathbf{X}\begin{bmatrix} v_l^*(1) \\ \vdots \\ v_l^*(N) \end{bmatrix} = \mathbf{X}\left(\mathbf{X}^{\mathrm{H}}\mathbf{w}_l - \mathbf{d}_l\right) \tag{2.129}$$

where

$$\mathbf{d}_l = \left[d_l(1), \quad d_l(2), \quad \ldots, \quad d_l(N)\right]^{\mathrm{H}} \tag{2.130}$$

By substituting (2.128) and (2.129) into (2.122), then

$$\begin{aligned} \mathbf{w}_l(k+1) &= \mathbf{w}_l(k) - \left[\mathbf{X}(k)\mathbf{X}^{\mathrm{H}}(k)\right]^{-1}\mathbf{X}(k)\left[\mathbf{X}^{\mathrm{H}}(k)\mathbf{w}_l(k) - \mathbf{d}_l(k)\right] \\ &= \left[\mathbf{X}(k)\mathbf{X}^{\mathrm{H}}(k)\right]^{-1}\mathbf{X}(k)\mathbf{d}_l(k) \end{aligned} \tag{2.131}$$

where

$$\mathbf{X}(k) = [\mathbf{x}(1+kN), \quad \mathbf{x}(2+kN), \quad \ldots, \quad \mathbf{x}(N+kN)]^{\mathrm{T}} \tag{2.132}$$

$$\mathbf{d}_l(k) = \hat{D}_l(k)[c_l(1+kN-\hat{\tau}_l), \quad c_l(2+kN-\hat{\tau}_l), \quad \ldots, \quad c_l(N+kN-\hat{\tau}_l)]^{\mathrm{T}} \tag{2.133}$$

where $\hat{D}_l(k)$ is the navigation message's estimate of the lth GNSS signal, i.e.

$$\hat{D}_l(k) = \mathrm{sgn}\left\{\mathrm{Re}\left[\sum_{k=1+kN}^{N+kN} y_l(k)c_l(k-\hat{\tau})\right]\right\} \tag{2.134}$$

where sgn is the signal function, then we have

$$\mathbf{y}_l(k) = [y_l(1+kN), \quad y_l(1+kN), \quad \ldots, \quad y_l(N+kN)]^{\mathrm{T}} = \left[\mathbf{w}_l^H(k)\mathbf{X}(k)\right]^{\mathrm{T}} \tag{2.135}$$

Based on the derivation steps of the above least squares de-spread re-spread algorithm, it can be seen that, to obtain $\hat{D}_l(k)$, we need to estimate the lth GNSS signal's $\hat{\tau}$. For a spread spectrum system, the estimation of this parameter can be regarded as a synchronization problem, and usually a correlator structure [35] is used to achieve synchronization and obtain the estimate on the τ, denoted by $\hat{\tau}$. During the synchronization process, the locally generated spread spectrum code

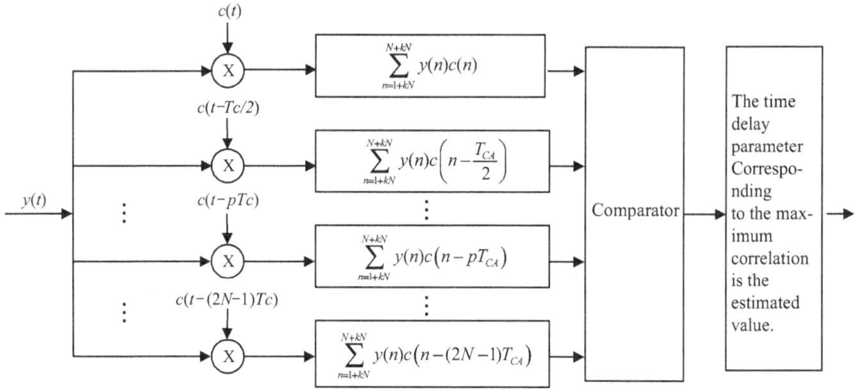

Fig. 2.20 Block diagram of estimating time delay parameters using correlators

$c_l(t)$ gradually delays by a half chip interval. Correlations are calculated between every delayed spread spectrum code sequence and the array output data $y_l(t)$. Then the outputs from every correlator are selected, and the maximum output is selected. At this point we can consider that the spread spectrum code on the branch corresponds to the maximum output and the transmitted signal achieves the coarse synchronization. The time delay of the corresponding spread spectrum chip is $\hat{\tau}$. The detailed block diagram of the correlator is shown in Fig. 2.20.

The least squares de-spread re-spread algorithm for the lth GNSS signal can be summarized as below:

(1) initialize weight vector $\mathbf{w}_l(0)$.
(2) calculate array output vector $\boldsymbol{y}_l(k)$ using (2.135).
(3) use the correlator shown in the Fig. 2.20 to estimate $\hat{\tau}$, and then use (2.134) to obtain the estimated navigation message $\hat{D}_l(k)$.
(4) use (2.133) to re-spread the navigation message, and obtain the reference signal $d_l\ (k)$.
(5) use (2.131) to update the weight vector $\mathbf{w}_l(k+1)$.
(6) repeat step (2)–step (5), until $F(\mathbf{w}_l(k+1))$ is smaller than the preset threshold, and the algorithm converges.

2.5.2 *New De-spread Re-spread Jamming Mitigation Algorithm*

The least squares de-spread re-spread algorithm described in Sect. 2.5.1 needs to have a suitable initial weight vector. The reason is that for a GNSS, when jamming exists, the correlators shown in Fig. 2.20 cannot estimate $\hat{\tau}$, so the GNSS signal cannot be reconstructed. In addition, when this algorithm re-spreads the signal, it only considers the time delay of the C/A code, but not the Doppler frequency. This greatly reduces the performance of the algorithm. Consequently we need to improve this algorithm.

It is well known that if no jamming exists, the acquisition and tracking modules of the GNSS can accurately obtain GNSS signal parameters such as chip delay, Doppler frequency, and navigation messages, i.e. it can de-spread the GNSS signal. To fully take advantage of the acquisition and tracking modules of the GNSS receiver, the jamming suppression has to be performed first. In this section we use array covariance matrix's inverse $\hat{\mathbf{R}}_x^{-1}$ to replace jamming orthogonal complement space (see Sect. 2.4.3 for details). By rewriting the signal after projection for (2.73), the signal becomes

$$\mathbf{y}(t) = \sum_{l=1}^{L} \bar{\mathbf{a}}(\theta_l)s_l(t) + \bar{\mathbf{e}}(t) \tag{2.136}$$

After removing the jamming, the receiver can acquire and track the satellite signal. Assuming that the time delay, Doppler frequency and navigation message for the lth satellite signal are $\hat{\tau}_l, \hat{\omega}_{dl}$ and $\hat{D}_l(t-\hat{\tau}_l)$ respectively, we can then obtain the reference signal after the re-spread as

$$d_l(t) = \hat{D}_l(t-\hat{\tau}_l)c_l(t-\hat{\tau}_l)\mathrm{e}^{\mathrm{j}\hat{w}_{dl}t} \tag{2.137}$$

Then, the cross-correlation between that signal and the projected array output is

$$\mathbf{r}_{yd} = E\{\mathbf{y}(t)d_l^*(t)\} \tag{2.138}$$

Since the C/A codes for GNSS signals are orthogonal to each other, and the GNSS signal and the noise are independent to each other, then we can substitute (2.136) into (2.138) to obtain

$$\mathbf{r}_{yd} = A_l\bar{\mathbf{a}}(\theta_l) \tag{2.139}$$

Obviously, $\mathbf{r}_{yd}$ is proportional to the projection direction vector $\bar{\mathbf{a}}(\theta_l)$ of the lth GNSS signal. Therefore $\mathbf{r}_{yd}$ can be used to enhance the lth GPS signal, then the array weight vector for signal after the projection is

$$\mathbf{w}_l = \mathbf{r}_{yd} \tag{2.140}$$

For actual weight vector calculation, the sample cross correlation vector $\hat{\mathbf{r}}_{yd}$ is usually used to substitute the theoretical cross correlation vector $\mathbf{r}_{yd}$, i.e.

$$\hat{\mathbf{r}}_{yd} = \sum_{n=1}^{N} \mathbf{y}(n)d_l^*(n) \tag{2.141}$$

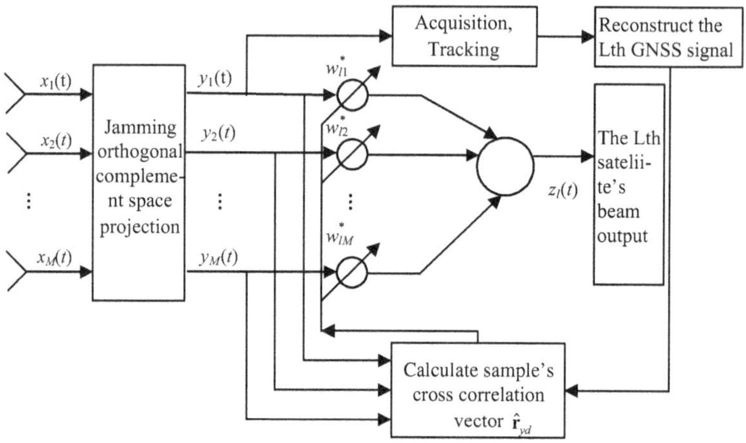

Fig. 2.21 Block diagram of the new de-spread re-spread algorithm

Since GNSS signals are weak, to fully take advantage of the GNSS signal's spread spectrum gain, the number of samples N is at least more than the number of samples within one C/A code period. The new de-spread and re-spread algorithm's block diagram is shown in Fig. 2.21.

As shown in Fig. 2.21, the new de-spread re-spread algorithm is a two-stage jamming mitigation processor. The stage one processor projects the signal received by the antenna array towards the jamming orthogonal complement space in order to eliminate the jamming signal; and the stage two processor acquires and tracks the output signal of the first antenna after projection and reconstructs the lth GPS signal. Then the sample cross correlation $\hat{\mathbf{r}}_{yd}$ between that signal and the signal after the projection is calculated, and the vector is used as the weight vector to enhance the lth GPS signal. Thereby, the array's overall weight vector is

$$\mathbf{w}_{lopt} = \hat{\mathbf{R}}_x^{-1}\hat{\mathbf{r}}_{yd} \tag{2.142}$$

Based on Fig. 2.21 and the derivation steps of the new algorithm, we know that compared with the least squares de-spread and re-spread algorithm, the new algorithm can accurately obtain GNSS signal information such as time delay, Doppler frequency and navigation messages. There is no need for iterative calculation, so the implementation is simple.

In Sects. 2.3–2.5, we discussed jamming mitigation algorithms. To facilitate readers to read and compare, Table 2.1 compares and summarizes the applicabilities of various algorithms mentioned before.

Table 2.1 Comparisons of applicabilities for various jamming mitigation algorithms

Applicability	Algorithm			
	Power minimization algorithm	Capon algorithm	Adaptive filtering algorithm using periodic repetition characteristic	Adaptive filtering algorithm using known spread spectrum code information
Provide signal processing gain	No	Yes	Yes	Yes
Need to know GNSS signal's direction of arrival	Not required	Required	Not required	Not required
Need to know array manifold	Not required	Required	Required	Not required
Array structure	Random array	Random array	Random array	Random array

2.5.3 Simulation Results

For simulation, the antenna array has a 10-element uniform linear array with an interval of half a wavelength. 4 GPS GNSS signals, PRN1, PRN2, PRN3 and PRN20 incident on the array from the directions of $0°$, $-55°$, $-20°$ and $30°$, and one jamming signal incidents on the array from the direction of $50°$. The jamming-to-noise ratio is 40 dB, and the signal-to-noise ratio is -20 dB. The digital IF frequency for the received signal by the array is 4.309 MHz, and the sampling rate is 5.714 MHz.

1. Comparisons of Adaptive Array Patterns

Figure 2.22 lists the adaptive array patterns obtained using three methods. Among them, Fig. 2.22a is the adaptive array pattern for the power minimization algorithm. Fig. 2.22b is for the SCORE algorithm, and Fig. 2.22c is for the new de-spread re-spread algorithm. The vertical solid lines in the subfigures represent the directions of the GNSS signals; the vertical dotted lines represent the directions of the jamming signals. By comparing Fig. 2.22a–c, we know that all three methods can suppress jamming. The SCORE algorithm generates the mainlobes towards the directions of the two GNSS signals (PRN1 and PRN3), and at the same time it attenuates the other two GNSS signals (PRN2 and PRN20). The power minimization method cannot provide gain on any satellite. But the new de-spread re-spread algorithm can generate multiple beams, and every beam's mainlobe aims

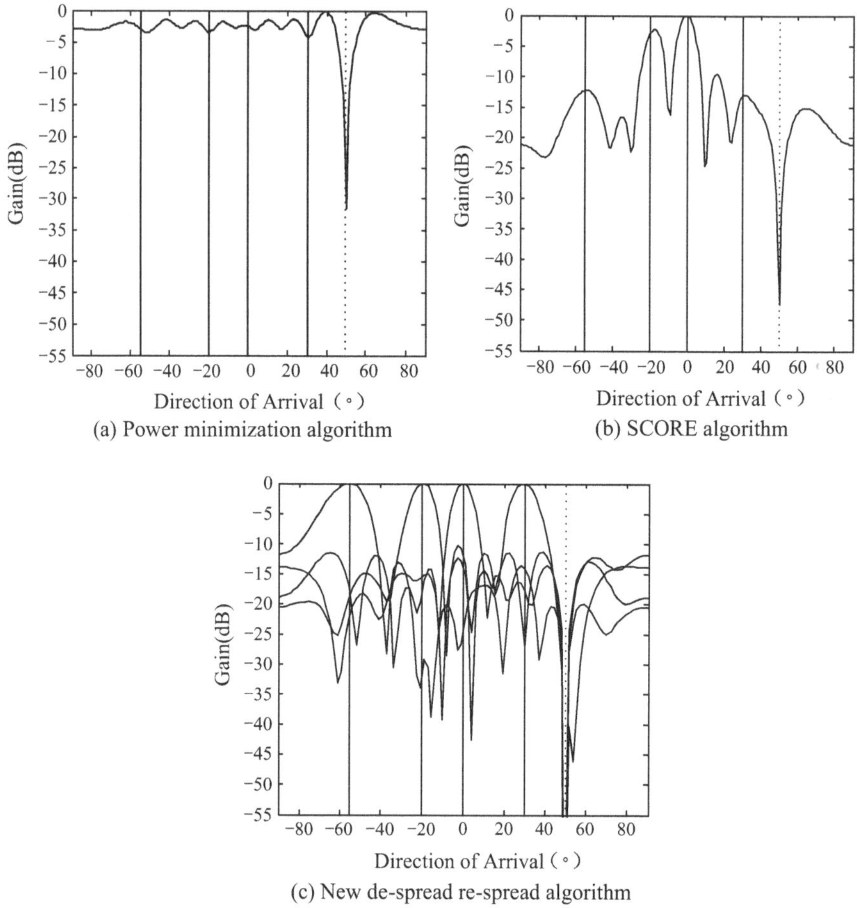

Fig. 2.22 Comparisons of adaptive array patterns for different algorithms

towards the GNSS signal, therefore the GNSS receivers can track signals from all satellites.

2. Comparisons of Acquisition Results

Figure 2.23 shows the acquisition results of using the SCORE algorithm for jamming mitigation. It can be seen in the figure that, since the SCORE algorithm attenuates the two GNSS signals (PRN2 and PRN20), the receiver only acquires two GNSS signals (PRN1 and PRN3).

Figure 2.24 shows the acquisition results using the new de-spread re-spread algorithm. It can be seen in Fig. 2.24 that every beam generated by the new algorithm aims towards a GNSS signal, thereby the receiver can acquire every GNSS signal.

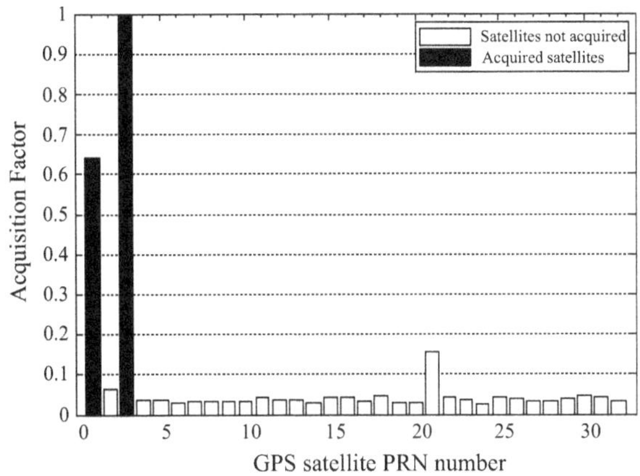

Fig. 2.23 Acquisition results of SCORE algorithm

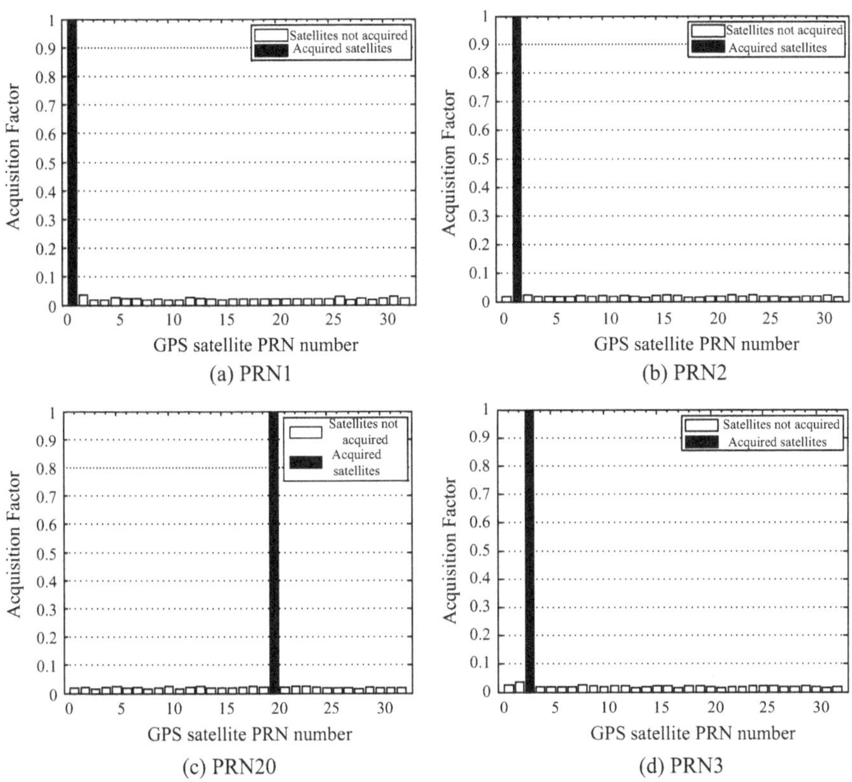

Fig. 2.24 Acquisition results of the new de-spread re-spread algorithm

Fig. 2.25 Comparisons of acquisition results

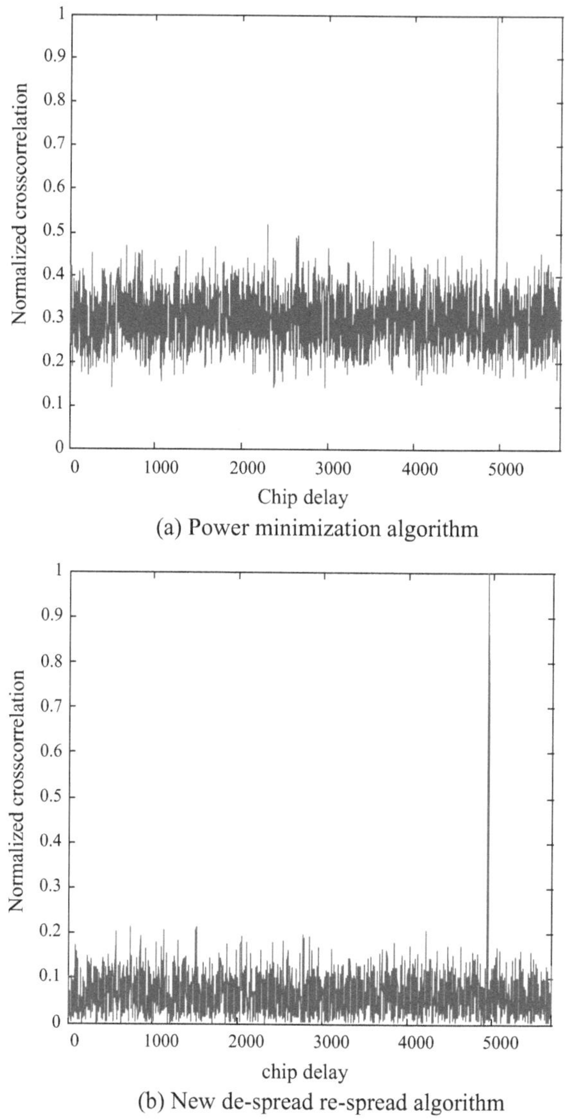

Figure 2.25 compares the acquisition results on the same GNSS signal using the new de-spread and re-spread algorithm and the power minimization algorithm. Among them, Fig. 2.25a is the result of the power minimization algorithm; Fig. 2.25b is the result for the new de-spread and re-spread algorithm. By comparing Fig. 2.25a, b, it can be seen that, after jamming suppression, the power minimization algorithm can acquire the GNSS signal, but can not provide the signal processing gain brought by the antenna array, thereby the normalized cross correlation number's sidelobe level is high. But the new de-spread and re-spread

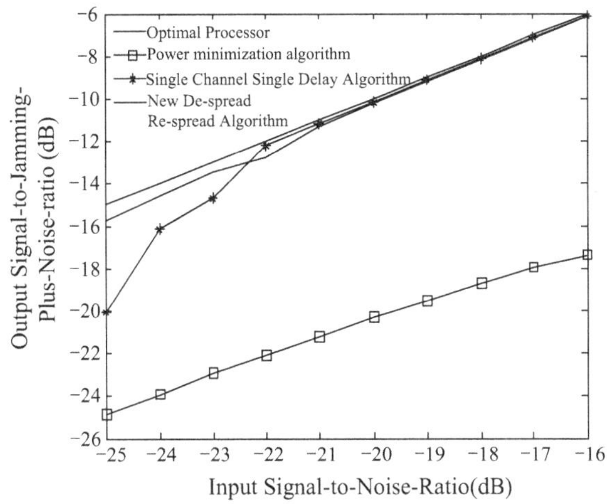

Fig. 2.26 Comparisons of signal-to-jamming-plus-noise ratio curves for different algorithms

algorithm can make the antenna mainlobe aim towards the direction of the GNSS signal, so that its acquisition result is better than that of the power minimization algorithm,

3. Comparisons of Output Signal-jamming-plus-noise Ratio

Figure 2.26 compares the output signal-jamming-plus-noise ratios for different algorithms. Among them, the solid line corresponds to the optimal processor; the dotted line corresponds to the new de-spread, re-spread algorithm; the "$\ast$" line corresponds to the single channel single delay cross correlation algorithm, and the "□" line corresponds to the power minimization algorithm. It can be seen in Fig. 2.26 that since the power minimization algorithm has no beam directions, it cannot improve a GNSS signal's carrier-to-noise ratio. Both the new de-spread re-spread algorithm and the single channel single delay cross correlation algorithm can improve the carrier-to-noise ratio of the GNSS signal.

2.5.4 Results on Hardware Platform Experiments

1. Brief Introduction on Hardware Platform Experiments

The new de-spread re-spread algorithm has the characteristics of high gain and high robustness. We have developed a digital multi-beam jamming mitigation real time processing system based on the algorithm. Based on the block diagram in the Fig. 2.21, the system can mainly be implemented using two main modules: the jamming mitigation module and the beamforming module. Figure 2.27 shows the experimental block diagram. In the Fig. 2.27, the experimental system is composed of two main parts: the signal transmitter and the receiver. The jamming signal in the transmitter is produced by a signal generator, and the GNSS signal

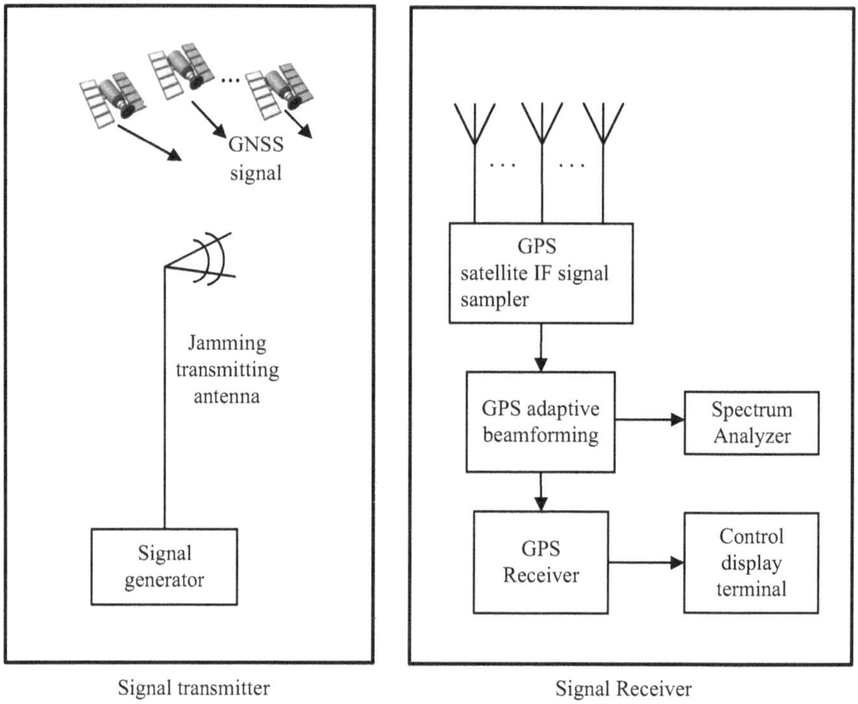

Fig. 2.27 Experimental block diagram for beam jamming mitigation real time processing system

refers to the signal transmitted by the actual GPS satellite. The receiver receives the signal using an array antenna first, then the received signal is converted to a digital IF signal using a down-converter and A/D sampling quantilization, and finally the digital IF signal is fed into the adaptive processing platform, After jamming suppression, GPS receiver acquisition, tracking and positioning, the results are transmitted to a personal computer control terminal for displaying the results of tracking and positioning. The effects of signal jamming can be observed by monitoring the frequency spectrum change before and after the jamming mitigation.

Figure 2.28 shows the experimental scene diagram for the digital multi-beam jamming mitigation real time processing system test. The left subfigure is the signal transmitter, with the top-left subfigure representing the GNSS signal and the bottom-left subfigure representing the signal source that generate the jamming signal to which a jamming transmitting antenna is connected. The right subfigure is the signal receiver, and in the order of processing there is a 7-element array antenna, IF signal sampler including an active downconverter, adaptive processing platform and computer display terminal. The sampling rate is 5.714286 MHz, and the intermediate frequency (IF) is 4.309 MHz.

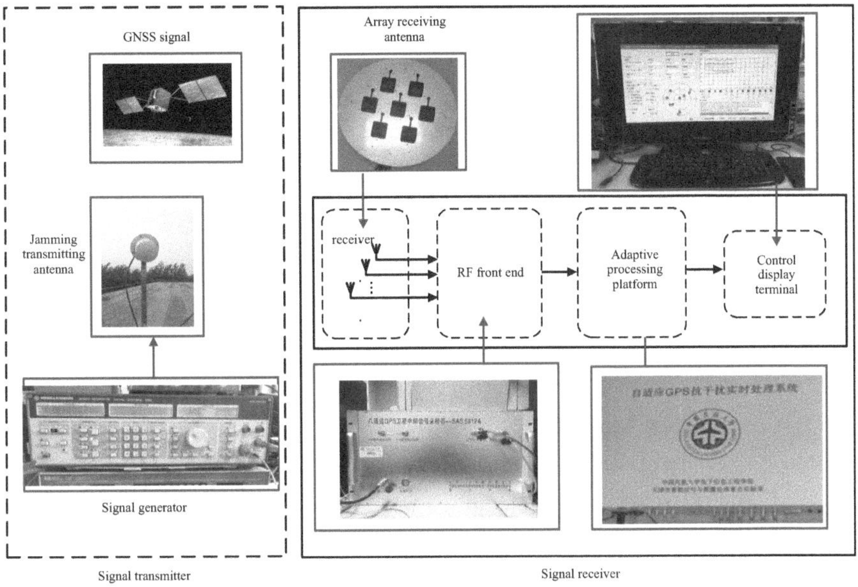

Fig. 2.28 Experimental scene diagrams for digital multi-beam jamming mitigation real time processing system test

2. Hardware Platform Testing Results

Figure 2.29 shows the IF signal frequency spectrum observed on the spectrum analyzer before and after the jamming suppression under the experiment conditions described in Fig. 2.27. The jamming signal produced by the signal generator has a frequency of 1575.42 MHz, and a jamming transmitting power of −10 dBm (jamming-to-noise ratio is 70 dB).

It can be seen in Fig. 2.29, after adaptive system processing, the peak of the signal spectrum is removed, so the jamming signal is effectively suppressed.

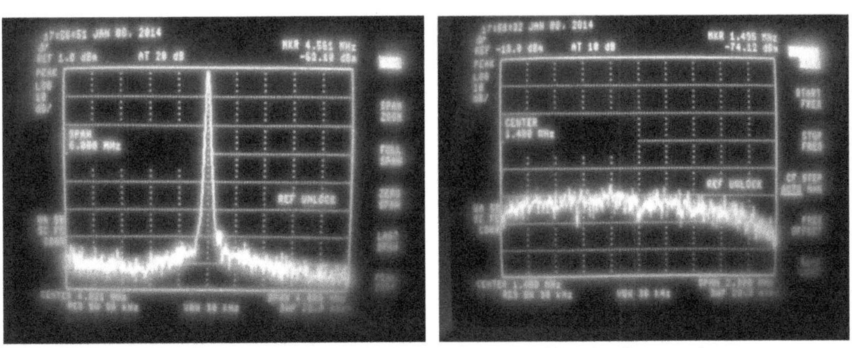

(a) Before Suppression (b) After Suppression

Fig. 2.29 Signal frequency spectrum comparisons before and after jamming suppression

Fig. 2.30 Tracking results obtained from the GPS receiver display terminal before and after adaptive beamforming processing

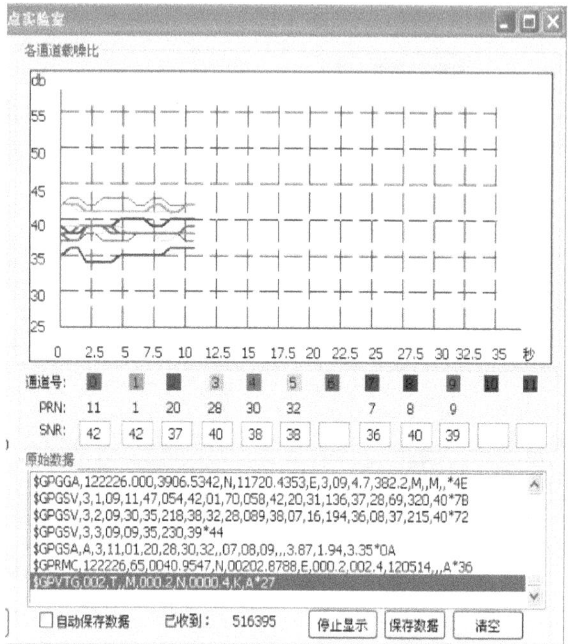

(a) No Jamming

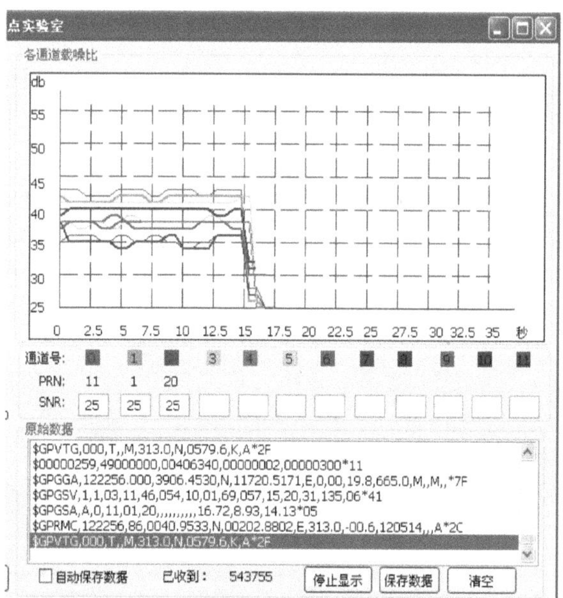

(b) With jamming

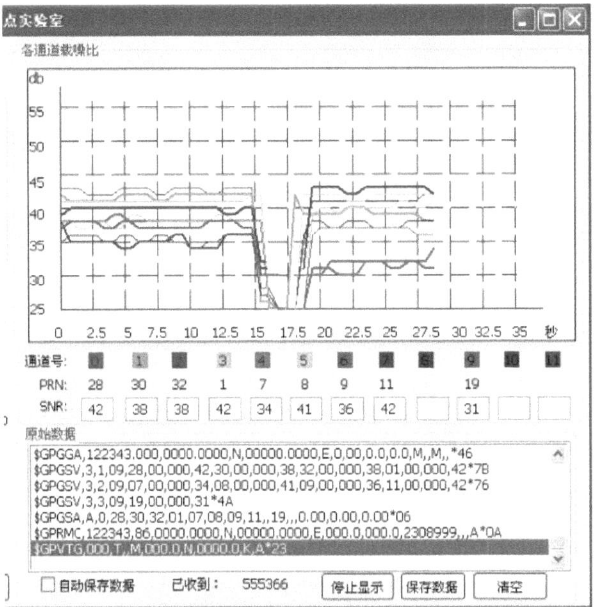

(c) After jamming suppression.

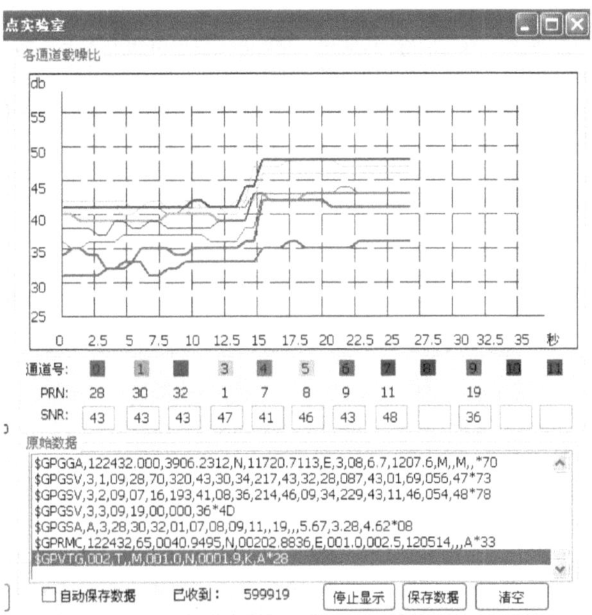

(d) After the beamforming is added.

Fig. 2.30 (continued)

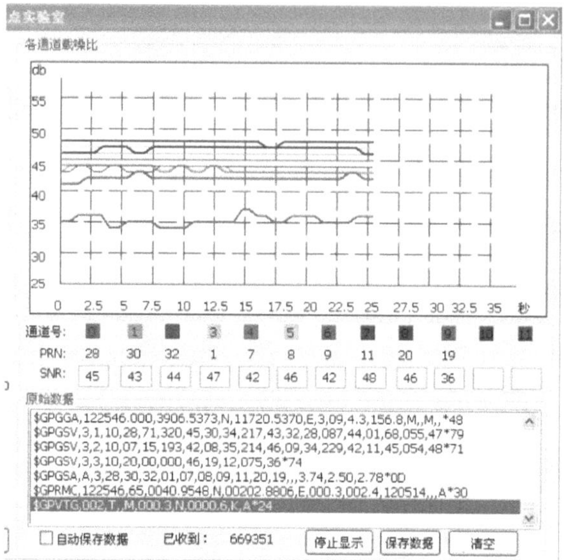

(e) After the beamforming is stabilized.

Fig. 2.30 (continued)

Figure 2.30 shows the tracking results obtained from the GPS receiver display terminal before and after adaptive beamforming processing. Every curve in Fig. 2.30 shows a GNSS signal's carrier-to-noise ratio change. Below every curve is the captured PRN, and the number inside the box under the PRN is the decibel number of the carrier-to-noise ratio. Figure 2.30a is the signal tracking diagram of the regular GPS receiver before adding the jamming signal; Fig. 2.30b is the instantaneous signal tracking diagram after adding the jamming signal; Fig. 2.30c is the signal tracking diagram when the jamming mitigation system is used; Fig. 2.30d is the instantaneous signal tracking diagram after adding the beamforming module; Fig. 2.30e is the tracking effect diagram after the beamforming module has been initiated and the signal becomes stable.

It can be seen in Fig. 2.30a that before adding the jamming signal, the GPS receiver can capture 9 satellites, and the average carrier-to-noise ratio is approximately 39.11 dB. In Fig. 2.30b, after adding a −50 dBm (the jamming-to-noise ratio is 30 dB) jamming signal, the receiver immediately loses the lock, and all signal tracking curves drop down. Then the satellite cannot perform normal acquisition and tracking. It can be seen in Fig. 2.30c that after initiating the GPS jamming mitigation system, the signal tracking curve immediately rises up. As the signal becomes stable, the carrier-to-noise ratio increases gradually until it becomes stabilized. As the signal gradually becomes stable, the carrier-to-noise ratio rises up gradually until it becomes stabilized. In Fig. 2.30d, after the beamforming module is initiated, the signal carrier-to-noise ratio rises up rapidly. In Fig. 2.30e, after the signal stabilizes, the average carrier-to-noise ratio improves by 5.68 dB compared with that of after the jamming mitigation due to the signal processing gain brought by the antenna array. This is because the GPS receiver module has acquisition function towards the low signal-to-noise visible satellites. In addition, at different moments, the visible satellites can change, so that after jamming mitigation, the system can track the PRN20 satellite. The experiment results show that while the system implemented suppresses the jamming signal, it forms a higher beam gain towards the direction of the satellites, so that effectively improves the signal's carrier-to-noise ratio.

2.6 Space Time Adaptive Filtering

Space-time Adaptive Processing (STAP) was proposed by Brennan et al. [48] in 1973 to solve the ground clutter suppression problem that shows a space-time two-dimensional coupled distribution for airborne radar. It essentially expands the one-dimensional spatial domain filtering technique to time and space two-dimensional domain, to form Time-Space two-dimensional processing structure. Reed et al. [49]. proposed the Sample Matrix Inverse (SMI) algorithm to compute adaptive weight. After this, to solve the problem of large computational load of two-dimensional optimal processing, Klemm proposed the auxiliary channel (ACR) [42] algorithm, Hong W. et al. proposed the joint domain localized processing (JDL) algorithm [50] and the $\sum - \Delta$ STAP algorithm [51], and Pao Zheng et al. proposed the method of "time first, space next" adaptive cascade processing method and partial joint adaptive processing method (abbreviated as two-dimensional Capon method and 3DT method) [47]. All these algorithms are computation-efficient. In addition, to promote the applicability of the STAP, scholars globally have studied problems such as reduced ranking processing and STAP under non-uniform environment.

For airborne and missile-borne receiver applications, the array antenna apertures are usually restricted. When there are various types of jamming (e.g. narrow band and wideband jamming, dispersion multipath interference etc.), only spatial domain processing cannot provide enough degrees of freedom to have good jamming suppression performance. Consequently, Fante et al. applied the STAP technique on GPS jamming mitigation, to greatly improve the degrees of freedom of the adaptive system. They also studied the criterion of the space-time jamming mitigation, the impact from multipath jamming, and the GPS signal distortion caused by STAP [16, 52–54] etc.

2.6.1 Space-Time Processing Data Model

Considering a uniform linear array composed of M array elements, and each array element has K taps with a tap delay of T. The STAP structure is shown in Fig. 2.31.

It can be seen on every array element channel that a FIR filter is constructed using delays at all levels that can be used in the temporal domain to remove jamming. From the perspectives of the nodes with the same time delays, different array elements constitute adaptive filter in the spatial domain, so it can identify spatial jamming sources, and suppress jamming by forming spatial domain nulls. Consequently, space-time processing has the ability to eliminate the jamming in the space-time domain.

We assume that after down-conversion, the received baseband signal $x(t)$ at the reference point at time t is represented as

$$x(t) = s(t) + \sum_{q=1}^{Q} j_q(t) + e(t) \tag{2.143}$$

where $s(t)$ denotes the satellite baseband signal; $j_q(t)$ denotes the baseband signal of the qth jamming signal; $e(t)$ is the additive Gaussian White noise. The data received by the array at time t can be written in the form of a $MK \times 1$ dimensional vector, i.e.

$$\mathbf{x}(t) = [x_{11}(t), x_{12}(t), \ldots, x_{1K}(t), \ldots, x_{M1}(t), \ldots, x_{MK}(t)]^{\mathrm{T}} \tag{2.144}$$

where $x_{mk}(t)$ is the received signal at time t for the mth array's kth delay unit, and it can be represented as

$$x_{mk}(t) = s_{mk}(t) + \sum_{q=1}^{Q} j_{qmk}(t) + e_{mk}(t) \tag{2.145}$$

For the GNSS signal in (2.145), the first array element of the uniform linear array is used as the reference point. By referring to the definition in (2.12), the GNSS signal received by the mth array's kth delay unit at time t can be written as

$$s_{mk}(t) = s(t - (k - 1)T)\mathrm{e}^{-\mathrm{j}\frac{2\pi(m-1)d\sin\theta}{\lambda}} \tag{2.146}$$

where θ denotes the angle of arrival of the satellite signal, and d denotes interval between array elements, and λ denotes the wavelength of the GNSS signal.

The GNSS signal received by the antenna array can be written at time t in the form of a $MK \times 1$ dimensional vector, i.e.

$$\bar{\mathbf{s}}(t) = [s_{11}(t),\ s_{12}(t),\ \ldots,\ s_{1K}(t),\ \ldots,\ s_{M1}(t),\ \ldots,\ s_{MK}(t)]^{\mathrm{T}} = \mathbf{A}\mathbf{s}(t) \tag{2.147}$$

where $\mathbf{s}(t) = [s(t), s(t-T), \ldots, s(t-(K-1)T)]^{\mathrm{T}}$ is the delayed GNSS signal at the Kth level, and $\mathbf{A} = \mathbf{I}_{K\times K} \otimes \mathbf{a}(\theta)$, $\mathbf{a}(\theta)$ is the spatial domain steering vector (reference (2.12)), and $\otimes$ denotes the Kronecker product.

In Fig. 2.31, $\{w_{mk}\}(m = 1, 2, \ldots, M; k = 1, 2, \ldots, K)$ is the space-time two-dimensional weight vector, and w_{mk} is the weight for the kth delay unit of

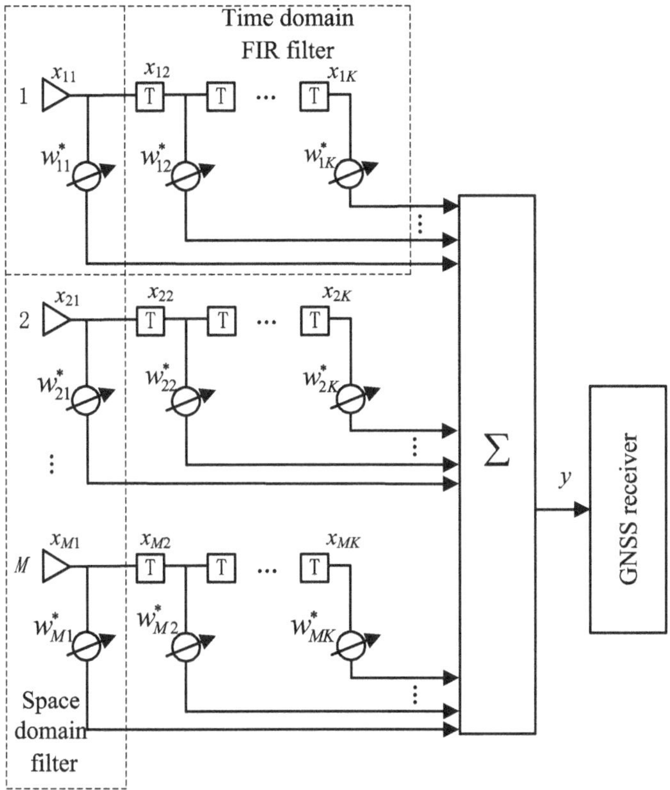

Fig. 2.31 STAP block diagram

the mth array element. The weight vector is written in the form of $MK \times 1$ dimensional vector, i.e.

$$\mathbf{w} = [w_{11}, w_{12}, \ldots, w_{1K}, \ldots, w_{M1}, \ldots, w_{MK}]^{\mathrm{T}} \tag{2.148}$$

Then, the output the space-time processor is

$$y(t) = \mathbf{w}^{\mathrm{H}}\mathbf{x}(t) \tag{2.149}$$

For the jamming component in (2.145), if multipath interference is not considered, its model is very similar to the model of the GNSS signal. If the jamming signal and the jamming signal's multipath signal incident on the array antenna simultaneously, as shown in Fig. 2.32, the received jamming signal becomes a combined jamming signal after the directly arrived wave and the reflected wave interfere with each other. This scenario happens more often when a GNSS receiver antenna is installed on a mobile platform. Using an airplane example, due to the reflections of the fuselage and the wings, the reflection waves of the jamming

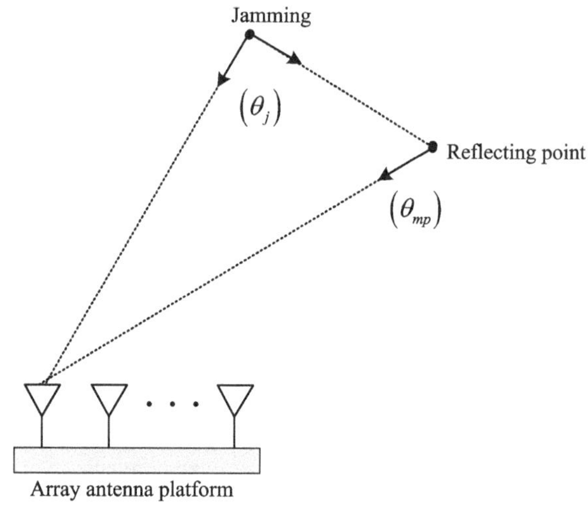

Fig. 2.32 Multipath jamming

among different antennas are only partially correlated, i.e. the array generates the dispersion phenomenon, also known as dispersion multipath jamming [55]. Below we use airborne dispersion multipath jamming as an example, to illustrate the necessity of the STAP processing.

For the qth jamming signal $j_q(t)$, if the array antenna receives another even larger reflected jamming signal relative to the LOS jamming signal, then the jamming signal received by the mth antenna becomes

$$J_{qm}(t) = j_{qm}(t) + \alpha j_{qm}(t - \tau_{mp}) \tag{2.150}$$

where the time delay of the multipath jamming relative to the LOS jamming is τ_{mp}; the multipath attenuation coefficient is α. The jamming signal received by the array can be written in the form of a $M \times 1$ dimension vector as $\mathbf{J}_q(t) = [J_{q1}(t), J_{q2}(t), \ldots, J_{qM}(t)]^{\mathrm{T}}$. Then the array covariance matrix for the jamming is

$$\begin{aligned}\mathbf{R}_j &= E\left\{\mathbf{J}_q(t)\mathbf{J}_q^{\mathrm{H}}(t)\right\} \\ &= \sigma_j^2 \left[\mathbf{a}(\theta_j) \quad \mathbf{a}(\theta_{mp})\right] \begin{bmatrix} 1 & \alpha r^*(\tau_{mp}) \\ \alpha^* r(\tau_{mp}) & |\alpha^2| \end{bmatrix} \left[\mathbf{a}(\theta_j) \quad \mathbf{a}(\theta_{mp})\right]^{\mathrm{H}}\end{aligned} \tag{2.151}$$

where σ_j^2 is the power of jamming signal; θ_j is the direction of LOS jamming; θ_{mp} is the direction of multipath multipath; $\mathbf{a}(\theta_j)$ is the array steering vector of LOS jamming; $\mathbf{a}(\theta_{mp})$ is the array steering vector of the multipath jamming; $r(\tau_{mp})$ is the correlation function between the LOS jamming and the multipath jamming.

Under the above given assumptions, if the correlation function between the LOS jamming and the multipath jamming $r(\tau_{mp})$ is not equal to 1, then the rank of the jamming covariance matrix is 2, and the array shows dispersion phenomenon. For this same reason, we can derive that, if one LOS jamming and P multipath jammings incident on the array simultaneously, since the received signals among different antennas are not correlated anymore, the rank of the jamming covariance matrix is larger than 1, and at maximum it can reach $P + 1$. Under the condition, if only spatial domain array process is used, multiple degrees of freedom are consumed to resist one jamming source. For M-element antenna array spatial domain processing, at maximum $M - 1$ nulls can be formed. When dispersion jamming exists, the reduction of degrees of freedom can degrade the performance of only spatial domain jamming suppression. STAP can greatly improve the degrees of freedom for adaptive processing, and furthermore it can improve performance by resisting dispersion jamming.

When wideband jammings exist, similar to the case above, the signals received by various array elements are not correlated with each other anymore. Therefore, STAP is also needed to improve the degrees of freedom for adaptive processing.

2.6.2 Space-Time Power Minimization Algorithm

The process used in space-time adaptive jamming mitigation uses different space-time algorithms to derive weight vectors for the array, so that the array output can be rendered immune from the impacts of jamming. This optimum criterion is the theoretical foundation in determining the space-time processing optimal weight vector. Different optimum weight vectors satisfying different requirements can be derived based on different criteria. Among them, the space-time power minimization algorithm [8] is widely applied on many GNSS receivers based on array antennas because it does not need any a priori information and the implementation is simple.

The power minimization algorithm, if the GNSS signal at the receiver is rather weak (around 20 dB lower than the noise level) and the jamming signal is very strong (often much larger than the noise level), determines the weight vector by minimizing the array output signal power. The method requires the weight of the first tap to be 1, and then a weight vector is selected to minimize the output power, i.e.

$$\begin{aligned} &\min_{w} \mathbf{w}^{\mathrm{H}}\mathbf{R}_x\mathbf{w} \\ &s.t.\ \ \mathbf{w}^{\mathrm{H}}\boldsymbol{\delta}_{MK} = 1 \end{aligned} \tag{2.152}$$

where $\mathbf{R}_x = E\{\mathbf{x}(t)\mathbf{x}^{\mathrm{H}}(t)\}$ is the theoretical covariance matrix for received signals by the array. In actual computation, it is usually substituted using the sample covariance matrix $\hat{\mathbf{R}}_x = \frac{1}{N}\sum_{n=1}^{N}\mathbf{x}(n)\mathbf{x}^{\mathrm{H}}(n)$; $\boldsymbol{\delta}_{MK} = [1, 0, \ldots, 0]^{\mathrm{T}}$ is a $MK \times 1$ dimension vector. Based on the Lagrange principle, the weight vector can be derived as

$$\mathbf{w}_{opt} = \frac{\mathbf{R}_x^{-1}\boldsymbol{\delta}_{MK}}{\boldsymbol{\delta}_{MK}^{\mathrm{H}}\mathbf{R}_x^{-1}\boldsymbol{\delta}_{MK}} \tag{2.153}$$

Even though the power minimization method does not need any a priori information and can be easily implemented, it cannot provide beam gain brought by array processing. So it cannot maximize the output signal to jamming-plus-noise ratio.

2.6.3 Space-Time De-spread Re-spread Algorithm

As described in Sect. 2.5.2, the blind adaptive algorithm based on de-spread re-spread algorithm takes advantage of the fact that the GNSS signal's spread spectrum code is known, so that it can simultaneously achieve array signal processing gain and jamming suppression with no need to know the directions of GNSS signals beforehand. Thereby, to expand the de-spread re-spread technique to the space-time domain, we can obtain space-time de-spread re-spread jamming suppression algorithm, and its block diagram is shown in Fig. 2.33.

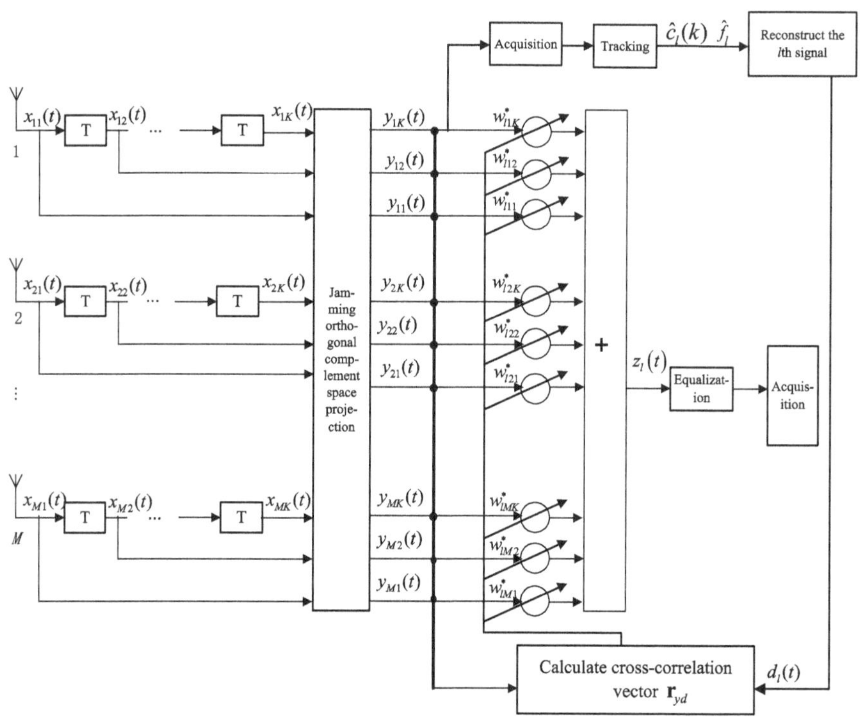

Fig. 2.33 Block diagram of space-time de-spread re-spread algorithm

First, we project the space-time data towards the inverse matrix of the covariance matrix, i.e.

$$\mathbf{y}(t) = \mathbf{R}_x^{-1}\mathbf{x}(t) \tag{2.154}$$

Then to perform acquisition and tracking on any one signal in (2.154), we can obtain the PRN code $c_l(t-\hat{\tau}_l)$, frequency $\hat{\omega}_l$ and navigation message $\hat{D}_l(t-\hat{\tau}_l)$ synchronized with the lth GNSS signal. From this information the lth GNSS signal can be reconstructed, i.e.

$$d_l(t) = \hat{D}_l(t)c_l(t-\hat{\tau}_l)\mathrm{e}^{\mathrm{j}\hat{\omega}_l(t-\hat{\tau}_l)} \tag{2.155}$$

We can reconstruct GNSS signals received from all L satellites or those satellites satisfying the position requirements using this method.

By performing correlation between reconstructed lth GNSS signal and the projected space-time data, the derived cross-correlation is

$$\hat{\mathbf{r}}_{yd} = \sum_{n=1}^{N}\mathbf{y}(n)d_l^*(n) \tag{2.156}$$

Based on this, similar to (2.142), we can obtain the space-time de-spread re-spread weight vector

$$\mathbf{w}_{lopt} = \mathbf{R}_x^{-1}\hat{\mathbf{r}}_{yd} \tag{2.157}$$

By repeating the steps described in (2.155)–(2.157), we can obtain the corresponding beams for L GNSS signals, and every beam points to one GNSS satellite.

2.6.4 Reduced Rank Space-Time Adaptive Filtering Algorithm

Good jamming suppression effects can be achieved by applying STAP technique on GNSS receivers. But, STAP processing involves the inverse of two-dimensional high order matrix. Not only does it require a large amount of computation, it also has a higher demand on the number of snapshots. Therefore, it is very important to perform reduced rank simplification on STAP [56]. It not only can greatly reduce the amount of computation, it also reduces the speed of convergence. In this section, we introduce some commonly used reduced rank STAP jamming suppression algorithms presently, including Principal Components (PC) [56], Cross Spectral Metric (CSM) [57], Multistage Nested Wiener Filter (MWF) [58–60], and Multistage Iterative Reduced Rank (MIRR).

Before introducing these commonly used reduced rank algorithms, we give the unified structure of reduced rank space-time adaptive jamming suppression algorithm based on the Wiener filter structure shown in Fig. 2.34.

Weiner-Hopf equation's solution is

$$\mathbf{w}_0 = \mathbf{R}_x^{-1}\mathbf{r}_{xd_0} \triangleq \left(E\{\mathbf{x}\mathbf{x}^{\mathrm{H}}\}\right)^{-1}E\{\mathbf{x}d_0^*\} \tag{2.158}$$

where d_0 denotes the desired signal; $\mathbf{x}$ is the two-dimension snapshot vector corresponding to the observed signal received by space-time processor; $\mathbf{w}_0$ is the weight vector of space-time processor; $\hat{d}_0 = \mathbf{w}_0^{\mathrm{H}}\mathbf{x}$ is the estimate on the desired signal; $\mathbf{R}_x = E\{\mathbf{x}\mathbf{x}^{\mathrm{H}}\}$ is the covariance matrix corresponding to the observed signal; $\mathbf{r}_{xd_0} = E\{\mathbf{x}d_0^*\}$ is the cross-correlation vector between the observed signal and the desired signal. The minimum mean square error of the output corresponding to (2.158) is

$$MMSE = \sigma_{d_0}^2 - \mathbf{r}_{xd_0}^{\mathrm{H}}\mathbf{R}_x^{-1}\mathbf{r}_{xd_0} \tag{2.159}$$

where $\sigma_{d_0}^2 = E\{d_0d_0^*\}$ is the average power of the desired signal. It can be seen in (2.158) that the weight vector $\mathbf{w}_0$ makes $\hat{d}_0$ approach the signal component of the desired signal d_0 that is correlated with the observed signal, i.e. $\hat{d}_0$ is the projection component of the observed signal $\mathbf{x}$ on the direction of cross-correlation $\mathbf{r}_{xd_0}$.

To solve for the classic Wiener filter's optimum weight vector, an inverse operation on the covariance matrix of the observed signal received by space-time processor is needed. The amount of computation is very large and the speed of convergence is slow (i.e. very high requirement on the number of snapshots). Thus further illustrates the necessity to perform reduced rank processing. By performing reduced rank processing, a smaller subspace can be created to approximate observation space that can greatly reduce the computation complexity and achieve faster speed of convergence. Theoretically, when the dimension of the reduced rank is reduced down to the space dimension of the jamming signal, the impacts on the processor's jamming suppression performance are not significant.

The reduced rank STAP jamming suppression algorithm can be expressed in a unified fashion: by using a $MK \times D$ dimension reduced rank transformation matrix $\mathbf{T} = [\mathbf{t}_1, \mathbf{t}_2, \ldots, \mathbf{t}_D]$ (where $D < MK$ represents the dimension after reducing the rank of the input observation signal), so that the dimension of the observed signal is reduced from MK dimension to D dimension, and the D-dimension adaptive

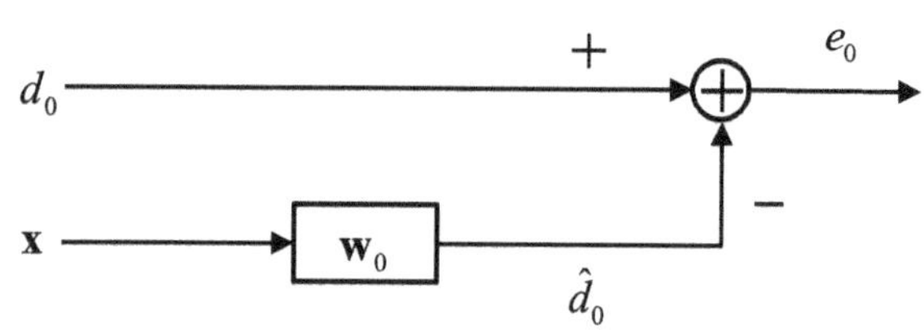

Fig. 2.34 Block diagram of a classical Wiener filter

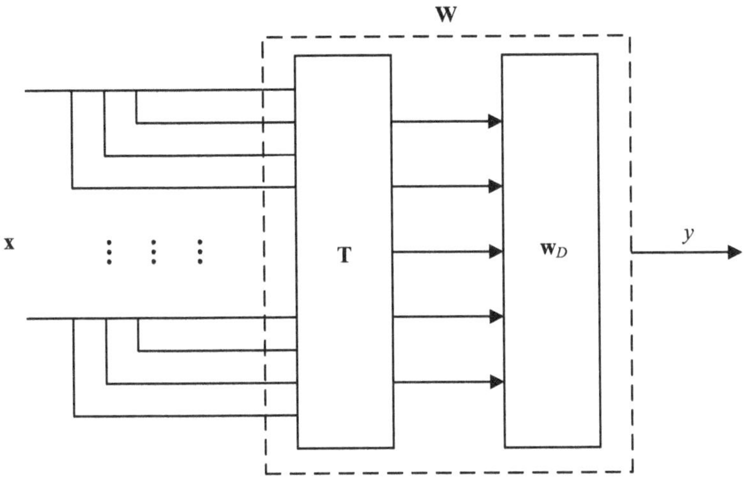

Fig. 2.35 Unified structure for reduced rank space-time adaptive jamming suppression algorithm

processing can be done in this reduced rank space. A $D \times 1$ dimension adaptive weight vector $\mathbf{w}_D$ can then be obtained, as shown in Fig. 2.35.

After performing **T** transformation reduced rank processing on the observed signal, the new input observed signal vector obtained is

$$\mathbf{x}_D = \mathbf{T}^{\mathrm{H}}\mathbf{x} \tag{2.160}$$

The covariance matrix corresponding to the observed data after rank reduction is

$$\mathbf{R}_D = \mathbf{T}^{\mathrm{H}}\mathbf{R}_x\mathbf{T} \tag{2.161}$$

It can be seen that the dimension of covariance matrix reduces from $MK \times MK$ to $D \times D$. Thereby, to solve the space-time adaptive weight vector after rank reduction $\mathbf{w}_D$, the amount of computation needed for covariance matrix inverse operation can be reduced significantly. The overall weight vector of the space-time adaptive processor is

$$\mathbf{w} = \mathbf{T}\mathbf{w}_D \tag{2.162}$$

The key for reduced rank processing is to select the proper reduced rank transformation matrix **T**, so that as the rank of the processor reduces, we still can achieve performance close to that of a full-dimensional optimum processor.

1. Principal Component Method

The main idea of the principal component method (PC) [56] is to perform eigenvalue decomposition on the observed data's covariance matrix $\mathbf{R}_x$, to obtain eigenvalue λ_i and the corresponding eigenvector $\mathbf{v}_i$, i.e.

$$\mathbf{R}_x = \mathbf{V}\Lambda\mathbf{V}^{\mathrm{H}} = \sum_{i=1}^{MK} \lambda_i \mathbf{v}_i \mathbf{v}_i^{\mathrm{H}} \tag{2.163}$$

where the diagonal element of the diagonal matrix Λ i is $\mathbf{R}_x$'s eigenvalue; $\mathbf{V}$ is the matrix composed of column vectors of $\mathbf{R}_x$'s eigenvector.

The GNSS signal reaching the receiver is much weaker than the noise level, while the jamming signal is fairly strong and is way above the noise level, thereby the space expanded from the eigenvector corresponding to the $\mathbf{R}_x$'s large eigenvalues can be regarded as the jamming subspace, while the space expanded from the eigenvector corresponding to $\mathbf{R}_x$'s small eigenvalues can be regarded as the noise subspace. By arranging $\mathbf{R}_x$'s eigenvalues in descending order (i.e. $\lambda_1 \geqslant \lambda_2 \geqslant \cdots \geqslant \lambda_{MK}$), the eigenvectors $\mathbf{v}_i$ corresponding to the first D largest eigenvalues λ_i can be used as the reduced rank matrix, i.e.

$$\mathbf{T} = \mathbf{V}_D = [\mathbf{v}_1,\ \mathbf{v}_2,\ \ldots,\ \mathbf{v}_D] \tag{2.164}$$

Now, the observed data's covariance matrix $\mathbf{R}_x$ can be approximated using the following formula

$$\mathbf{R}_x \approx \mathbf{R}_{PC} = \mathbf{V}_D \mathbf{\Lambda}_D \mathbf{V}_D^{\mathrm{H}} = \sum_{i=1}^{D} \lambda_i \mathbf{v}_i \mathbf{v}_i^{\mathrm{H}} \tag{2.165}$$

where the diagonal matrix $\mathbf{\Lambda}_D = \mathrm{diag}\{\lambda_1, \lambda_2, \ldots, \lambda_D\}$, and is a $D \times D$ dimensional matrix; the reduced rank eigenvector $\mathbf{V}_D$ is a $MK \times D$ dimension matrix, and the reduced rank dimension $D \leq MK - 1$. If D is bigger than the dimension of the jamming subspace, the jamming information can be preserved. Then, the principal component method's weight vector is

$$\mathbf{w}_{PC} = \mathbf{R}_{PC}^{-1} \mathbf{r}_{xd_0} = \sum_{i=1}^{D} \frac{1}{\lambda_i} \mathbf{v}_i \mathbf{v}_i^{\mathrm{H}} \mathbf{r}_{xd_0} \tag{2.166}$$

If all jamming components that need to be suppressed are all contained in the reduced rank matrix $\mathbf{T}$, the principal component method has very good reduced rank performance. But the PC method approximates the auto-correlation matrix based on the eigenvalues' amplitudes only. It does not consider the space-time two-dimension steering vector of the desired signal. Consequently, the space compression is not complete. When there is characteristic spectrum expansion and subspace leakage, the performance of the PC method's compression usually experiences sharp degradation. Aiming towards the above questions, we introduce the cross spectral metric to improve the principal component method.

2. Cross Spectral Metric

The main idea of Cross spectral metric (CSM) [57] is to use the amplitude of cross spectral energy as the criterion to select reduced rank matrix. By substituting the observed data covariance matrix's eigenvector decomposition expansion formula (2.163) into the Wiener filter output's minimum mean square error expression (2.159), we can obtain

$$\begin{aligned} MMSE &= \sigma_{d_0}^2 - \mathbf{r}_{xd_0}^{\mathrm{H}} \mathbf{R}_x^{-1} \mathbf{r}_{xd_0} = \sigma_{d_0}^2 - \mathbf{r}_{xd_0}^{\mathrm{H}} \sum_{i=1}^{MK} \frac{\mathbf{v}_i \mathbf{v}_i^{\mathrm{H}}}{\lambda_i} \mathbf{r}_{xd_0} \\ &= \sigma_{d_0}^2 - \sum_{i=1}^{MK} \frac{\left|\mathbf{v}_i^{\mathrm{H}} \mathbf{r}_{xd_0}\right|^2}{\lambda_i} \triangleq \sigma_{d_0}^2 - \sum_{i=1}^{MK} \frac{\rho_i^2}{\lambda_i} \end{aligned} \tag{2.167}$$

where $\rho_i = \left|\mathbf{v}_i^{\mathrm{H}} \mathbf{r}_{xd_0}\right|$ denotes the space correlation between the ith eigenvector and the desired signal's space-time two-dimension steering vector. By defining ρ_i^2/λ_i as cross spectral energy, we maximize the overall cross spectral energy by minimizing the output MMSE in formula (2.167). Therefore, the cross spectral metric method selects the eigenvectors $\tilde{\mathbf{v}}_i$ corresponding to the first D number of eigenvalues $\tilde{\lambda}_i$ that maximize cross spectrum energy as the reduced rank matrix, i.e.

$$\mathbf{T} = \tilde{\mathbf{V}}_D = [\tilde{\mathbf{v}}_1, \tilde{\mathbf{v}}_2, \ldots, \tilde{\mathbf{v}}_D] \tag{2.168}$$

Then, the observed data's covariance matrix $\mathbf{R}_x$ can be approximated as

$$\mathbf{R}_x \approx \mathbf{R}_{CSM} = \tilde{\mathbf{V}}_D \tilde{\mathbf{\Lambda}}_D \tilde{\mathbf{V}}_D^{\mathrm{H}} = \sum_{i=1}^{D} \tilde{\lambda}_i \tilde{\mathbf{v}}_i \tilde{\mathbf{v}}_i^{\mathrm{H}} \tag{2.169}$$

Then, we can obtain the cross spectral metric method's weight vector

$$\mathbf{w}_{CSM} = \mathbf{R}_{CSM}^{-1} \mathbf{r}_{xd_0} = \sum_{i=1}^{D} \frac{1}{\tilde{\lambda}_i} \tilde{\mathbf{v}}_i \tilde{\mathbf{v}}_i^{\mathrm{H}} \mathbf{r}_{xd_0} \tag{2.170}$$

The data compression effect of the cross spectrum measure method is better than that of the principal component method. But, both the cross spectrum measure method and the principal component method need to perform eigenvalue decomposition on the observed data's covariance matrix to obtain reduced rank subspace, so that the amount of computation required is still large. To some degree, this constraint limits the application of these methods in actual engineering practices. The multistage nested wiener filter method introduced below can avoid matrix inverse and eigenvalue decomposition so it can reduce the computation complexity for the above two problems.

3. Multistage Nested Wiener Filter

The multistage Nested Wiener Filter (MWF) [58] is a multistage equivalent realization of the Wiener filter. It uses a series of orthogonal projection to perform multistage decomposition on observed input signal vector, then performs multistage scalar Wiener filtering to synthesize the output error signal of the Wiener filter. Compared with the PC and CSM methods that are based on feature decomposition, the MWF method avoids complex operations such as matrix inverse and feature decomposition.

Before we introduce multistage Wiener filter, we must first give the following theorems [38].

Theorem 2.1 *Using full rank matrix* $\mathbf{T} \in \mathbb{C}^{MK \times MK}$, *we perform pre-filtering on observed data* $\mathbf{x}_0$ *(originally* $\mathbf{x}$*) to obtain new observed data* $\mathbf{z}_1 = \mathbf{T}\mathbf{x}_0$. *The Wiener filter weight vector* $\mathbf{w}_{z_1}$ *and estimated signal* $\hat{d}_0 = \mathbf{w}_{z_1}^{\mathrm{H}}\mathbf{z}_1$ *are also derived based on this data, and the MMSE derived is the same as that of no prefiltering.*

Below we give proof on Theorem 2.1, to derive the Weiner filtering performance on full rank matrix pre-filtering. Figure 2.36 shows the block diagram of using full rank matrix $\mathbf{T} \in \mathbb{C}^{MK \times MK}$ to perform pre-filtering processing on observed data $\mathbf{x}_0$.

New observed data after pre-filtering can be written as

$$\mathbf{z}_1 = \mathbf{T}\mathbf{x}_0 \tag{2.171}$$

The corresponding new mean square error is

$$MSE_{z_1} = \sigma_{d_0}^2 - \mathbf{w}_{z_1}^{\mathrm{H}}\mathbf{r}_{z_1 d_0} - \mathbf{w}_{z_1}\mathbf{r}_{z_1 d_0}^{\mathrm{H}} + \mathbf{w}_{z_1}^{\mathrm{H}}\mathbf{R}_{z_1}\mathbf{w}_{z_1} \tag{2.172}$$

where $\mathbf{R}_{z_1} = E\{\mathbf{z}_1\mathbf{z}_1^{\mathrm{H}}\} = E\{\mathbf{T}\mathbf{x}_0\mathbf{x}_0^{\mathrm{H}}\mathbf{T}^{\mathrm{H}}\} = \mathbf{T}\mathbf{R}_{x_0}\mathbf{T}^{\mathrm{H}}$ is the covariance matrix for the new observed data; $\mathbf{r}_{z_1 d_0} = E\{\mathbf{z}_1 d_0^*\} = E\{\mathbf{T}\mathbf{x}_0 d_0^*\} = \mathbf{T}\mathbf{r}_{x_0 d_0}$ is the cross-correlation between new observed data and expected data. The derived Wiener filtering weight vector $\mathbf{w}_{\mathbf{z}_1}$ can be represented as

$$\begin{aligned}\mathbf{w}_{z_1} &= \mathbf{R}_{z_1}^{-1}\mathbf{r}_{z_1 d_0} \\ &= \left[\mathbf{T}^{-1}\right]^{\mathrm{H}}\mathbf{R}_{x_0}^{-1}\mathbf{T}^{-1}\mathbf{T}\mathbf{r}_{x_0 d_0} = \left[\mathbf{T}^{-1}\right]^{\mathrm{H}}\mathbf{R}_{x_0}^{-1}\mathbf{r}_{x_0 d_0} \\ &= \left[\mathbf{T}^{-1}\right]^{\mathrm{H}}\mathbf{w}_{x_0}\end{aligned} \tag{2.173}$$

The new desired signal estimated based on new observed data is

$$\hat{d}_0 = \mathbf{w}_{z_1}^{\mathrm{H}}\mathbf{z}_1 = \mathbf{w}_{x_0}^{\mathrm{H}}\mathbf{T}^{-1}\mathbf{T}\mathbf{x}_0 = \mathbf{w}_{x_0}^{\mathrm{H}}\mathbf{x}_0 \tag{2.174}$$

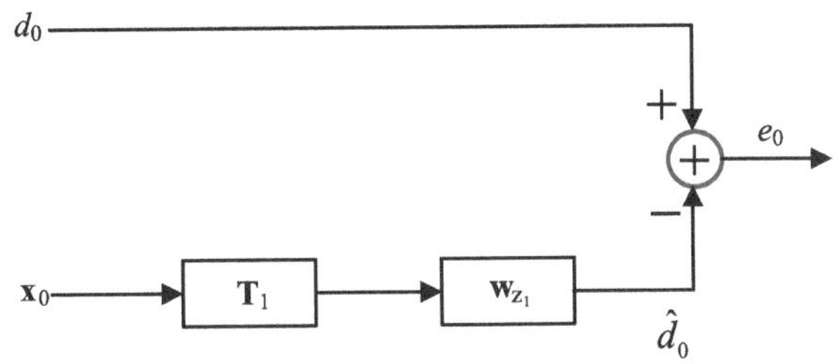

Fig. 2.36 Wiener filter including pre-filtering

It can be seen that the estimated results are the same as those before pre-filtering. Now the new MMSE can be recalculated. Under the minimum mean square error criterion, we can derive using (2.172)

$$\begin{aligned} MMSE_{z_1} &= \sigma_{d_0}^2 - \mathbf{w}_{z_1}^{\mathrm{H}}\mathbf{R}_{z_1}\mathbf{w}_{z_1} \\ &= \sigma_{d_0}^2 - \mathbf{w}_{x_0}^{\mathrm{H}}\mathbf{T}^{-1}\mathbf{T}\mathbf{R}_{x_0}\mathbf{T}^{\mathrm{H}}\left[\mathbf{T}^{-1}\right]^{\mathrm{H}}\mathbf{w}_{x_0} \\ &= \sigma_{d_0}^2 - \mathbf{w}_{x_0}^{\mathrm{H}}\mathbf{R}_{x_0}\mathbf{w}_{x_0} \\ &= MMSE_{x_0} \end{aligned} \tag{2.175}$$

It can be seen that the output MMSE is the same as before pre-filtering. Based on Theorem 2.1, we can use the full rank pre-filtering matrix

$$\mathbf{T}_1 = \begin{bmatrix} \mathbf{h}_1^{\mathrm{H}} \\ \mathbf{B}_1 \end{bmatrix} \in \mathbb{C}^{MK \times MK} \tag{2.176}$$

where $\mathbf{h}_1 = \mathbf{r}_{x_0d_0}/\|\mathbf{r}_{x_0d_0}\| \in \mathbb{C}^{MK\times 1}$ is a normalized cross-correlation vector; $\mathbf{B}_1 = null(\mathbf{h}_1) \in \mathbb{C}^{MK-1\times MK}$ is a $(MK-1)\times MK$ dimension matrix orthogonal to $\mathbf{h}_1$, and is regarded as the blocking matrix, and its expanded space is $\mathbf{h}_1$'s null space, i.e. $\mathbf{B}_1\mathbf{h}_1 = 0$. Then, Fig. 2.37 shows the structure of a Wiener filter. Where $d_1 = \mathbf{h}_1^{\mathrm{H}}\mathbf{x}_0$ is $\mathbf{x}_0$'s projection on the projection direction of the cross-correlation vector $\mathbf{r}_{x_0d_0}$; $\mathbf{h}_1$ ensures that d_1 and d_0 have the largest correlation; $\mathbf{x}_1 = \mathbf{B}_1\mathbf{x}_0$ is a $MK-1$

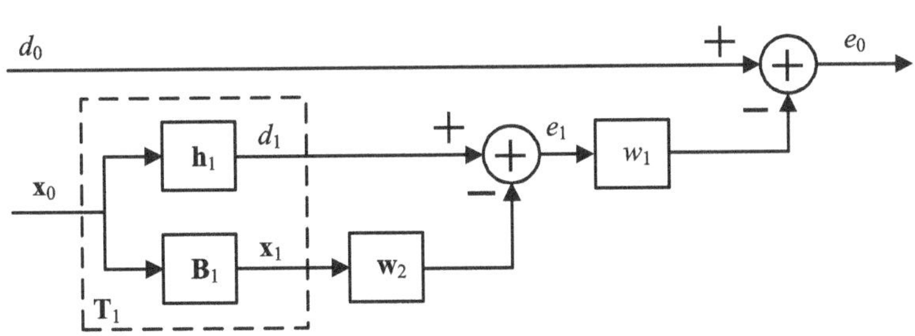

Fig. 2.37 Two-stage Wiener filter structure

dimension vector, and it corresponds to the projection $\mathbf{x}_0$'s projection on its $MK-1$ dimension subspace orthogonal to cross-correlation vector $\mathbf{r}_{x_0 d_0}$, not containing any information in d_0.

The new observed data after pre-filtering the observed data $\mathbf{x}_0$ can be represented as

$$\mathbf{z}_1 = \mathbf{T}_1\mathbf{x}_0 = \begin{bmatrix} \mathbf{h}_1^{\mathrm{H}}\mathbf{x}_0 \\ \mathbf{B}_1\mathbf{x}_0 \end{bmatrix} = \begin{bmatrix} d_1 \\ \mathbf{x}_1 \end{bmatrix} \tag{2.177}$$

Its corresponding covariance matrix is

$$\mathbf{R}_{z_1} = \begin{bmatrix} \sigma_{d_1}^2 & \mathbf{r}_{x_1 d_1}^{\mathrm{H}} \\ \mathbf{r}_{x_1 d_1} & \mathbf{R}_{x_1} \end{bmatrix} \tag{2.178}$$

where

$$\begin{aligned} \sigma_{d_1}^2 &= E\{d_1 d_1^{\mathrm{H}}\} = \mathbf{h}_1^{\mathrm{H}}\mathbf{R}_{x_0}\mathbf{h}_1 \\ \mathbf{r}_{x_1 d_1} &= E\{\mathbf{x}_1 d_1^*\} = \mathbf{B}_1\mathbf{R}_{x_0}\mathbf{h}_1 \\ \mathbf{R}_{x_1} &= E\{\mathbf{x}_1\mathbf{x}_1^{\mathrm{H}}\} = \mathbf{B}_1\mathbf{R}_{x_0}\mathbf{B}_1^{\mathrm{H}} \end{aligned} \tag{2.179}$$

The $\mathbf{R}_{x0}$ in the above formula is the original $\mathbf{R}_x$. The cross-correlation vector between the new observed data and the desired data is

$$\mathbf{r}_{z_1 d_0} = E\{\mathbf{z}_1 d_0^*\} = \mathbf{T}_1\mathbf{r}_{x_0 d_0} = \begin{bmatrix} \|\mathbf{r}_{x_0 d_0}\| \\ \mathbf{0}_{(MK-1)\times 1} \end{bmatrix} \tag{2.180}$$

Therefore, the Wiener filtering weight vector corresponds to the data in (2.177) is

$$\mathbf{w}_{z_1} = \mathbf{R}_{z_1}^{-1}\mathbf{r}_{z_1 d_0} \tag{2.181}$$

By applying the block Hermite matrix inverse formula, we can derive

$$\mathbf{R}_{z_1}^{-1} = \xi_1^{-1}\begin{bmatrix} 1 & -\mathbf{r}_{x_1 d_1}^{\mathrm{H}}\mathbf{R}_{x_1}^{-1} \\ -\mathbf{R}_{x_1}^{-1}\mathbf{r}_{x_1 d_1} & \mathbf{R}_{x_1}^{-1}\left(\xi_1\mathbf{I} + \mathbf{r}_{x_1 d_1}\mathbf{r}_{x_1 d_1}^{\mathrm{H}}\mathbf{R}_{x_1}^{-1}\right) \end{bmatrix} \tag{2.182}$$

where $\xi_1 = E\left\{|e_1|^2\right\} = \sigma_{d_1}^2 - \mathbf{r}_{x_1 d_1}^{\mathrm{H}}\mathbf{R}_{x_1}^{-1}\mathbf{r}_{x_1 d_1}$. By substituting (2.182) into (2.181), we can derive the Wiener filtering weight vector after pre-filtering [59]

$$
\begin{aligned}
\mathbf{w}_{z_1} &= \xi_1^{-1}\begin{bmatrix} 1 & -\mathbf{r}_{x_1d_1}^{\mathrm{H}}\mathbf{R}_{x_1}^{-1} \\ -\mathbf{R}_{x_1}^{-1}\mathbf{r}_{x_1d_1} & \mathbf{R}_{x_1}^{-1}\left(\xi_1\mathbf{I}+\mathbf{r}_{x_1d_1}\mathbf{r}_{x_1d_1}^{\mathrm{H}}\mathbf{R}_{x_1}^{-1}\right)\end{bmatrix}\begin{bmatrix}\|\mathbf{r}_{x_0d_0}\| \\ \mathbf{0}_{(MK-1)\times 1}\end{bmatrix} \\
&= \xi_1^{-1}\|\mathbf{r}_{x_0d_0}\|\begin{bmatrix} 1 \\ -\mathbf{R}_{x_1}^{-1}\mathbf{r}_{x_1d_1}\end{bmatrix} \\
&= w_1\begin{bmatrix} 1 \\ -\mathbf{w}_2\end{bmatrix}
\end{aligned} \tag{2.183}
$$

It can be seen that, an MK dimension Wiener weight vector can be decomposed into a form of one scalar and an MK − 1 dimension weight vector. Among them, the scalar $w_1 = \xi_1^{-1}\|\mathbf{r}_{x_0d_0}\|$, the $MK - 1$ dimension weight vector $\mathbf{w}_2 = \mathbf{R}_{x_1}^{-1}\mathbf{r}_{x_1d_1}$ is the Wiener solution of the estimated scalar d_1 based on $MK - 1$ dimension data $\mathbf{x}_1$.

Using the same method we can decompose the Wiener filter stage by stage, as shown in Fig. 2.38, until it can be decomposed to the (MK − 1)th stage. Then, we can simplify the matrix inverse as the reciprocal of multiple scalars. $\mathbf{h}_r$ of every decomposition is the cross-correlation between the previous upper branch's desired signal and the lower branch's observed data $\mathbf{x}_{r-1}$, in order to preserve the information from the previous step as much as possible. The block matrix of every decomposition $\mathbf{B}_r$ can ensure orthogonalities among ever reduced rank components. After Multistage decomposition, we can still obtain the same output MMSE as the original Wiener filtering.

Actually, there is no need to completely decompose the filter. Only D-step decompositions ($D \le MK - 1$) can obtain almost all useful information. Then, the observed data $\mathbf{x}_D$ contains almost no jamming components, and its covariance matrix tends to whiten, so that decomposition can stop [60]. As shown in Fig. 2.38, a multistage Winder filter is composed of an analysis filter and a synthetic filter. The analysis filter uses an orthogonal projection transformation to perform decomposition, and the synthetic filter uses a set of iterative Wiener filters for synthesis, and its input is the output of the analysis filter. The detailed steps of the original multistage Wiener reduced rank algorithm's iterative process are listed in Table 2.2. In the table, $\mathbf{B}_{ptran} = \mathbf{B}_p^{\mathrm{H}}$.

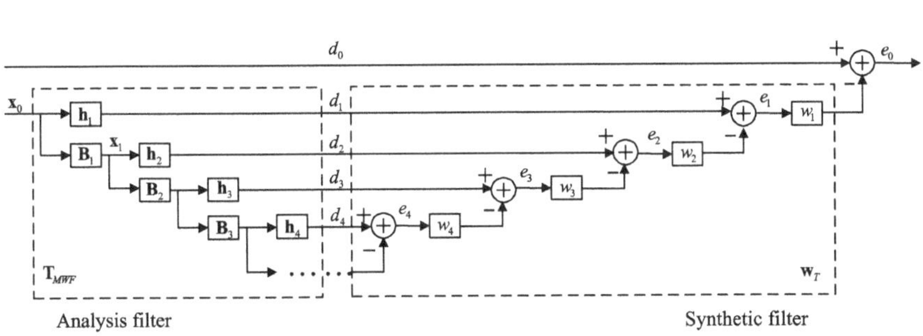

Fig. 2.38 Multistage Wiener filter structure

Table 2.2 Detailed steps of original multistage Wiener filtering algorithm

Analysis filter
Initialization: $\mathbf{R}_x = \mathbf{R}_{x_0} = E\{\mathbf{x}_0\mathbf{x}_0^{\mathrm{H}}\}$, $\mathbf{r}_{xd} = \mathbf{r}_{x_0d_0} = E\{\mathbf{x}_0 d_0^{\mathrm{H}}\}$, $\mathbf{B}_p = \mathbf{I}_{MK-1}$, $\mathbf{B}_{ptran} = \mathbf{I}_{MK-1}$
Iterative computation: for $i = 1, 2, \ldots, D$ $\delta_i = \sqrt{\mathbf{r}_{xd}^{\mathrm{H}}\mathbf{r}_{xd}}$, $\mathbf{h}_i = \mathbf{r}_{xd}/\delta_i$, $\sigma_i^2 = \mathbf{h}_i^{\mathrm{H}}\mathbf{R}_x\mathbf{h}_i$, $\mathbf{B}_i = null(\mathbf{h}_i)$, $\mathbf{t}_i = \mathbf{B}_i^{\mathrm{H}}\mathbf{h}_i$, $\mathbf{B}_{ptran} = \mathbf{B}_p^{\mathrm{H}}$, $\mathbf{B}_p = \mathbf{B}_i\mathbf{B}_p$, $\mathbf{r}_{xd} = \mathbf{B}_p\mathbf{R}_{x_0}\mathbf{B}_{ptran}\mathbf{h}_i$, $\mathbf{R}_x = \mathbf{B}_p\mathbf{R}_{x_0}\mathbf{B}_p^{\mathrm{H}}$
Synthetic filter
Initialization: $\xi_D = \sigma_D^2$, $w_D = \delta_D/\xi_D$
Iterative computation: for $i = D-1, D-2, \ldots, 2, 1$ $\xi_i = \sigma_i^2 - \delta_{i+1}^2/\xi_i$, $w_i = \delta_i^2/\xi_i$
Weight computation
Initialization: $\mathbf{w}_r = 0$, $w_p = 1$
Iterative computation: for $i = 1, 2, \ldots, D$ $w_p = -w_p w_i^*$, $\mathbf{w}_r = \mathbf{w}_r + \mathbf{t}_i w_p$ When iteration completes, we can obtain: $\mathbf{w}_{MWF} = \mathbf{w}_r$

Based on the above iterative solution process for the original multistage Wiener filter, we can implement the reduced rank filter for the input observed signal. Below we give the explicit expression of the method, i.e. (2.184)–(2.186). After completing the decomposition, the equivalent reduced rank matrix can be expressed as

$$\mathbf{T}_{MWF} = [\mathbf{t}_1, \mathbf{t}_2, \ldots, \mathbf{t}_D] = \left[\mathbf{h}_1, \mathbf{B}_1^{\mathrm{H}}\mathbf{h}_2, \ldots, \left(\prod_{i=1}^{D-1}\mathbf{B}_i^{\mathrm{H}}\right)\mathbf{h}_D\right] \tag{2.184}$$

$[\mathbf{t}_1, \mathbf{t}_2, \ldots, \mathbf{t}_D]$ is the basis vector of the reduced rank subspace. The equivalent weight vector after rank reduction can be expressed as

$$\mathbf{w}_D = \left[w_1^*, -w_1^* w_2^*, \ldots, (-1)^{D-1}\prod_{i=1}^{D-1} w_i^*\right]^{\mathrm{T}} = \left(\mathbf{T}_{MWF}^{\mathrm{H}}\mathbf{R}_{x_0}\mathbf{T}_{MWF}\right)^{-1}\mathbf{T}_{MWF}^{\mathrm{H}}\mathbf{r}_{x_0d_0} \tag{2.185}$$

The overall weight matrix can be denoted as

$$\mathbf{w}_{MWF} = \mathbf{T}_{MWF}\mathbf{w}_D \tag{2.186}$$

Different block matrices can enable different implementations for multistage Wiener filter. Goldstein, Reed and Scharf, who originally proposed the multistage Wiener filter, provided a method to compute a block matrix in the appendix of

Ref. [58]. The implementation method is known as GRS-MWF. The vector is represented as

$$\mathbf{h}_i = \left[h_{i1}, h_{i2}, \ldots, h_{i(MK-i)}\right] \tag{2.187}$$

Then GRS-MWF's block matrix is

$$\mathbf{B}_i = \begin{bmatrix} \frac{1}{h_{i1}} & \frac{-1}{h_{i2}} & 0 & \cdots & 0 & 0 & 0 \\ 0 & \frac{1}{h_{i2}} & \frac{-1}{h_{i3}} & \cdots & 0 & 0 & 0 \\ \vdots & \vdots & \vdots & \ddots & \vdots & \vdots & \vdots \\ 0 & 0 & 0 & \cdots & \frac{1}{h_{i(MK-i-2)}} & \frac{-1}{h_{i(MK-i-1)}} & 0 \\ 0 & 0 & 0 & \cdots & 0 & \frac{1}{h_{i(MK-i-1)}} & \frac{-1}{h_{i(MK-i)}} \end{bmatrix} \tag{2.188}$$

It can be seen that $\mathbf{B}_i$ is a $(MK - i - 1) \times (MK - i)$ rectangular array. Since $\mathbf{x}_i = \mathbf{B}_i\mathbf{x}_{i-1}$, we know that the dimension of $\mathbf{x}_i$ decrements by 1 as i increases, and that it is very beneficial for reducing computation load and storage requirement. By truncating the multistage Wiener filter after the Dth step ($D < MK - 1$), we can obtain a *D-order* reduced rank multistage filter, and this is also known as a multistage Wiener filter with rank D.

Multistage Wiener filter can also be implemented using the correlation subtraction structure called CSA-MWF, as shown in Fig. 2.39. The structure is equivalent to a block matrix as

$$\mathbf{B}_i = \mathbf{I}_i - \frac{\mathbf{h}_i\mathbf{h}_i^{\mathrm{H}}}{\|\mathbf{h}_i\|} \tag{2.189}$$

Furthermore the analysis filter in Fig. 2.37 can also be simplified. As shown in Fig. 2.39, the CSA-MWF structure is based on a data field, so it does not need to explicitly compute the input observation signal's covariance matrix and block matrix. Therefore, the amount of computation can be greatly reduced compared with the GRS-MWF structure.

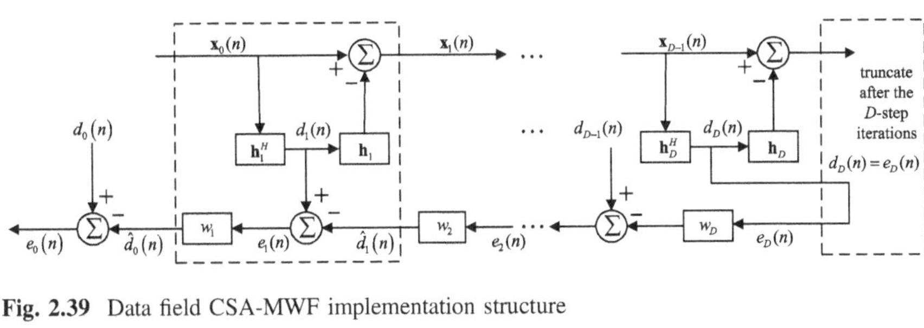

Fig. 2.39 Data field CSA-MWF implementation structure

Table 2.3 Detailed steps for data field CSA-MWF algorithm

Synthetic filter
Initialization: $d_0(\mathrm{n}) = \boldsymbol{\delta}_{MK}^{\mathrm{H}}\mathbf{x}(\mathrm{n})$, $\mathbf{x}_0(\mathrm{n}) = \mathbf{x}(\mathrm{n}) - \boldsymbol{\delta}_{MK} d_0(\mathrm{n})$
Iterative computation: for $i = 1, 2, \ldots, D$ $\mathbf{h}_i = E\{\mathbf{x}_{i-1}(\mathrm{n})d_{i-1}^*(\mathrm{n})\}/\|E\{\mathbf{x}_{i-1}(\mathrm{n})d_{i-1}^*(\mathrm{n})\}\|$, $d_i(\mathrm{n}) = \mathbf{h}_i^{\mathrm{H}}\mathbf{x}_{i-1}(\mathrm{n})$, $\mathbf{x}_i(\mathrm{n}) = \mathbf{x}_{i-1}(\mathrm{n}) - \mathbf{h}_i(\mathrm{n})d_i(\mathrm{n})$
Synthetic filter
Initialization: $e_D(\mathrm{n}) = d_D(\mathrm{n})$
Synthetic filter
Iterative computation: for $i = D, D-1, \ldots, 2, 1$ $w_i = E\{e_i(\mathrm{n})d_{i-1}^*(\mathrm{n})\}/E\{\|e_i(\mathrm{n})\|^2\}$, $e_{i-1}(\mathrm{n}) = d_{i-1}(\mathrm{n}) - w_i^* e_i(\mathrm{n})$
Weight calculation
Initialization: $w_p = 1$
Iterative computation: for $i = 1, 2, \ldots, D$ $w_p = -w_p w_i^*$, $w_i = w_p$

The detailed steps for the data field CSA-MWF algorithm iterative process are shown in Table 2.3.

Below we give the explicit expression for the method, i.e. (2.190)–(2.192). After completing the decomposition, the equivalent reduced rank matrix can be expressed as

$$\mathbf{T}_{MWF} = [\boldsymbol{\delta}_{MK}, \mathbf{h}_1, \ldots, \mathbf{h}_D] \tag{2.190}$$

After rank reduction, the equivalent weight vector can be expressed as

$$\mathbf{w}_D = [1, w_1, \ldots, w_D]^{\mathrm{T}} \tag{2.191}$$

The overall weight vector can be denoted as

$$\mathbf{w}_{MWF} = \mathbf{T}_{MWF}\mathbf{w}_D \tag{2.192}$$

4. Multistage Iterative Reduced Rank Method

It can be discovered from the Multistage Nested Wiener Filter introduced above that these methods use iterative computation to avoid the processes of covariance matrix inverse and feature decomposition so the amount of computation can be reduced. Among them, the data field's CSA-MWF implementation structure improves the computational complexity on the analysis filter by not calculating the covariance matrix, as well as reduces every iteration's computation complexity by calculating cross correlations among vectors to derive new desired signal and observed data. This method then iterates backwards in the synthetic filter to determine the weight vector. Even though it is a scalar Weiner filter structure, but they nest with each other so that they can not be realized using parallel computation, thereby the overall real time performance of the algorithm suffers. The synthetic filter structure can be improved by making it execute one iteration for every added step, e.g. compute the weight vector in parallel. This parallel multistage iterative structure after improving both the analysis filter and synthetic filter is known as the Multistage Iterative Reduced Rank method.

From (2.185) and (2.186), the overall weight vector can be represented as

$$\mathbf{w}_{MWF} = \mathbf{T}\left(\mathbf{T}^{\mathrm{H}}\mathbf{R}_{x_0}\mathbf{T}\right)^{-1}\mathbf{T}^{\mathrm{H}}\mathbf{r}_{x_0 d_0} \tag{2.193}$$

where we can simplify so that $\mathbf{T} = \mathbf{T}_{MWF}$. For the multistage Wiener filter structure shown in Fig. 2.36 with a rank of D, the ith desired signal after analysis filter decomposition can be written as

$$d_i = \left(\prod_{j=1}^{i-1}\mathbf{B}_j^{\mathrm{H}}\right)\mathbf{h}_i\mathbf{x}_0 = \mathbf{t}_i^{\mathrm{H}}\mathbf{x}_0 \tag{2.194}$$

By writing various stages of desired signals into vector form representation, we have

$$\mathbf{d}^{(D)} = [d_1,\ d_2,\ \ldots,\ d_D]^{\mathrm{T}} = \left[\mathbf{T}^{(D)}\right]^{\mathrm{H}}\mathbf{x}_0 \tag{2.195}$$

By substituting (2.195) into (2.193), we can obtain the overall weight vector as

$$\mathbf{w}_{MWF} = \mathbf{T}^{(D)}\left[\mathbf{R}_d^{(D)}\right]^{-1}\mathbf{r}_{dd_0} \tag{2.196}$$

where $\mathbf{R}_d^{(D)} = E\{\mathbf{d}^{(D)}[\mathbf{d}^{(D)}]^{\mathrm{H}}\} = [\mathbf{T}^{(D)}]^{\mathrm{H}}\mathbf{R}_{x_0}\mathbf{T}^{(D)}$ is a transformation covariance matrix, and we derive

$$\begin{aligned}
&\mathbf{R}_d^{(D)}(i,i) = E\left\{d_i d_i^*\right\} = \sigma_{d_i}^2 \\
&\mathbf{R}_d^{(D)}(i,i-1) = E\left\{d_i d_{i-1}^*\right\} = E\left\{\mathbf{h}_i^{\mathrm{H}}\mathbf{x}_{i-1}d_{i-1}^*\right\} = \mathbf{h}_i^{\mathrm{H}}\mathbf{r}_{x_{i-1}d_{i-1}} = \delta_i \\
&\mathbf{R}_d^{(D)}(i-1,i) = E\left\{d_{i-1}d_i^*\right\} = E\left\{d_{i-1}\mathbf{x}_{i-1}^{\mathrm{H}}\mathbf{h}_i\right\} = \mathbf{r}_{\mathbf{x}_{i-1}d_{i-1}}^{\mathrm{H}}\mathbf{h}_i = \delta_i, \quad i = 2,3,\ldots,D
\end{aligned} \tag{2.197}$$

Therefore, $\mathbf{R}_d^{(D)}$ is a real number matrix in the form of three diagonal matrices

$$\mathbf{R}_d^{(D)} = \begin{bmatrix} \sigma_{d_1}^2 & \delta_2 & 0 & \cdots & 0 & 0 \\ \delta_2 & \sigma_{d_2}^2 & \delta_3 & \cdots & 0 & 0 \\ 0 & \delta_3 & \sigma_{d_3}^2 & \ddots & 0 & 0 \\ \vdots & \vdots & \vdots & \ddots & \delta_{D-1} & 0 \\ 0 & 0 & 0 & \delta_{D-1} & \sigma_{d_{D-1}}^2 & \delta_D \\ 0 & 0 & 0 & 0 & \delta_D & \sigma_{d_D}^2 \end{bmatrix} = \begin{bmatrix} \mathbf{R}_d^{(D-1)} & & \mathbf{0} \\ & & \delta_D \\ \mathbf{0} & \delta_D & \sigma_{d_D}^2 \end{bmatrix} \tag{2.198}$$

Also in (2.196)

$$\mathbf{r}_{dd_0} = E\left\{\mathbf{d}^{(D)} d_0^*\right\} = E\left\{\left[\mathbf{T}^{(D)}\right]^{\mathrm{H}} \mathbf{x}_0 d_0^*\right\} = \left[\mathbf{T}^{(D)}\right]^{\mathrm{H}} \mathbf{r}_{x_0 d_0} = \left[\mathbf{T}^{(D)}\right]^{\mathrm{H}} \mathbf{h}_1 \delta_1 = \begin{bmatrix} \delta_1 \\ \mathbf{0} \end{bmatrix} \tag{2.199}$$

It can be known from (2.199) that the weight vector in (2.196) only needs to be used in the first row of $[\mathbf{R}_d^{(D)}]^{-1}$, and multiply it by δ_1. By defining

$$\mathbf{C}^{(D)} = \left[\mathbf{C}_1^{(D)},\ \mathbf{C}_2^{(D)},\ \ldots,\ \mathbf{C}_D^{(D)}\right] = \left[\mathbf{R}_d^{(D)}\right]^{-1} \tag{2.200}$$

and further expanding (2.200) based on the block Hermite matrix inversion formula, we can obtain

$$\mathbf{C}^{(D)} = \left[\mathbf{R}_d^{(D)}\right]^{-1} = \begin{bmatrix} \mathbf{C}^{(D-1)} & \mathbf{0} \\ \mathbf{0} & 0 \end{bmatrix} + \beta_D^{-1} \mathbf{b}^{(D)} \left[\mathbf{b}^{(D)}\right]^{\mathrm{H}} \tag{2.201}$$

$$\beta_D = \sigma_{d_D}^2 - \delta_D^2 \mathbf{C}_{D-1,D-1}^{(D-1)} \tag{2.202}$$

$$\mathbf{b}^{(D)} = \begin{bmatrix} -\delta_D \mathbf{C}_{D-1}^{(D-1)} \\ 1 \end{bmatrix} \tag{2.203}$$

where $\mathbf{C}_{i,j}^{(D)}$ denotes the ith element of the jth column $\mathbf{C}_j^{(D)}$. We only need the first row $\mathbf{C}_1^{(D)}$ of $\left[\mathbf{R}_d^{(D)}\right]^{-1}$ to compute the weight vector in (2.196).

$$\mathbf{C}_1^{(D)} = \begin{bmatrix} \mathbf{C}_1^{(D-1)} \\ 0 \end{bmatrix} + \beta_D^{-1} \mathbf{C}_{1,D-1}^{(D-1)} \begin{bmatrix} \delta_D^2 \mathbf{C}_{D-1}^{(D-1)} \\ -\delta_D \end{bmatrix} \tag{2.204}$$

where $\mathbf{C}_D^{(D)} = \beta_D^{-1} \begin{bmatrix} -\delta_D \mathbf{C}_{D-1}^{(D-1)} \\ 1 \end{bmatrix}$, $\mathbf{C}_{D,D}^{(D)} = \beta_D^{-1}$, $\beta_D = \sigma_{d_D}^2 - \delta_D^2 \beta_{D-1}^{-1}$. Since $\mathbf{C}_1^{(D)}$ can be solved iteratively based on (2.204), the overall weight vector is

$$\mathbf{w}_{MWF} = \delta_1 \mathbf{T}^{(D)} \mathbf{C}_1^{(D)} \tag{2.205}$$

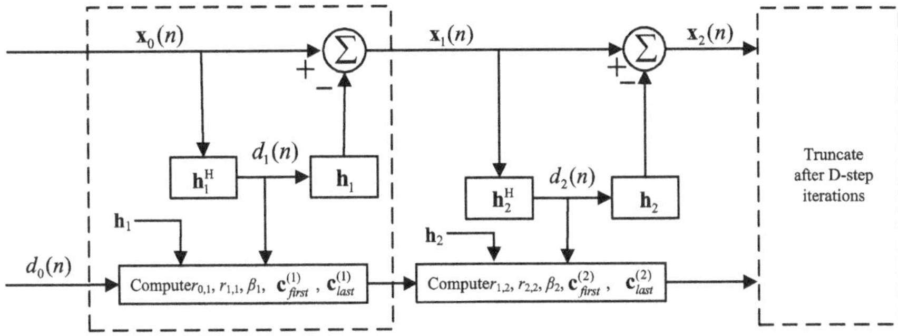

Fig. 2.40 Implementation structure of multi-step iteration

From (2.205), we extrapolate that for every incremental stage of the filter, only one iteration needs to be added, so that the recursive updating for backward iteration can be avoided. If the analysis filter adopts the CSA-MWF implementation structure, and at the same time the synthetic filter adopts the backward iterative algorithm, the constructed multistage iterative implementation structure is shown in Fig. 2.40.

Detailed steps of multi-step iterative algorithm are listed in Table 2.4.

We can summarize the various reduced rank methods introduced above: When the jamming subspace components are all contained in the reduced rank matrix **T**, the principal component method has very good reduced rank performance; When there exist a characteristic spectrum expansion and subspace leakage, the cross spectral metric method has better data compression effect; The multistage nested

Table 2.4 Detailed steps of the multi-step iteration algorithm

Initialization
$d_0(\mathrm{n}) = \boldsymbol{\delta}_{MK}^H \mathbf{x}(\mathrm{n})$, $\mathbf{x}_0(\mathrm{n}) = \mathbf{x}(\mathrm{n}) - \boldsymbol{\delta}_{MK} d_0(\mathrm{n})$, $\mathbf{h}_1 = E\{\mathbf{x}_0(\mathrm{n}) d_0^*(\mathrm{n})\} / \left\| E\{\mathbf{x}_0(\mathrm{n}) d_0^*(\mathrm{n})\}\right\|$, $d_1(\mathrm{n}) = \mathbf{h}_1^H \mathbf{x}_0(\mathrm{n})$, $\mathbf{x}_1(\mathrm{n}) = \mathbf{x}_0(\mathrm{n}) - \mathbf{h}_1(\mathrm{n}) d_1(\mathrm{n})$, $r_{0,1} = 0$, $r_{1,1} = E\{d_1(\mathrm{n}) d_1^*(\mathrm{n})\}$, $\mathbf{c}_{\mathrm{first}}^{(1)} = r_{1,1}^{-1}$, $\mathbf{c}_{\mathrm{last}}^{(1)} = r_{1,1}^{-1}$
Iterative computation
for $i = 2 : MK - 1$ $\mathbf{h}_i = E\{\mathbf{x}_{i-1}(\mathrm{n}) d_{i-1}^*(\mathrm{n})\} / \left\| E\{\mathbf{x}_{i-1}(\mathrm{n}) d_{i-1}^*(\mathrm{n})\}\right\|$, $d_i(\mathrm{n}) = \mathbf{h}_i^H \mathbf{x}_{i-1}(\mathrm{n})$, $\mathbf{x}_i(\mathrm{n}) = \mathbf{x}_{i-1}(\mathrm{n}) - \mathbf{h}_i(\mathrm{n}) d_i(\mathrm{n})$, $r_{i,i} = E\{d_i(\mathrm{n}) d_i^*(\mathrm{n})\}$, $r_{i-1,i} = E\{d_{i-1}(\mathrm{n}) d_i^*(\mathrm{n})\}$, $\beta_i = r_{i,i} - \left\|r_{i-1,i}\right\|^2 \mathbf{c}_{\mathrm{last},i-1}^{(i-1)}$, $\mathbf{c}_{\mathrm{first}}^{(i)} = \begin{bmatrix} \mathbf{c}_{\mathrm{first}}^{(i-1)} \\ 0 \end{bmatrix} + \beta_i^{-1} \mathbf{c}_{\mathrm{last},1}^{(i-1),*} \begin{bmatrix} \left\|r_{i-1,i}\right\|^2 \mathbf{c}_{\mathrm{last}}^{(i-1)} \\ -r_{i-1,i}^* \end{bmatrix}$, $\mathbf{c}_{\mathrm{last}}^{(i)} = \beta_i^{-1} \begin{bmatrix} -r_{i-1,i} \mathbf{c}_{\mathrm{last}}^{(i-1)} \\ 1 \end{bmatrix}$, $MSE^{(i)} = \sigma_{d_0}^2 - \left\|\mathbf{r}_{x_0 d_0}\right\|_2^2 \mathbf{c}_{\mathrm{first},1}^{(i)}$ If $MSE^{(D)} < MSE_{\mathrm{threshold}}$, Then $D = i$, jump out
Weight calculation
$\mathbf{T}_D = [\mathbf{t}_1, \mathbf{t}_2, \ldots, \mathbf{t}_D] = [\mathbf{h}_1, \ldots, \mathbf{h}_D]$, $\mathbf{w}_{MWF} = \left\|\mathbf{r}_{x_0 d_0}\right\|_2 \mathbf{T}_D \mathbf{c}_{\mathrm{first}}^{(D)}$

Wiener filter method avoids performing matrix inverse and eignevalue decomposition; The multistage Wiener filter with correlated subtractive structure does not need to explicitly compute observed data's covariance matrix, and does not need to compute the block matrix, therefore it can further reduce dynamic range and amount of computation. The multi-step iterative reduced rank method achieves parallel computation of the weight vector, so the real-time performance of the reduced rank algorithm improves significantly.

2.6.5 *Impacts on GNSS Signal by Space-Time Processing and Equalization Technique*

Since a space-time processing method performs jamming suppression in the joint space-time domain, frequency responses over the whole processing bandwidth are not consistent, inevitably leading to distortions on GNSS signals, as a result cross correlation functions between array outputs and local signals have phenomena such as mainlobe broadening, cross correlation peak shift etc. Those phenomena have impacts on GPS signal acquisition, pseudo-range measurement, and determination of user positions [61]. This section analyzes the impacts of space-time processing on GNSS signals, and introduces some existing main STAP compensation techniques, including weight constraint [61], temporal-domain filter [61], least squares inverse filtering and equalization method based on homomorphic filtering principle [62].

Array output signal after the space-time adaptive jamming mitigation is

$$y(n) = \mathbf{w}^{\mathrm{H}}\mathbf{x}(n) = \mathbf{w}^{\mathrm{H}}\mathbf{A}\mathbf{s}(n) + \mathbf{w}^{\mathrm{H}}\mathbf{j}(n) + \mathbf{w}^{\mathrm{H}}\mathbf{e}(n) \tag{2.206}$$

where $\mathbf{x}(n)$ denotes the space-time two-dimension snapshot; $\mathbf{w}$ denotes the adaptive weight vector corresponding to the STAP algorithm; $\mathbf{A} = \mathbf{I}_{K\times K} \otimes \mathbf{a}(\theta)$, $\mathbf{a}(\theta) = \left[1, \mathrm{e}^{-\mathrm{j}\frac{2\pi d \sin\theta}{\lambda}}, \ldots, \mathrm{e}^{-\mathrm{j}\frac{2\pi(M-1)d\sin\theta}{\lambda}}\right]^{\mathrm{T}}$ is related to the actual direction of arrival θ for GNSS signals; $\mathbf{s}(n)$ is the *K-stage* delayed GNSS signal; $\mathbf{j}(n)$ and $\mathbf{e}(n)$, respectively, are the jamming and noise components of space-time two dimensional snapshot.

To suppress and eliminate the jamming, it is equivalent to

$$\mathbf{w}^{\mathrm{H}}\mathbf{j}(n) \approx 0 \tag{2.207}$$

Then the output of the array is

$$y(n) \approx \mathbf{w}^{\mathrm{H}}\mathbf{A}\mathbf{s}(n) + \mathbf{w}^{\mathrm{H}}\mathbf{e}(n) = \mathbf{w}^{\mathrm{H}}\mathbf{A}\mathbf{s}(n) + \bar{e}(n) \tag{2.208}$$

where $\bar{e}(n) = \mathbf{w}^{\mathrm{H}}\mathbf{e}(n)$ represents noise output.

After the STAP processing, uniform linear array response coefficients for different time delays can be represented using a $K \times 1$ dimension vector $\mathbf{h} = (\mathbf{w}^{\mathrm{H}}\mathbf{A})^{\mathrm{T}}$, i.e.

$$\mathbf{h} = [h(0),\ h(1),\ \ldots,\ h(K-1)]^{\mathrm{T}} \tag{2.209}$$

where

$$h(k-1) = \sum_{m=1}^{M} w_{mk} \mathrm{e}^{-\mathrm{j}\frac{2\pi(M-1)d\sin\theta}{\lambda}} \tag{2.210}$$

By transforming (2.210) to frequency domain, the GNSS signal's frequency response can be obtained using space-time processing

$$\begin{aligned} H(\omega) &= \sum_{k=1}^{K} h(k-1)\mathrm{e}^{-\mathrm{j}\omega(k-1)T} \\ &= \sum_{k=1}^{K}\sum_{m=1}^{M} w_{mk}\mathrm{e}^{\left[-\mathrm{j}\frac{2\pi(M-1)d\sin\theta}{\lambda} - \mathrm{j}\omega(k-1)T\right]} \end{aligned} \tag{2.211}$$

If it is assumed that the frequency response of a satellite transmitting signal $s(t)$ is $S(\omega)$, then the frequency response for the output after the STAP processing is $S(\omega)H(\omega)$. Therefore, we can obtain the output signal after the STAP for the GNSS signal, and it has the following format

$$y(n) = \frac{1}{2\pi}\int_{-\pi}^{\pi} S(\omega)H(\omega)\mathrm{e}^{\mathrm{j}\omega n}\mathrm{d}\omega \tag{2.212}$$

It can be known from distortion-free transmission conditions $H(\omega) = K\mathrm{e}^{-\mathrm{j}\omega t_0}$ (where K and t_0 are constants) that when the STAP's frequency response to the GNSS signal $H(\omega)$ does not satisfy the above conditions, the GNSS signal has distortion.

To compensate for the impacts on the GNSS signal by the STAP process, scholars all over the world have performed related research studies. Among them, the weight constraint method proposed by professor Hatke from MIT [61] performs equalization on all distortions of the GNSS signal, to make their impacts on the distortions the same. This method adds an orthogonal constraint condition to solve new weights for different jamming mitigation algorithms. The MITRE Corporation in the US proposed an equalization algorithm based on a temporal-domain filter [61] to make different GNSS beams' STAP responses consistent, i.e. let every GNSS signal's beam output convolute with STAP responses of other GNSS signals using an added temporal-domain filter. The University of Chinese National Defense proposed the method of adding an inverse filter. The method, based on the principle of deterministic least squares inverse filter, solves for inverse filter coefficients for every GNSS signal, and adds array outputs, to recover the desired GPS signal. Since linear filter has difficulty separating the product signal or convolution signal, the China Civil Aviation University proposed an equalization algorithm based on homomorphic filtering [62]. The method transforms a convolution type combined signal to an additive signal, then designs a filter to separate signals in it, so that responses of STAP at different delay taps are consistent. Below we introduce the basic principles of these methods one by one.

1. Weighted Constraint Equalization Algorithm

Weighted Constraint Equalization Algorithm [61] directly adds constraint conditions on jamming mitigation weights of the STAP process, and re-solves for new weights. From (2.211) we can obtain the response of the space-time processing on the GNSS signal as: $\mathbf{h} = [h(0), h(1), h(2), \ldots, h(K-1)]^{\mathrm{T}}$, $\mathbf{h} = (\mathbf{w}^{\mathrm{H}}\mathbf{A})^{\mathrm{T}}$ is a $K \times 1$ dimension vector. The weighted constraint method, in terms of a specific satellite, means that the group delay of the constraint $\mathbf{h}$ is a fixed value. And the constraint can be achieved using the following orthogonal constraint condition

$$\mathbf{h}^{\mathrm{T}}\partial\mathbf{c} = 0 \tag{2.213}$$

where

$$\partial\mathbf{c} = \begin{bmatrix} \partial R_s(\Delta T(D-1)) \\ \cdots \\ \partial R_s(\Delta T(D-K)) \end{bmatrix} \Big/ \partial\tau \tag{2.214}$$

$D\Delta T$ is the fixed time delay; $R_s(\tau)$ is the auto-correlation of the GNSS signal; $\partial(\bullet)$ represents the derivative operation.

This method, for different space-time adaptive jamming mitigation processing algorithms, needs to have additional orthogonal constraint conditions. For example, by adding a constraint condition for the Capon jamming mitigation algorithm $\mathbf{w}^{\mathrm{H}}\mathbf{A}\partial\mathbf{c} = 0$, the solution of weights can be achieved using the formula below

$$\begin{aligned} &\min \ \mathbf{w}^{\mathrm{H}}\mathbf{R}_x\mathbf{w} \\ &\text{s.t.} \ \begin{cases} \mathbf{A}^{\mathrm{H}}\mathbf{w} = \boldsymbol{\beta}_{K\times 1} \\ \mathbf{w}^{\mathrm{H}}\mathbf{A}\partial\mathbf{c} = 0 \end{cases} \end{aligned} \tag{2.215}$$

where $\boldsymbol{\beta}_{K\times 1}$ is a constant vector. Let $\mathbf{B} = [\mathbf{A}, \ \ \mathbf{A}\partial\mathbf{c}]$, $\mathbf{C} = [\boldsymbol{\beta}^{\mathrm{H}}, \ \ 0]$. By writing (2.215) in the form of augmented matrix, weights can be recalculated as

$$\begin{aligned} &\min \ \mathbf{w}^{\mathrm{H}}\mathbf{R}_x\mathbf{w} \\ &\text{s.t.} \ \ \mathbf{w}^{\mathrm{H}}\mathbf{B} = \mathbf{C} \end{aligned} \tag{2.216}$$

Using the Lagrange method to solve weights, the new weights can be obtained as

$$\mathrm{w} = \mathbf{R}_x^{-1}\mathbf{B}\left(\mathbf{B}^{\mathrm{H}}\mathbf{R}_x^{-1}\mathbf{B}\right)^{-1}\mathbf{C}^{\mathrm{H}} \tag{2.217}$$

2. Temporal domain filter equalization algorithm

Temporal domain filter equalization algorithm [61] makes different GNSS' STAP processing responses consistent by adding temporal domain filters. The goal of this method is to convolute one GNSS signal's beam output with other GNSS signals' STAP responses. Thereby, the minimum order of filters for this equalization algorithm is $3K$–2.

Assuming signals from four GNSS satellites incident onto the array, the output signals after array jamming mitigation for every satellite are y_1, y_2, y_3 and y_4 respectively, and the space-time responses on the four satellites are $\mathbf{h}_1$, $\mathbf{h}_2$, $\mathbf{h}_3$ and $\mathbf{h}_4$ respectively. To make the responses of STAP processes the same for all four GNSS signals, we perform the computation below, to obtain four new output signals $y_{1\Delta}(n)$, $y_{2\Delta}(n)$, $y_{3\Delta}(n)$ and $y_{4\Delta}(n)$.

$$y_{1\Delta}(n) = y_1(n) * \mathbf{h}_2 * \mathbf{h}_3 * \mathbf{h}_4 = \left\{\mathbf{w}_1^{\mathrm{H}}\mathbf{x}(n)\right\} * \mathbf{h}_2 * \mathbf{h}_3 * \mathbf{h}_4 \tag{2.218}$$

$$y_{2\Delta}(n) = y_2(n) * \mathbf{h}_1 * \mathbf{h}_3 * \mathbf{h}_4 = \left\{\mathbf{w}_2^{H}\mathbf{x}(n)\right\} * \mathbf{h}_1 * \mathbf{h}_3 * \mathbf{h}_4 \tag{2.219}$$

$$y_{3\Delta}(n) = y_3(n) * \mathbf{h}_1 * \mathbf{h}_2 * \mathbf{h}_4 = \left\{\mathbf{w}_3^{\mathrm{H}}\mathbf{x}(n)\right\} * \mathbf{h}_1 * \mathbf{h}_2 * \mathbf{h}_4 \tag{2.220}$$

$$y_{4\Delta}(n) = y_4(n) * \mathbf{h}_1 * \mathbf{h}_2 * \mathbf{h}_3 = \left\{\mathbf{w}_4^{\mathrm{H}}\mathbf{x}(n)\right\} * \mathbf{h}_1 * \mathbf{h}_2 * \mathbf{h}_3 \tag{2.221}$$

where $*$ denotes the convolution operation.

After using this method of adding temporal domain filter, we can make the impacts of STAP processes the same for different GNSS signals. The four output signals can have correlation processing with different local spread spectrum codes, and the obtained cross-correlations are consistent. But this method can narrow the signal spread spectrum. For example, the frequency spectrum of the first GNSS becomes

$$Y_l(\omega) = H_1(\omega)S(\omega)H_2(\omega)H_3(\omega)H_4(\omega) \tag{2.222}$$

This also leads to the expansion of correlation function.

3. Least Squares Inverse Filter Equalization Algorithm

The equalization compensation method based on deterministic least squares inverse filter solves an inverse filter

$$\mathbf{g} = [g(0),\ g(1),\ \ldots,\ g(K)]^{\mathrm{T}} \tag{2.223}$$

which minimizes the error energy $V(\mathbf{g}) = \|\delta - \mathbf{h} * \mathbf{g}\|$, δ denotes an unit impulse function. To solve this inverse filter, we need to solve the derivative of $V(\mathbf{g})$ related to $\mathbf{g}$, and make that derivative as zero. Based on this, we can derive

$$\sum_{k=0}^{K} g(k)r(k-l) = q(l), \quad l = 0,\ 1,\ \ldots,\ K \tag{2.224}$$

where

$$\begin{aligned} r(k-l) &= \sum_{i=-\infty}^{+\infty} h(i-l)h^*(i-k) \\ &= \sum_{i=0}^{+\infty} h(i)h^*(i-(l-k)), \quad l = 0, 1, \ldots, K \end{aligned} \tag{2.225}$$

$$\begin{aligned} q(l) &= \sum_{k=-\infty}^{+\infty} \delta(k)h(k-l) = h(-l), \quad l = 0, 1, \ldots, K \\ &= \begin{cases} h(0) & l = 0 \\ 0 & l > 0 \end{cases} \end{aligned} \tag{2.226}$$

In matrix form, it is

$$\begin{bmatrix} r(0) & r(1) & \cdots & r(K) \\ r(-1) & r(0) & \cdots & r(K-1) \\ \vdots & \vdots & & \vdots \\ r(-K) & r(-(K-1)) & \cdots & r(0) \end{bmatrix} \begin{bmatrix} g(0) \\ g(1) \\ \vdots \\ g(K) \end{bmatrix} = \begin{bmatrix} h(0) \\ 0 \\ \vdots \\ 0 \end{bmatrix} \tag{2.227}$$

Then it becomes convenient to solve for the inverse filter's coefficients $\mathbf{g}$.

4. Equalization Algorithm Based on Homomorphic Filtering

Homomorphic filtering [63] can be used to separate or process non-additive combinational signals (e.g. product-type signal and convolution-type signal). The basic approach of the method is to perform Fourier transform or Z-transform and logarithmic transformation on mixed signals, so that the product-type signal and convolution-type signal can be converted to additive signals. After performing signal separation using linear systems, the signals can be converted back.

It can be known from (2.208) that the array output after the STAP jamming mitigation processing can be represented as

$$
\begin{aligned}
y(n) &\approx \mathbf{w}^{\mathrm{H}}\mathbf{A}\mathbf{s}(n) + \bar{e}(n) = \mathbf{h}^{\mathrm{T}}\mathbf{s}(n) + \bar{e}(n) \\
&= [h(0), h(1), \ldots, h(K-1)] \begin{bmatrix} s(n) \\ s(n-1) \\ \vdots \\ s(n-(K-1)) \end{bmatrix} + \bar{e}(n) \\
&= \sum_{k=0}^{K-1} h(k)s(n-k) + \bar{e}(n)
\end{aligned} \tag{2.228}
$$

It can be seen that unlike the expected receiving GNSS signal $s(n)$, array output is equivalent to the process of convoluting the array response and the signal, i.e.

$$
y(n) = h * s(n) + \bar{e}(n) \tag{2.229}
$$

It can be seen in (2.229) that the impact on the signal by the STAP process can be seen as a convolution distortion. Therefore, the correlation function between the derived array output signal and the local spread spectrum code has distortion, leading to errors in positioning results.

Below this method of homomorphic filtering can be used to isolate the GNSS signals. First, Fourier transformations are performed on both sides of (2.229) to convert the convolution distortion to a product-type distortion.

$$
\begin{aligned}
Y(\omega) &= H(\omega)S(\omega) + \bar{E}(\omega) \\
&= H(\omega)[S(\omega) + E_1(\omega)]
\end{aligned} \tag{2.230}
$$

where $\bar{E}(\omega) = H(\omega)E_1(\omega)$. Then let $S_1(\omega) = S(\omega) + E_1(\omega)$, then (2.230) can be written as

$$
Y(\omega) = H(\omega)S_1(\omega) \tag{2.231}
$$

After taking the logarithm, the product-type distortion can be converted to an additive distortion, and we have

$$
\ln(Y(\omega)) = \ln(H(\omega)S_1(\omega)) = \ln(H(\omega)) + \ln(S_1(\omega)) \tag{2.232}
$$

We then subtract the additive distortion after the transformation

$$
\ln(S_1(\omega)) = \ln(Y(\omega)) - \ln(H(\omega)) \tag{2.233}
$$

By taking exponential operations on both sides of (2.23), we can obtain

$$
S_1(\omega) = \exp[\ln(Y(\omega)) - \ln(H(\omega))] \tag{2.234}
$$

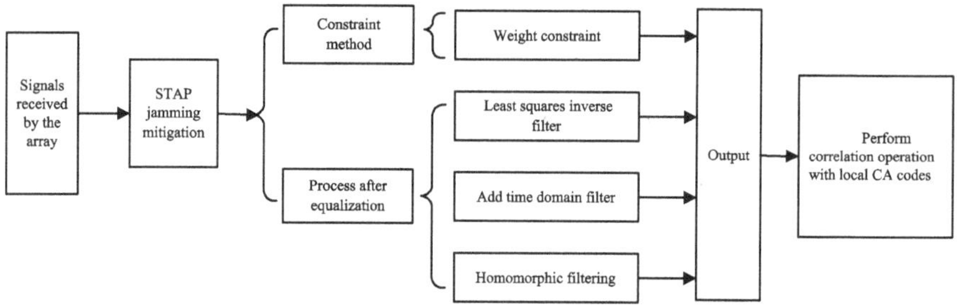

Fig. 2.41 STAP jamming mitigation and equalization compensation process

It can be known from (2.230) that $S_1(\omega)$ is the GNSS signal frequency response under the influence of noise. By performing inverse Fourier transformation on (2.234), the obtained signal is $s_1(n) = s(n) + e_1(n)$, we then can recover the desired GNSS signal, but the obtained signal still contains the noise component $e_1(n)$.

The four methods described above compensate for STAP process, and all need to use the GNSS signal steering vector. Among them, the weight constraint method needs to re-add orthogonal constraint conditions for different jamming mitigation algorithms to solve for the new weights, but the added computation is mainly due to matrix multiplication, so it adds relatively small loads. Every satellite beam output for the temporal domain filter equalization algorithm needs to convolute other GNSS signals' response coefficients with different time delays, to let the correlation function have certain degrees of expansion and time delay, so that the added computation is for multiple convolution operations. When it is used to solve the inverse filter coefficients for every GNSS signal, the least squares inverse filter algorithm needs to have the covariance matrix of a GNSS signal's STAP response, so it increases computational load. Equalization algorithm based on homomorphic filtering uses the homomorphic filtering to separate out the GNSS signal, so that it is compatible with algorithms for estimating GNSS signal's steering vector such as the space-time de-spread re-spread method. The additional amount of computation is mainly for fast Fourier transformation, logarithmic operation, and exponential operation, so the amount of computation is still relatively small. Figure 2.41 shows the STAP jamming mitigation and equalization compensation process.

2.6.6 Simulation Results

Below we perform simulations to compare the performances of various algorithms described in this section. For simulations, we use an equidistant linear array with the number of array elements being $M = 4$, and the interval between array elements as half a wavelength. The signals received by the array can be down-converted to the intermediate frequency 4.309 MHz, and the sampling rate is 5.714 MHz. After every array element, three time taps are connected to form a space-time processing structure, and every time tap has a one-sample point delay. We set the following simulation environment: five satellites PRN2, PRN6, PRN14, PRN20 and PRN25, and their signals incident onto the antenna array with angles of −15°, 20°, −5°, 5°, and 15°. The signal-to-noise ratio is −20 dB. Two single-frequency jamming signals with frequencies of 2.5948 MHz and 3.7376 MHz, incident onto the antenna array with angles of −50° and 70°, and the jamming-to-noise ratio is 40 dB. Two wideband jamming signals incident on the antenna array with angles of −70° and 50°, and the jamming-to-noise ratio is 40 dB.

1. Simulation Result Comparisons Between Spatial Domain Processing and Space-Time Processing

When both wideband jamming and single frequency jamming exist, the simulation results of spatial domain jamming mitigation and space-time jamming mitigation are compared in Fig. 2.42. Spatial domain processing uses the spatial domain power minimization algorithm, and space-time processing can use the space-time power minimization algorithm and space-time de-spread and re-spread algorithm described in this chapter. Figure 2.42a shows the antenna pattern of the spatial domain processing, where the dotted position represents the direction of the jamming signal. It can be seen in the figure that not enough nulls are formed towards the jamming directions for the spatial domain processing due to the constraints on the degrees of freedom. Figure 2.42b–e show the space-time domain processing's space-time two-dimension response graphs and the corresponding top-view graphs. In these graphs, we mark the GNSS signal (PRN2) and single-frequency jamming's space-time two-dimension coordinates on the top view graph. The nulls of wideband jamming aim towards the directions of −70° and 50°. It can be seen from the subfigures that the space-time power minimization algorithm and the space-time de-spread re-spread algorithm can form nulls towards the jamming directions since they have more degrees of freedom. The space-time de-spread re-spread algorithm can use more a priori information, so that the obtained space-time two-dimension response graph has its mainlobe aiming towards the GNSS signal's space-time two-dimension coordinates, to achieve array signal processing gain.

Figure 2.43 further explains the performances of spatial domain jamming mitigation and space-time jamming mitigation processes by acquiring PRN2. In the figure, the x-axis denotes the chip delay, the y-axis denotes the Doppler frequency, and the z-axis denotes the accumulated correlation. Figure 2.43a is the acquisition result after spatial domain processing. It can be seen that since the spatial domain processing can not effectively eliminate jamming impacts, the correlation output does not have obvious peaks, so that it can not acquire the GNSS. Figure 2.43b, c show acquisition results using the space-time power minimization algorithm and the space-time de-spread re-spread algorithm. It can be seen that both subfigures have obvious peaks, so they show that the GNSS signal PRN2 can be successfully

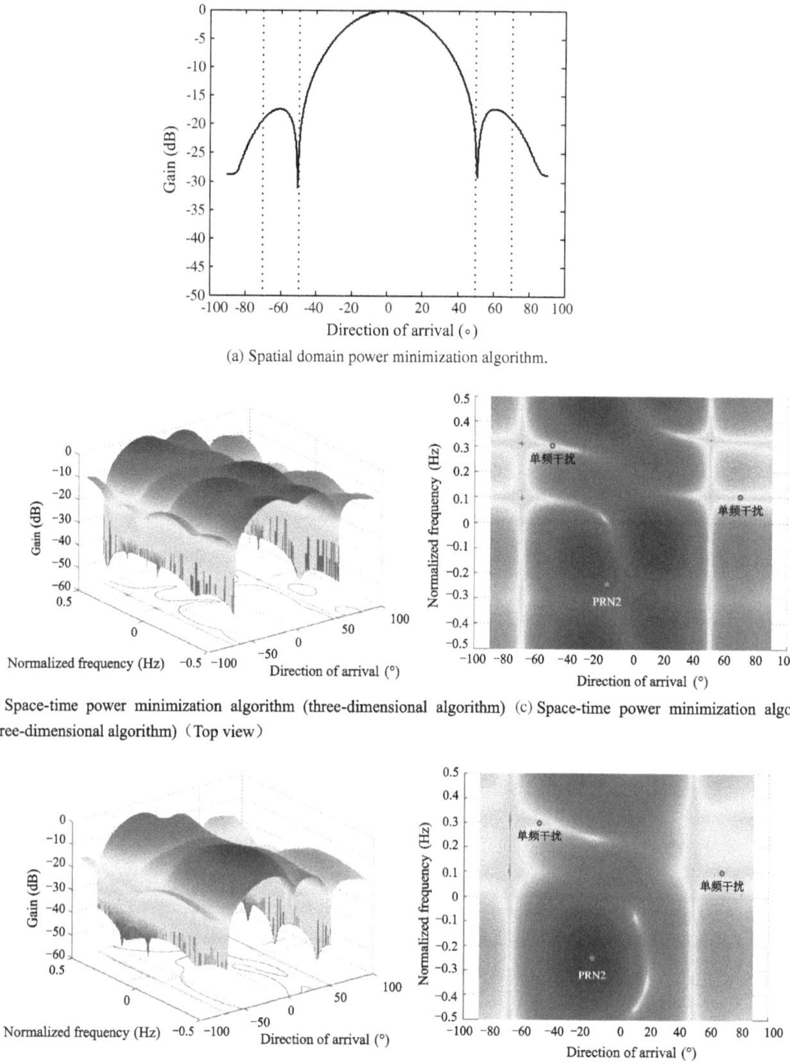

(a) Spatial domain power minimization algorithm.

(b) Space-time power minimization algorithm (three-dimensional algorithm) (c) Space-time power minimization algorithm (three-dimensional algorithm) (Top view)

(d) Space-time de-spread re-spread algorithm PRN2 (three-dimensional algorithm. (e) Space-time de-spread re-spread algorithm PRN2 (Top view)

Fig. 2.42 Comparisons of adaptive patterns

acquired. The jamming mitigation performance of space-time processing is better than that of spatial domain processing. By comparing Fig. 2.43b, c further, we find that the space-time de-spread re-spread algorithm can provide signal processing gains, so that its correlation output peaks of acquiring GNSS are greater. This is beneficial for the follow-up GNSS signal tracking.

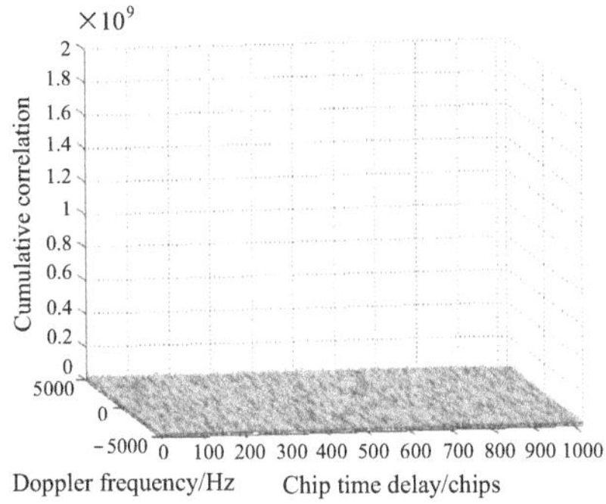

(a) Spatial domain power minimization algorithm.

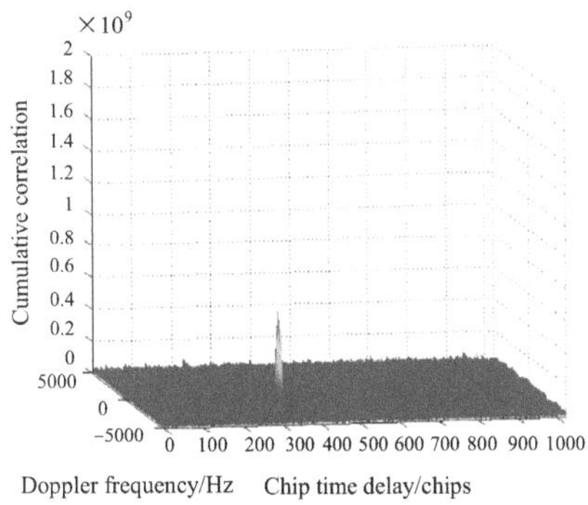

(b) Space-time power minimization algorithm

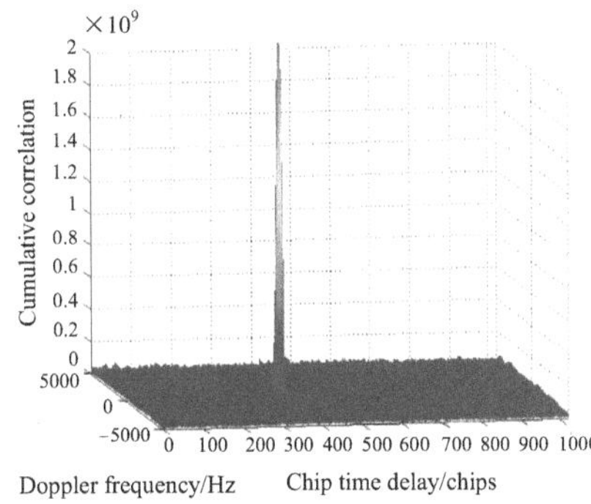

(c) Space-time de-spread re-spread algorithm

Fig. 2.43 Comparisons on PRN2's acquisition results after jamming mitigation processing

2. Simulation Result Comparisons of Several Reduced Rank STAP Algorithms

Below we discuss and compare the rank reduction performances of applying several rank reduction algorithms on the space-time power minimization algorithm. Figure 2.44 shows the simulation results. For every set of subfigures, the left side is the space-time two-dimension response, where the x-axis denotes the direction of arrival; the y-axis denotes the normalization frequency; and the z-axis denotes the space-time processing gain; the right side shows the corresponding top view. To make it convenient to compare, we redraw Fig. 2.42b, which is the result before rank reduction, as Fig. 2.44a, b. Due to simulation conditions, the dimension of jamming subspace is 10, so the number of dimensions for several rank reduction algorithms in the simulations are all set as $D = 10$. From the space-time two-dimension responses and the corresponding top views, it can be seen that various rank reduction methods can all form deep nulls towards the jammings.

3. Simulation Result Comparisons of Several STAP Equalization Algorithms

Figures 2.45 and 2.46 list comparative simulation results for several different STAP equalization algorithms. In Fig. 2.45, the solid line denotes the normalized

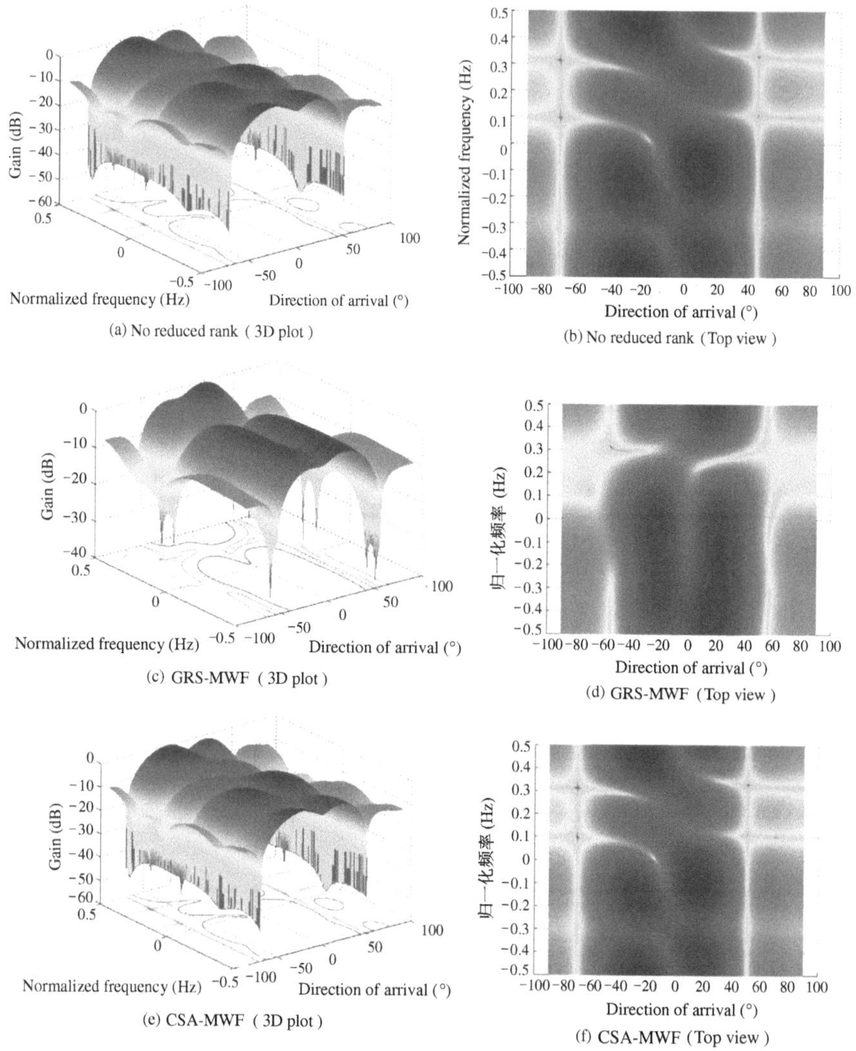

Fig. 2.44 Comparisons of STAP algorithm adaptive response diagrams for different reduced ranks

cross correlation between the array output signal without equalization processing and the local PRN codes; the "○" line denotes the normalized cross correlation results after the STAP jamming mitigation and the additional homomorphic filtering processing; the "*" line corresponds to the least squares inverse filtering method; and the "◇" line corresponds to the weight constraint method. It can be seen that after the STAP jamming mitigation the peak values of correlation functions have some offsets and broading, and this results in the measured PRN code's initial

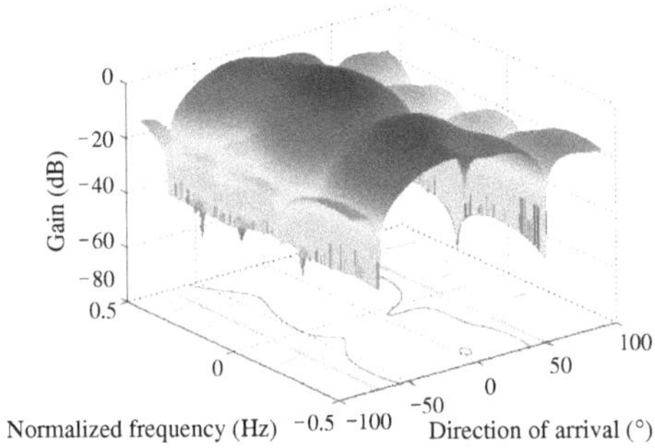

(g) PC (3D plot)

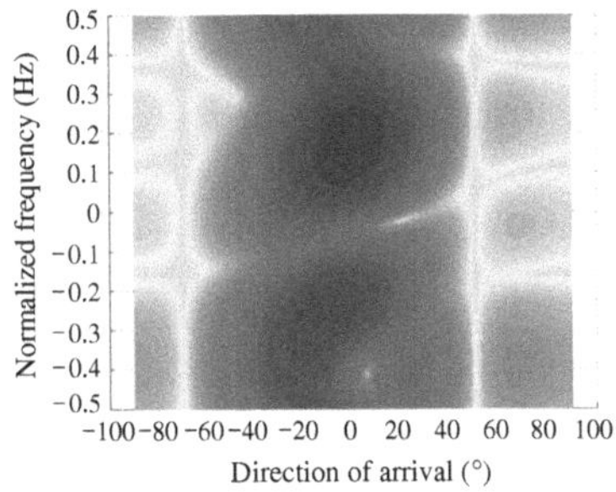

(h) PC (Top view)

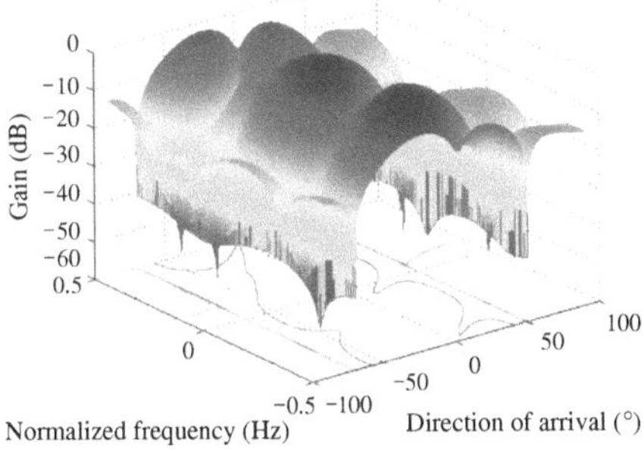

(i) CSM (3D plot)

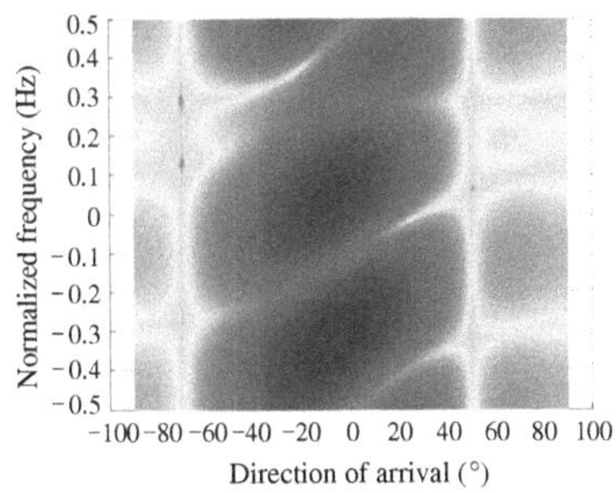

(j) CSM (Top view)

Fig. 2.44 (continued)

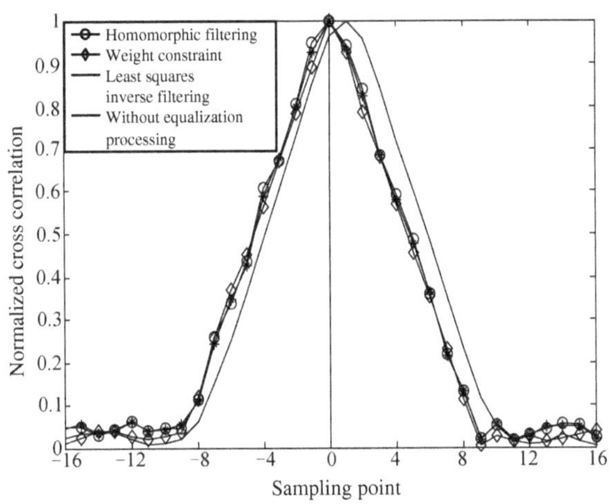

Fig. 2.45 Comparison of normalized cross-correlations

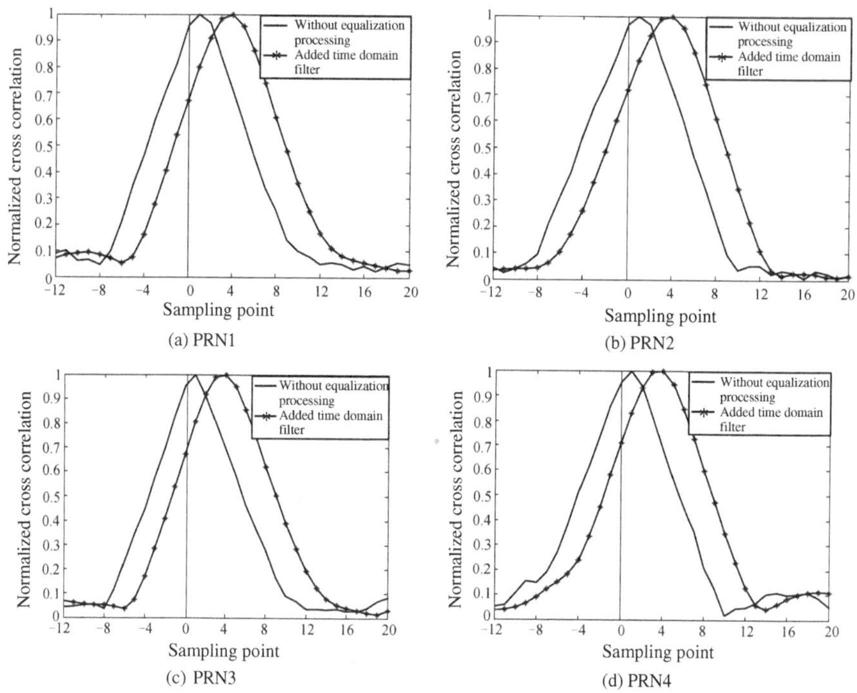

Fig. 2.46 Different satellites' cross correlation graphs with added temporal domain filtering

position errors, thereby acquisition and tracking are impacted. Several equalization methods can all correct this type of distortion.

Since the added temporal domain filter method is different from other methods, its results are listed in Fig. 2.46. When 4 GNSS signals incident on the array, it can be seen from Fig. 2.46 that, for every satellite, when there is no added temporal domain filter, peak offsets of cross correlation are different. When the temporal domain filter is added, even though the correlation function has certain broadening, the delays towards every GNSS signals are consistent, i.e. the offset amount is the same, so it achieves the objective of equalization.

2.7 Summary

In this chapter, we discuss jamming suppression techniques from the perspectives of temporal domain, spatial domain, and space-time domain. Temporal domain processing is low cost and easy to implement, so it is suitable for suppressing single frequency jamming with a low number of jamming sources. Spatial domain processing is a fairly mature jamming suppression technique, so it is suitable in environments of multiple narrowband jamming sources, and it has many corresponding adaptive filtering algorithms. In this chapter, we also analyze in detail various spatial domain algorithms' features and applicabilities. We proposed a high-gain and high-robustness jamming mitigation algorithm for hardware implementation. The space-time process represents a developing trend of jamming mitigation, and can be used under various complex jamming environments. But the amount of computation needed is large, and it is easier to trigger signal distortion. Therefore we need to consider rank reduction processing and signal distortion compensation techniques.

References

1. Upton DM, Upadhyay TN, Marchese J. Commercial-off-the-shelf (COTS) GPS interference canceller and test results. In: ION Navigation conference. 1998. p. 319–25.
2. Dimos G, Upadhyay T, Jenkins T. Low-cost solution to narrowband GPS interference problem. Process IEEE Natl Conf Aerosp Electron. 1995;1:45–153.
3. Rifkin R, Vaccaro JJ. Comparisons of narrowband adaptive filter technologies for GPS. MITRE technique report. 2000. p. 125–31.
4. Milstein LB. Interference rejection technique in spread spectrum communications. Process IEEE. 1988;76(6):657–71.
5. Ma W, Mao W, Chang F. Design of adaptive all-pass based notch filter for narrowband anti-jamming GPS system. In: Proceedings of 2005 international symposium on intelligent signal processing and communication system, Hong Kong. 2003. p. 305–8.
6. Soderstrand MA, Johnson TG, Strandberg RH, et al. Suppression of multiple narrow-band interference using real-time adaptive notch filters. IEEE Trans Circuits and Syst II: Analog Digit Signal Process. 1997;44(3):217–25.
7. Landry RJ, Calmettes V, Bousquet M. Impact of interference on a generic GPS receiver and assessment of mitigation techniques. In: IEEE 5th international symposium on spread spectrum techniques and applications. 1998. vol 1. p. 87–91.
8. Zoltowski MD, Haber F. Advanced adaptive null steering concepts for GPS. In: Processing MILCOM conference, universal communications, San Diego, CA. 1995. vol 3. p. 1214–8.
9. Gecan A, Zoltowski MD. Power minimization technique for GPS null steering antennas. In: Institute of Navigation conference, Palm Springs, CA. 1995.vol 12, no 1, p. 13–5.
10. Wang WY, Du QR, Wu RB et al. Interference suppression with flat gain constraint for satellite navigation systems. IET Radar Sonar Navig. 2015;9(7):852–6.
11. Moelker DJ, Pol EV, Yeheskel BN. Adaptive antenna arrays for interference cancellation in GPS and GLONASS receivers. In: Proceeding of the IEEE position location and navigation symposium. 1996. p. 191–8.
12. Jay RS. Interference mitigation approaches for the global positioning system. Lincoln Lab J. 2003;14(2):168–80.
13. Kaplan ED, Hegarty CJ. Understanding GPS principles and applications, 2nd ed. Artech House; 2006.
14. Sun W, Amin MG. A self-coherence anti-jamming GPS receiver. IEEE Trans Signal Process. 2005;53(10):3910–5.
15. Li P, Lu D, Wu R, et al. Adaptive anti-jamming algorithm based on the characteristics of the GPS signal. In: Proceedings of 2008 international symposium on intelligent signal processing and communication systems. 2008. p. 181–4.
16. Fante RL, Vaccaro JJ. Wideband cancellation of interference in a GPS receiver array. IEEE Trans Aerosp Electron Syst. 2000;36(2):549–64.

17. Myrick WL, Goldstein JS, Zoltowski MD. Anti-jam space-time preprocessor for GPS based on multistage nested wiener filter. IEEE Mil Commun (Atlantic NJ). 1999;1:675–81.
18. Lu D, Wu R. Global positioning system anti-jamming algorithm based on period repetitive CLEAN. IET Radar Sonar Navig. 2013;7(2):164–9.
19. Poor HV, Rusch LA. Narrowband interference suppression in spread spectrum CDMA. IEEE Pers Commun Mag. 1994;3:14–27.
20. Haykin S. Adaptive filter theory. 4th ed. New Jersey: Prentice Hall; 2002.
21. Young JA, Lehnert J, Lehnert S. Analysis of DFT-based frequency excision algorithm for direct sequence spread spectrum communications. IEEE Trans Commun. 1998;46:1076–87.
22. Capozza PT, Holland BJ, Hopkinson TM, et al. A single-chip narrow-band frequency-domain excisor for a global positioning system (GPS) receiver. IEEE J Solid-State Circuits. 2000;35 (3):51–5.
23. Hlawatsch F, Boudreaux-Bartels GF. Linear and quadratic time-frequency signal representations. IEEE Signal Process. 1992;9(2):21–67.
24. Badke B, Spanias AS. Partial band interference excision for GPS using frequency-domain exponents. In: Proceedings of the 2002 IEEE international conference on acoustic, speech, and signal processing. 2002. vol 4. p. 3936–9.
25. Ouyang X, Amin MG. Short-time fourier transform receiver for nonstationary interference excision in direct sequence spread spectrum communications. IEEE Trans Signal Process. 1988;184–213.
26. Zhao L, Amin MG, Lindsey AR. Subspace projection techniques for anti-FM jamming GPS receivers. IEEE SSAP; 2000. p. 529–33.
27. Zhang Y, Amin MG, Lindsey AR. Anti-jamming GPS receivers based on bilinear signal distributions. In: Proceedings of the MILCOM conference, universal Communications. 2001. vol 2. p. 1070–4.
28. Carlson BD. Covariance matrix estimation errors and diagonal loading in adaptive arrays. IEEE Trans Aerosp Electron Syst. 1988;24(4):397–401.
29. Misra, Pratap. Global positioning system: signals, measurements, and performance. Ganga-Jamuna Press; 2006.
30. Agee BG, Schell SV, Gardner WA. Spectral self-coherence restoral: a new approach to blind adaptive signal extraction using antenna arrays. Proc IEEE. 1990;78(4):753–67.
31. Amin MG, Sun W. A novel interference suppression scheme for global navigation satellite systems using antenna array. IEEE J Sel Areas Commun. 2005;23(5):999–1012.
32. Gershman AB, Nickel U, Bohme JF. Adaptive beamforming algorithms with robustness against jammer motion. IEEE Trans Signal Process. 1997;45(7):1878–85.
33. Shmidt RO. Multiple emitter location and signal parameter estimation. IEEE Trans Antennas Propag. 1986;34(3):276–80.
34. Roy R, Paulraj A, Kailath T. ESPRIT—a subspace rotation approach to estimation of parameter of cissoids in noise. IEEE Trans Acoust Speech Signal Process. 1986;34:1340–2.
35. Pickholtz RL, Schilling DL, Milstein LB. Theory of spread-spectrum communications: a tutorial. IEEE Trans Commun. 1982;30(5):855–84.
36. Hogbom JA. Aperture synthesis with a non-regular distribution of interferometer baselines. Astron Astrophys Suppl Ser. 1974;15:417–26.
37. Gough PT. A fast spectral estimation algorithm based on the FFT. IEEE Trans Signal Process. 1994;42(6):1317–22.
38. Tsao J, Steinberg BD. Reduction of sidelobe and speckle artifacts in microwave imaging: the CLEAN technique. IEEE Trans Antennas Propag. 1988;36(4):543–56.
39. Li J, Stioica P. Efficient mixed-spectrum estimation with applications to target feature extraction. IEEE Trans Signal Process. 1996;44(2):281–95.
40. Li J, Zheng D, Stoica P. Angle and waveform estimation via RELAX. IEEE Trans Aerosp Electron Syst. 1997;33(3):1077–87.
41. Liu Z, Li J. Implementation of the RELAX algorithm. IEEE Trans Aerosp Electron Syst. 1998;34(2):657–64.

42. Klemm R. Adaptive airborne MTI: an auxiliary channel approach. IEE Proc Commun Radar Signal Process. 1987;134(3):269–76.
43. Rong Z. Simulations of adaptive array algorithm for CDMA system. M.S. thesis. Blacksburg: Virginia Tech; 1996.
44. Rong Z, Rappaport TS. Simulation of multitarget adaptive algorithms for wireless CDMA systems. In: Proceeding of IEEE vehicular technology conference. 1997. p. 1–5.
45. Thanh VD, Hung NL. New adaptive beamforming algorithms for smart antennas in DS-CDMA mobile communication systems. In: 2004 3th international conference on computational electromagnetics and its applications proceedings. 2004. p. 165–8.
46. Du Z, Gong P, Wu W. Block based RLS de-spread re-spread multitarget array: algorithm and performance. In: Proceeding of IEEE vehicular technology conference. 2001. p. 190–3.
47. Wu R, Bao Z. Array pattern distortion and remedies in space-time adaptive processing for airborne radar. IEEE Trans Antennas Propag. 1998;46(7):963–70.
48. Brennan LE, Mallett JD, Reed IS. Adaptive arrays in airborne MTI radar. IEEE Trans Antennas Propag. 1976;24(5):607–15.
49. Reed IS, Mallett JD, Brennan LE. Rapid convergences rate in adaptive arrays. IEEE Trans Aerosp Electron Syst. 1974;10(6):853–63.
50. Wang H, Cai LJ. On adaptive spatial-temporal processing for airborne surveillance radar systems. IEEE Trans Aerosp Electron Syst. 1994;30(3):660–9.
51. Wang H. An overview of space-time adaptive processing for airborne radars. In: CIE international conference of radar. 1996. p. 789–94.
52. Fante RL, Vaccaro JJ. Cancellation of jammers and jammer multipath in a GPS receiver. IEEE Aerosp Electron Syst Mag. 1998;13(11):25–8.
53. Fante RL, Vaccaro JJ. Evaluation of adaptive space-time-polarization cancellation of broadband interference. In: IEEE position location and navigation symposium. 2002. p. 1–3.
54. David S. Navigation accuracy and interference reject for GPS adaptive antenna arrays. Doctor degree thesis. Stanford University; 2007.
55. Fante RL, Torres JA. Cancellation of diffuse jammer multipath by airborne adaptive radar. IEEE Trans Aerosp Electron Syst. 1995;31(2):805–20.
56. Tufts DW, Kumaresan R, Kirsteins I. Data adaptive signal estimation by signal value decomposition of a data matrix. Proc IEEE. 1982;70(6):684–5.
57. Goldstein JS, Reed IS. Redueed-rank adaptive filtering. IEEE Trans Signal Proeess. 1997;45 (2):492–6.
58. Goldstein JS, Reed IS, Seharf LL. A multistage representation of the wiener filter based on orthogonal projections. IEEE Trans Inf Theory. 1998;44(7):2943–59.
59. Honig ML, Xiao WM. Performance of reduced-rank linear interference suppression. IEEE Trans Inf Theory. 2001;47(5):1928–46.
60. Ricks DC, Goldstein JS. Efficient architectures for implementing adaptive algorithms. In: Proceedings of the 2000 antenna applications symposium. USA: Monticello. p. 29–41.
61. Hatke GF. Adaptive array processing for wideband nulling in GPS systems. Signals, systems & computers. In: Conference record of the thirty-second Asilomar conference. 1998. p. 1332–6.
62. Wu R, Xu R, Lu D, et al. STAP equalization technique based on homomorphic filtering in GPS. In: 2010 IEEE international symposium on phased array systems and technology. 2010. p. 841–5.
63. Vaseghi SV. Advanced digital signal processing and noise reduction. 3rd ed. USA: Wliey; 2006. p. 27–269.

Chapter 3
High-Dynamic GNSS Jamming Suppression Techniques

3.1 Introduction

All the jamming mitigation techniques introduced in Chap. 2 assume that receivers are in a stationary state or in a state of low to medium dynamic motion, which means that relative to GNSS receivers, the direction of jamming does not change or changes slowly, as a result there is enough snapshots to perform jamming covariance matrix estimation, so adaptive jamming suppression can be achieved. But, in military application scenarios, carriers of receivers have greater variation rates on motion velocity, acceleration, and jerk, meaning that they are in high-dynamic motion states. For example, a military aircraft performs air combat manoeuvring (dogfight) or a missile tracks targets in the process of hitting a target, etc. In these instances, relative to receivers, jamming DOAs change dynamically so conventional adaptive anti-jamming techniques do not apply anymore. Since the nulls are very narrow, interferences can easily move out of the beamforming null zones.

One effective method to solve this problem of fast changing jamming direction is the so-called null-widen technique [1–4], which ensures that during the processing time interval, the jamming DOA is always located in a wider null zone. Conventional null-widen techniques can be categorized into two types: one type is the derivative constraint method [1], which needs to know the jamming direction, And it is not flexible in terms of controlling the null width. The other type is the Covariance Matrix Taper (CMT) method [5], which includes the Mailloux method [3] and the Zatman method [4]. For uniform linear arrays, these two CMT methods do not need to know the DOAs of jamming signals. However these two null-widen algorithms were mainly proposed in the context of dynamic jamming environments, not aiming towards high-dynamic GNSS jamming mitigation.

In this chapter, we use null-widen techniques for high-dynamic GNSS anti-jamming. First, we analyze signal models for high-dynamic receivers in detail, and derive derivative constraint power minimization algorithm [6]. Then we analyze applicability of the Mailloux method and the Zatman method, and derive a

R. Wu et al., *Adaptive Interference Mitigation in GNSS*, Navigation: Science and Technology, https://doi.org/10.1007/978-981-10-5571-3_3

null-widen algorithm based on the Laplace statistical model [7, 8]. Subsequently, we expand the above two methods to the space-time domain, and then we discuss the reduced rank space-time null-widen algorithm [9]. At last, we use the sparse representation model to solve jamming DOA estimation problem using a single snapshot.

3.2 Definition of High-Dynamic and Signal Model

In 1988, S. Hinedi, in his report submitted to GPS Ranging Application Office, proposed definitions on two types of high-dynamic motions [10]: the first definition prescribes that within 1 s (second) there must be more than 70 g of acceleration rise, i.e. 70 g/s jerk sustaining more than 1 s; the second definition on dynamic stress is stricter than the first definition, which allows 50 g acceleration rise within 0.5 s, i.e. a sustained 100 g/s jerk lasting for 0.5 s, and the ensuing 50 g acceleration lasts for 2 s, then within 0.5 s there is a 50 g acceleration decrease. For the second type of high-dynamic motion definition, Figs. 3.1 and 3.2 show an acceleration trajectory and a jerk trajectory for a high-dynamic receiver.

Below we simulate the jamming DOA under a high-dynamic environment as time changes. Assuming the receiver carrier moves in a straight line, the relative position between the jamming source and the receiver carrier is given in Fig. 3.3 where r denotes the distance between the jamming source and the receiver's motion trajectory. Within 3 s, the receiver carrier moves from position s_0 to s_1 with the initial velocity 1 km/s (assuming that s_0 and s_1 are symmetric related to the origin).

Figure 3.4 shows the DOA curve as time changes while r takes on different values. In Fig. 3.4, it can be seen that, when jamming source is very close to the receiver, the DOA of jamming changes fast in relation to time.

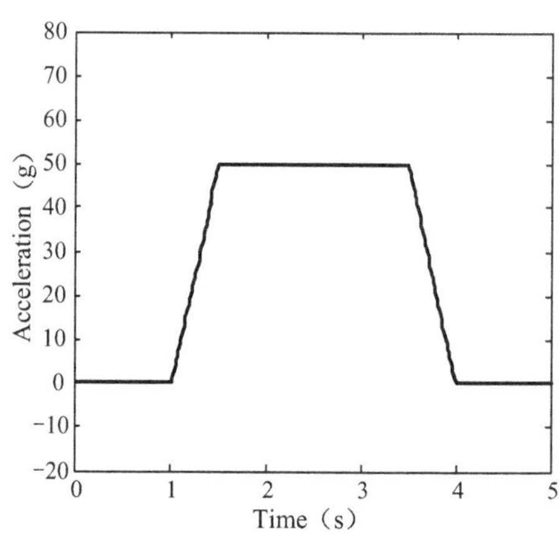

Fig. 3.1 Acceleration trajectory for a high-dynamic receiver carrier

Fig. 3.2 Jerk trajectory for a high-dynamic receiver carrier

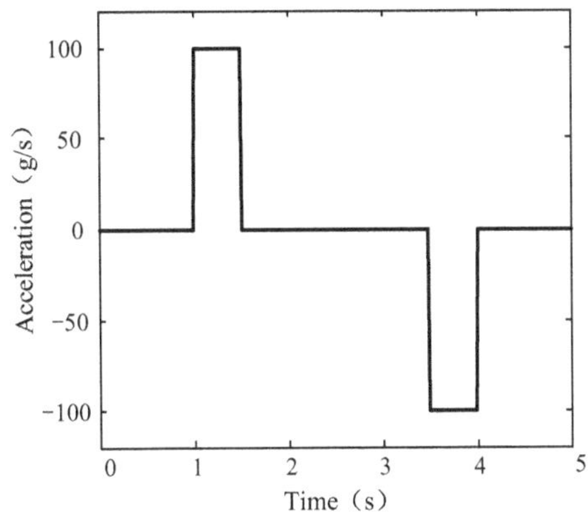

Fig. 3.3 Relative positions between jamming source and receiver's motion trajectory

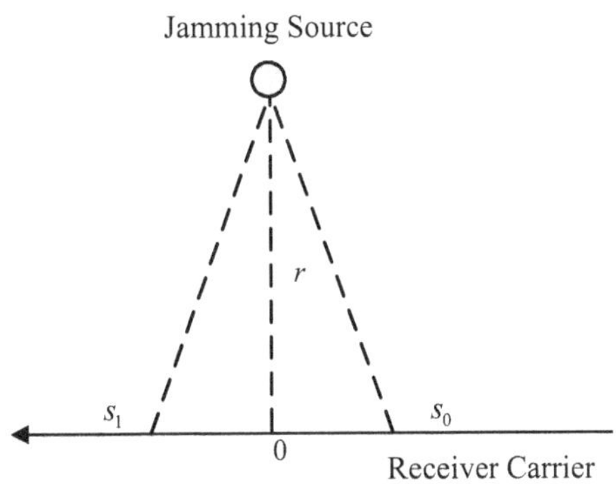

Fig. 3.4 Jamming direction curves as time changes

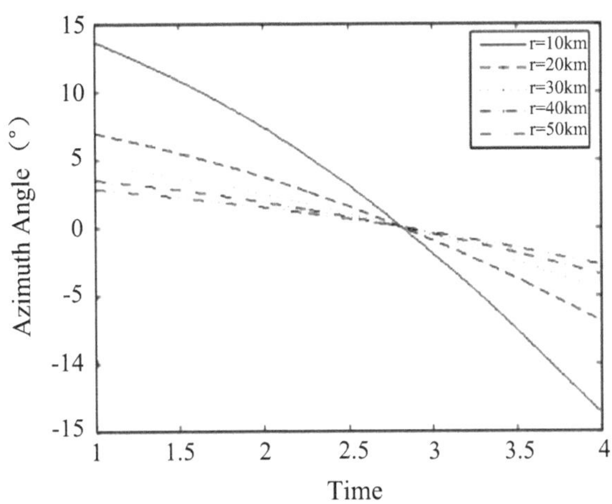

It can be known from the above definition on high-dynamic and the change of the simulated jamming direction, the data received by the GNSS receivers has the following characteristics:

(1) The carrier's Doppler shift increases.
(2) The direction of jamming signal, relative to the receiver, changes rapidly over time.
(3) For a stable receiver platform (corresponding to particle high-dynamic), due to far distance between a satellite and the receiver, the direction of GNSS signal can be approximated as a constant within the processing time interval.

Therefore, in a high-dynamic environment, for a stable receiver carrier platform containing M array elements, the received signal can be represented as:

$$\mathbf{x}(t) = \sum_{l=1}^{L} \mathbf{a}(\theta_l, \phi_l)\, s_l(t) + \sum_{q=1}^{Q} \mathbf{a}\big(\theta_q(t), \phi_q(t)\big)\, j_q(t) + \mathbf{e}(t) \tag{3.1}$$

where $s_l(t)$ denotes the lth GNSS signal $(l = 1, 2, \ldots, L)$; θ_l and ϕ_l are pitch and azimuth of the incident signal; since within the processing time interval, the DOA of the GNSS signal can be approximated as a constant, the steering vector $\mathbf{a}(\theta_l, \phi_l)$ (see (2.9)) does not change over time; $j_q(t)$ denotes the qth jamming signal $(q = 1, 2, \ldots, Q)$; $\mathbf{a}\big(\theta_q(t), \phi_q(t)\big)$ is the steering vector of the qth jamming signal, and this value changes rapidly over time; $\mathbf{e}(t)$ denotes the thermal noise vector of the receiver, which can be assumed as an additive Gaussian white noise vector with zero mean and σ_e^2 covariance. We can assume that the GNSS signal, jamming signal, and noise are not correlated with each other.

Since the GNSS signal power is far below the noise power level, the array covariance matrix $\mathbf{R}_x$ is mainly dominated by jamming and noise covariance matrices, and its element located at row m and column n can be approximated as:

$$\mathbf{R}_x(m, n) \approx \sum_{q=1}^{Q} \sigma_q^2\, \mathrm{e}^{-\mathrm{j}\mathbf{u}^{\mathrm{T}}(\mathbf{p}_m - \mathbf{p}_n)} + \sigma_e^2 \delta_{mn} \tag{3.2}$$

where $\mathbf{u}$ is the wave number vector (as expressed in (2.10)); $\mathbf{p}_m$ denotes the three dimensional position vector for the mth array element (as expressed in (2.11)); σ_q^2 is the qth jamming signal's power; δ_{mn} is the Kronecker δ function.

Based on the power minimization algorithm discussed in Chap. 2 [11], the array's weight vector is

$$\mathbf{w} = \frac{\mathbf{R}_x^{-1} \boldsymbol{\delta}_M}{\boldsymbol{\delta}_M^{\mathrm{H}} \mathbf{R}_x^{-1} \boldsymbol{\delta}_M} \tag{3.3}$$

where $\mathbf{w} = [w_1, w_2, \ldots, w_M]^{\mathrm{T}}$ denotes the array's weight vector, and $\boldsymbol{\delta}_M = [1, 0, \ldots, 0]^{\mathrm{T}}$ is the $M \times 1$ dimensional vector. In actual computation, the array covariance matrix $\mathbf{R}_x$ can be replaced using the sample covariance matrix $\widehat{\mathbf{R}}_x$ (see (2.32)). It can be known from the model described by (3.1), in the case of high-dynamic scenario, the DOA of jamming always changes dynamically, and as a result only few snapshot of data can be used to estimate sample covariance matrix. Therefore, unless adaptive jamming suppression algorithm has a very fast convergence speed to track the dynamic DOA change, jamming can move out of null zone and cannot be effectively cancelled. In the following sections we discuss high-dynamic null-widen algorithms from the perspectives of derivative constraint and covariance matrix taper methods, in order to solve the above problems.

3.3 Null-Widen Spatial Domain Adaptive Filtering Based on Derivative Constraints

3.3.1 Derivative Constraint Power Minimization Algorithm for Uniform Linear Array

For an uniform linear array, (3.1) can be simplified as

$$\mathbf{x}(t) = \sum_{l=1}^{L} \mathbf{a}(\theta_l)s_l(t) + \sum_{q=1}^{Q} \mathbf{a}\left(\theta_q(t)\right)j_q(t) + \mathbf{e}(t) \tag{3.4}$$

Let θ_q denotes the initial direction of the qth jamming signal during the computation interval for array weight vector. Based on the derivative constraint null-widen algorithm's principle, the p-order derivative constraint power minimization method's cost function [6] is:

$$\begin{aligned} &\min\ \mathbf{w}^{\mathrm{H}}\mathbf{R}_x\mathbf{w} \\ &\text{s.t. } \mathbf{w}^{\mathrm{H}}\boldsymbol{\delta}_M = 1 \\ &\quad \left.\frac{\partial^p\left(\mathbf{w}^{\mathrm{H}}\mathbf{a}(\theta)\right)}{\partial\xi^p}\right|_{\theta=\theta_q} = 0, \quad p = 1, 2, \ldots, P \end{aligned} \tag{3.5}$$

where $\xi = -\pi\sin\theta$; and p is the derivative order. To ensure that the degrees of freedom needed for the array to suppress the jamming are available, it needs to satisfy $PQ < M$.

It can be known from (2.12) that for a uniform linear array with an array element interval of half a wavelength, we have

$$\mathbf{a}(\theta_q) = \left[1, \quad \mathrm{e}^{-\mathrm{j}\pi\sin\theta_q}, \quad \ldots, \quad \mathrm{e}^{-\mathrm{j}(M-1)\pi\sin\theta_q}\right]^{\mathrm{T}} \tag{3.6}$$

By substituting it into (3.5)'s derivative constraint expression, we can have

$$\mathbf{w}^{\mathrm{H}}\mathbf{B}^{p}\mathbf{a}(\theta_q) = 0, \quad p = 1, 2, \ldots, P \tag{3.7}$$

where

$$\mathbf{B} = diag\{0, \quad 1, \quad \ldots, \quad M-1\} \tag{3.8}$$

is a $M \times M$ dimensional diagonal matrix.

By defining the $M \times Q$ dimensional matrix formed by Q jamming signals' steering vectors as

$$\mathbf{A} = \left[\mathbf{a}(\theta_1), \quad \mathbf{a}(\theta_2), \quad \ldots, \quad \mathbf{a}(\theta_q)\right] \tag{3.9}$$

The $P \times Q$ dimensional subspace expanded based on $\mathbf{B}^{p}\mathbf{a}(\theta_q)$ is an $M \times QP$ dimensional matrix, which can be denoted as

$$\mathbf{F} = \left[\mathbf{BA}, \quad \mathbf{B}^2\mathbf{A}, \quad \ldots, \quad \mathbf{B}^P\mathbf{A}\right] \tag{3.10}$$

Therefore, (3.7) can be written as

$$\mathbf{w}^{\mathrm{H}}\mathbf{F} = \mathbf{0}_{1\times QP} \tag{3.11}$$

By substituting (3.11) into (3.5), we have

$$\begin{aligned} &\min \mathbf{w}^{\mathrm{H}}\mathbf{R}_x \mathrm{w} \\ &\text{s.t. } \mathbf{w}^{\mathrm{H}}\boldsymbol{\delta}_M = 1 \\ &\quad\ \mathbf{w}^{\mathrm{H}}\mathbf{F} = \mathbf{0}_{1\times QP} \end{aligned} \tag{3.12}$$

Obviously, to directly solve (3.12), we need DOA information on jamming signals, and it becomes a linear multiple constraints problem. Below we solve (3.12) using another solution where the optimization problem of (3.12) can be transformed to

$$\begin{aligned} &\min \mathbf{w}^{\mathrm{H}}\mathbf{R}_x\mathbf{w} + \zeta\mathbf{w}^{\mathrm{H}}\mathbf{F}\mathbf{F}^{\mathrm{H}}\mathbf{w} \\ &\text{s.t. } \mathbf{w}^{\mathrm{H}}\boldsymbol{\delta}_M = 1 \end{aligned} \tag{3.13}$$

In (3.13)

$$\mathbf{F}\mathbf{F}^{\mathrm{H}} = \sum_{p=1}^{P}\mathbf{B}^{p}\mathbf{A}\mathbf{A}^{\mathrm{H}}\mathbf{B}^{p} \tag{3.14}$$

The $\mathbf{AA}^{\mathrm{H}}$ in (3.14) needs DOA information of jamming signals. Since jamming power is very large, we can use $\mathbf{R}_x$ to approximate the $\mathbf{AA}^{\mathrm{H}}$ term. By defining

$$\widetilde{\mathbf{R}}_x = \mathbf{R}_x + \zeta \sum_{p=1}^{P} \mathbf{B}^p \mathbf{R}_x \mathbf{B}^p \tag{3.15}$$

Therefore, (3.13) becomes

$$\begin{aligned} &\min \mathbf{w}^{\mathrm{H}} \widetilde{\mathbf{R}}_x \mathrm{w} \\ &\text{s.t. } \mathbf{w}^{\mathrm{H}} \boldsymbol{\delta}_M = 1 \end{aligned} \tag{3.16}$$

Based on the Lagrange multiplier method, the solution of (3.16) is

$$\mathbf{w}_{opt} = (\boldsymbol{\delta}_M{}^{\mathrm{H}} \widetilde{\mathbf{R}}_x^{-1} \boldsymbol{\delta}_M)^{-1} \widetilde{\mathbf{R}}_x^{-1} \boldsymbol{\delta}_M \tag{3.17}$$

It can be seen from (3.17) that the derivative constraint power minimization algorithm used on a uniform linear array does not need to know the information on jamming signal DOAs. The detailed steps of the algorithm can be summarized as below:

(1) Compute the received data's sample covariance matrix $\widehat{\mathbf{R}}_x$.
(2) Use $\widehat{\mathbf{R}}_x$ to replace the $\mathbf{R}_x$ in (3.15) and compute $\widehat{\widetilde{\mathbf{R}}}_x$.
(3) Use $\widehat{\widetilde{\mathbf{R}}}_x$ to substitute $\widetilde{\mathbf{R}}_x$ in (3.17) and compute antenna array's weight vector $\mathbf{w}_{opt}$.

3.3.2 *Derivative Constraint Power Minimization Algorithm for a Uniform Circular Array*

Usually, conformal circle arrays are used on carriers moving with high-speed such as jets and missiles [12]. Figure 3.5 shows a schematic of an M-element uniform circle array, where the center of the circle is selected as the array's reference point, and the radius of the circle array is d. Using (2.9), the mth element of the steering vector in (3.1) is

$$\mathbf{a}_m(\theta, \phi) = \mathrm{e}^{-\mathrm{j}\mathrm{u}^T(\theta,\phi)\mathbf{p}_m} \tag{3.18}$$

Fig. 3.5 Illustration of a uniform circle array

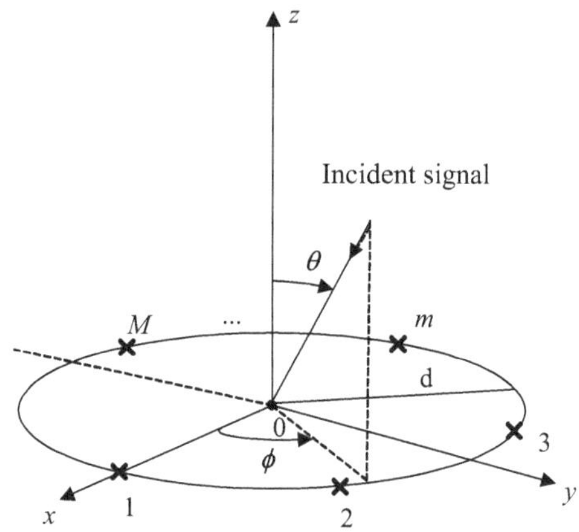

where

$$\mathbf{u}(\theta,\phi)=\frac{2\pi}{\lambda}\begin{bmatrix}\sin(\theta)\cos(\phi)\\ \sin(\theta)\sin(\phi)\end{bmatrix} \tag{3.19}$$

is the wave number vector, and

$$\mathbf{p}_m=d[\cos r_m,\ \sin r_m]^{\mathrm{T}} \tag{3.20}$$

is the mth array element's position vector, where

$$r_m=\frac{2\pi(m-1)}{M} \tag{3.21}$$

Usually for a uniform linear array, only one angle needs to be considered for the signal DOA. Thereby when adding derivative constraint on the jamming DOA, we only need to aim towards one angle. For a uniform circle array, the signal DOA contains two angles: pitch and azimuth, so derivative constraints must be imposed simultaneously on the two angles. Therefore, the cost function of the p-order derivative constraint power minimization method applied on the uniform circle array can be written as

$$\left.\begin{array}{c}\min\ \mathbf{w}^{\mathrm{H}}\mathbf{R}_x\mathbf{w}\\ \text{s.t.}\ \ \mathbf{w}^{\mathrm{H}}\boldsymbol{\delta}_M=1\\ \left.\frac{\partial^p(\mathbf{w}^{\mathrm{H}}\mathbf{a}(\theta,\phi))}{\partial\theta^p}\right|_{\theta=\theta_q}=0,\quad p=1,2,\ldots,P\\ \left.\frac{\partial^p(\mathbf{w}^{\mathrm{H}}\mathbf{a}(\theta,\phi))}{\partial\phi^p}\right|_{\phi=\phi_q}=0,\quad p=1,2,\ldots,P\end{array}\right\} \tag{3.22}$$

To make it simple, we can select the derivative order $p = 1$, then (3.22) can be simplified to

$$\begin{aligned} &\min \mathbf{w}^{\mathrm{H}}\mathbf{R}_x\mathbf{w} \\ &\text{s.t. } \mathbf{w}^{\mathrm{H}}\mathbf{C} = \mathbf{G} \end{aligned} \tag{3.23}$$

where

$$\mathbf{C} = \left[\boldsymbol{\delta}_M, \left.\frac{\partial \mathbf{a}(\theta,\phi)}{\partial\theta}\right|_{\substack{\theta=\theta_1\\ \phi=\phi_1}}, \left.\frac{\partial \mathbf{a}(\theta,\phi)}{\partial\phi}\right|_{\substack{\theta=\theta_1\\ \phi=\phi_1}}, \ldots, \left.\frac{\partial \mathbf{a}(\theta,\phi)}{\partial\theta}\right|_{\substack{\theta=\theta_Q\\ \phi=\phi_Q}}, \left.\frac{\partial \mathbf{a}(\theta,\phi)}{\partial\phi}\right|_{\substack{\theta=\theta_Q\\ \phi=\phi_Q}}\right]^{\mathrm{T}} \tag{3.24}$$

$$\mathbf{G} = [1, \quad 0, \quad \ldots, \quad 0]^{\mathrm{T}}_{1\times(2Q+1)} \tag{3.25}$$

and

$$\frac{\partial \mathbf{a}(\theta,\phi)}{\partial\theta} = \mathrm{j}\frac{2\pi}{\lambda}\mathbf{p}^T\frac{\partial \mathbf{u}(\theta,\phi)}{\partial\theta} \odot \mathbf{a}(\theta,\phi) \tag{3.26}$$

$$\frac{\partial \mathbf{a}(\theta,\phi)}{\partial\phi} = \mathrm{j}\frac{2\pi}{\lambda}\mathbf{p}^T\frac{\partial \mathbf{u}(\theta,\phi)}{\partial\phi} \odot \mathbf{a}(\theta,\phi) \tag{3.27}$$

where, $\odot$ denotes matrix's Hadamard product. The solution of (3.23) is

$$\mathbf{w} = \mathbf{R}_x^{-1}\mathbf{C}(\mathbf{C}^{\mathrm{H}}\mathbf{R}_x^{-1}\mathbf{C})^{-1}\mathbf{G}^{\mathrm{H}} \tag{3.28}$$

It can be seen from (3.23) to (3.28) that, to solve weight vector, we need to know the DOAs of the jamming signals. Professor E.A. Mohamed proposed to use the power minimization method to estimate the jamming directions [13]. The method first uses (3.29) to derive weight vector, and then searches for angles to make (3.30) smaller than a critical value (which can be determined based on specific situations), and the angles found are the jamming DOAs.

$$\mathbf{w} = \frac{\mathbf{R}_x^{-1}\boldsymbol{\delta}_M}{\boldsymbol{\delta}_M^{\mathrm{H}}\mathbf{R}_x^{-1}\boldsymbol{\delta}_M} \tag{3.29}$$

$$P(\theta,\phi) = \left|\mathbf{w}^{\mathrm{H}}\mathbf{a}(\theta,\phi)\right|^2 \tag{3.30}$$

In (3.29), the computation load is relatively heavy for solving $\mathbf{R}_x^{-1}$, but (3.28) which uses a new method to solve weights also uses $\mathbf{R}_x^{-1}$, thereby its computation does not need to be counted as an additional load. What has been added to the new method, compared with the original power minimization method, is the two dimensional search.

The detailed implementation steps used for applying the derivative constraint algorithm on a uniform circle array are listed below

(1) Compute the received data's sample covariance matrix $\widehat{\mathbf{R}}_x$.
(2) Compute weight vector using the power minimization method.
(3) Substitute the derived weight vector from step (2) into (3.30) to set a threshold, and search for (θ, ϕ) values that can make $P(\theta, \phi)$ smaller than the threshold value, then the (θ, ϕ) found can be identified as the jamming DOA.
(4) Substitute the estimated (θ, ϕ) in step (3) into (3.26) and (3.27), and compute the derivative constraints $\frac{\partial \mathbf{a}(\theta,\phi)}{\partial \theta}$ and $\frac{\partial \mathbf{a}(\theta,\phi)}{\partial \phi}$.
(5) Substitute the array covariance matrix $\widehat{\mathbf{R}}_x$ derived in step (1), and $\frac{\partial \mathbf{a}(\theta,\phi)}{\partial \theta}$ and $\frac{\partial \mathbf{a}(\theta,\phi)}{\partial \phi}$ derived in step (4) into (3.28), in order to derive an antenna array's weight vector $\mathbf{w}$.

3.3.3 *Simulation Results*

1. Uniform Linear Array Simulation Results

For simulation, a 7-element uniform linear array with half a wavelength interval is used. The DOA of the GNSS signal is $\theta = 50°$. When a jamming signal incidents on an array, the initial DOA of jamming is 14°, and the jamming DOA gradually changes from 14° to 12°. The jamming-to-noise ratio is 35 dB and the signal-to-noise ratio is –20 dB. For the signal received by the array, the digital IF is 4.309 MHz and the sampling rate is 5.714 MHz.

We consider a first-order derivative constraint, i.e. $p = 1$, and use the constant coefficient $\zeta = 1$. Figure 3.6 compares the beam patterns of the derivative constraint power minimization algorithm and the power minimization algorithm. To observe the null zone clearly, Fig. 3.6b shows the partially enlarged view of the array pattern. It can be clearly seen from Fig. 3.6b that, compared with the power minimization method, the null beam of the derivative constraint power minimization method becomes wider, but at the same time the null becomes a bit shallower.

To further compare the suppression performances on high-dynamic GPS jamming between the derivative constraint algorithm and the power minimization algorithm, Fig. 3.7 shows the results of comparing output signal-jamming-plus-noise ratios. It can be seen from Fig. 3.7 that the output signal-jamming-plus-noise ratio by the derivative constraint power minimization algorithm is clearly better than that of the power minimization algorithm.

Figure 3.8 shows the acquisition effects of GPS receivers using the derivative power minimization algorithm and the power minimization algorithm for jamming mitigation in high-dynamic environments. Here we need to notice that for the data used to solve for the array weight vector, the DOA of jamming is 14°, but due to the

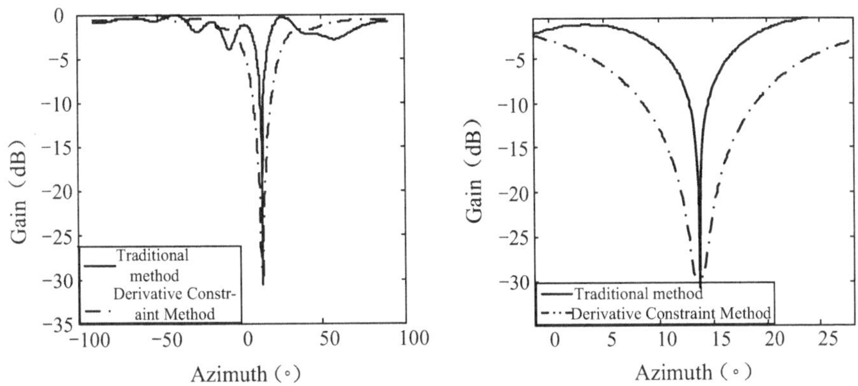

(a) Complete pattern (b) Partially enlarged pattern

Fig. 3.6 Comparisons of adaptive array patterns between the derivative constraint algorithm and the power minimization algorithm

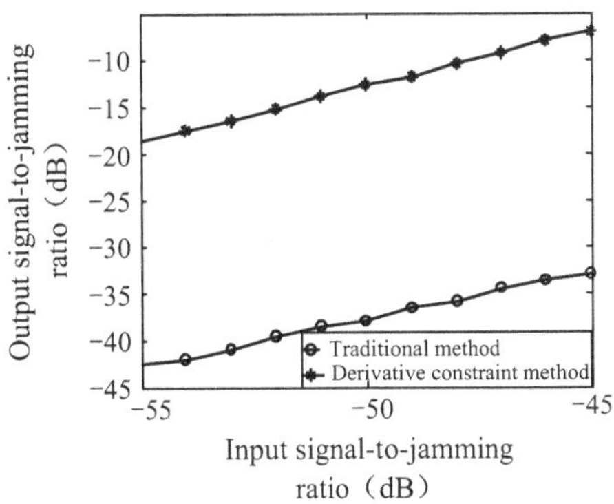

Fig. 3.7 Comparisons of output signal-to-jamming ratios for derivative constraint algorithm and power minimization algorithms

fast motion of high-dynamic GPS receivers, the DOA of jamming in the data used to acquire has changed to 12°.

By comparing Fig. 3.8a, b, it is obvious that the power minimization algorithm cannot maintain the normal operations of dynamic GPS receivers; while the derivative constraint power minimization algorithm can guarantee the correct acquisition of GNSS signals by using high-dynamic GPS receivers, thereby maintaining the normal operation of dynamic GPS receivers. This is because the null zone formed by the traditional jamming mitigation algorithm is too narrow. When the jamming DOA changes very fast relative to the GPS receiver, and the convergence speed of the algorithm is not fast enough (i.e. the convergence of the algorithm cannot keep up with the speed of the jamming DOA change), the null zone of the antenna beam easily deviates from the jamming DOA and consequently it cannot effectively suppress the jamming. When the jamming power is fairly large, it cannot be suppressed effectively by relying on the de-spreading functionality of

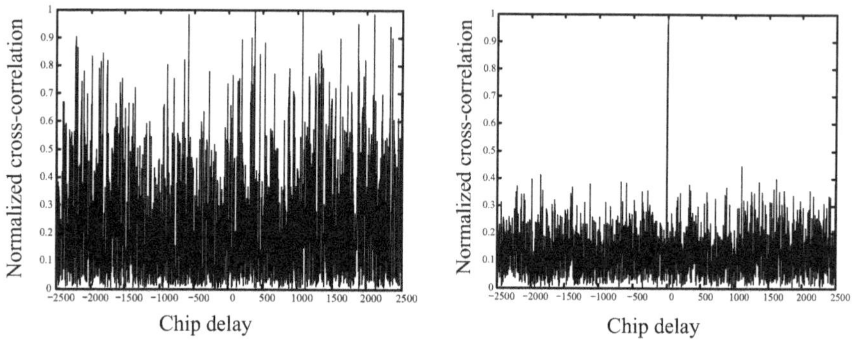

(a) Power minimization algorithm (b) Derivative constraint power minimization algorithm

Fig. 3.8 Comparisons of acquisition results

GPS receivers, and consequently the GPS receivers cannot acquire GNSS signals properly.

2. Uniform Circular Array Simulation Results

For simulation, a 7-element uniform circular array is used. We assume that the GNSS signal incidents onto the circular array from the direction of $(20°, 80°)$, the jamming signal's initial incident angle is $(50°, 150°)$, and the jamming DOA gradually changes from $(50°, 150°)$ to $(53°, 154°)$. The jamming-to-noise ratio is 35 dB, the signal-to-noise ratio is –20 dB, and the derivative constraint order is $p = 1$. Figure 3.9 compares the beam patterns for the derivative constraint method and the power minimization method, where subfigures (a), (c) and (e) show the beam patterns formed by the power minimization method, the top view and side view of the beam pattern. Subfigures (b), (d) and (f) are the beam pattern, top view and side view of the beam pattern formed by the first order derivative constraint power minimization method. It can be seen by comparing subfigures (a), (b), (c), (d), (e), and (f) in Fig. 3.9 that the null zone formed by the derivative constraint power minimization method is wider than the null zone formed by the power minimization, but shallower.

Figure 3.10 compares acquisition results of using the first order derivative constraint method and the power minimization method for jamming mitigation. The jamming DOA of the data used for the array's weighted vector is $(50°, 150°)$, and the jamming DOA for the data used for acquisition changes to $(53°, 154°)$ as time changes. By comparing Fig. 3.10a, b, it can be seen that, the acquisition performance of the derivative constraint power minimization method is significantly better than that of the power minimization method.

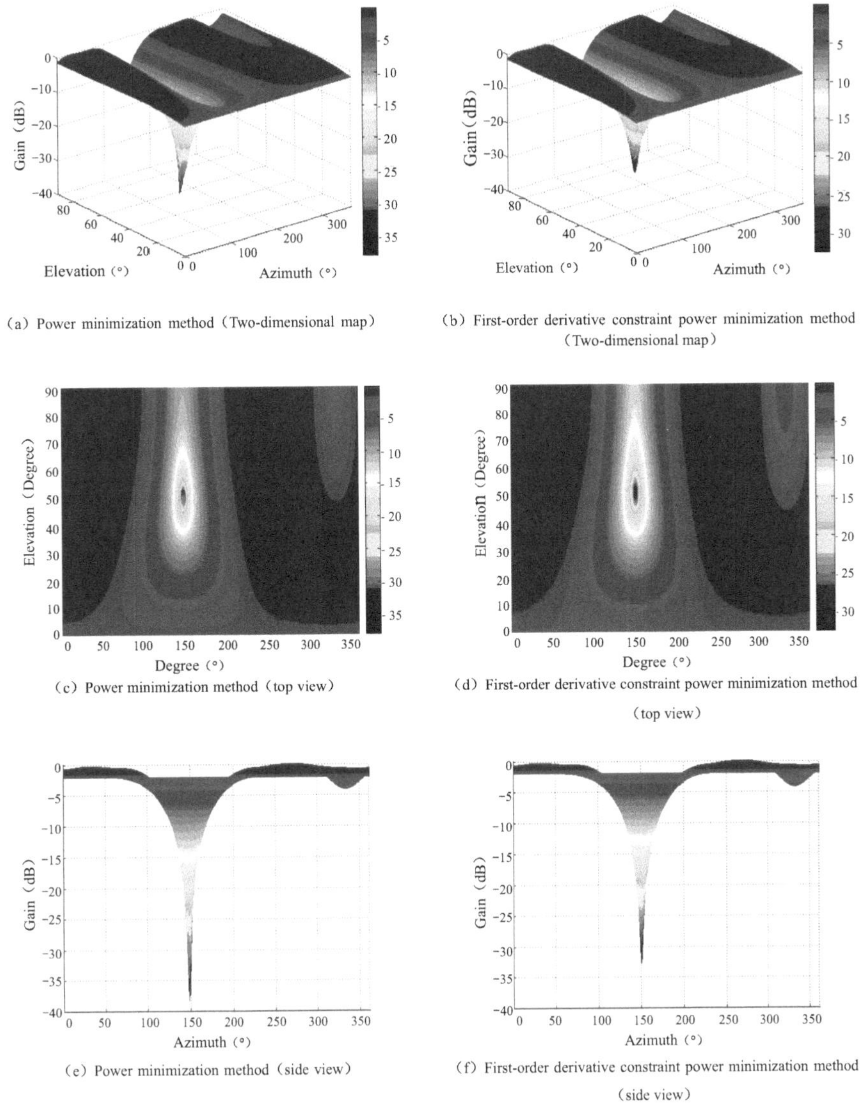

(a) Power minimization method (Two-dimensional map)

(b) First-order derivative constraint power minimization method (Two-dimensional map)

(c) Power minimization method (top view)

(d) First-order derivative constraint power minimization method (top view)

(e) Power minimization method (side view)

(f) First-order derivative constraint power minimization method (side view)

Fig. 3.9 Comparisons of adaptive array patterns

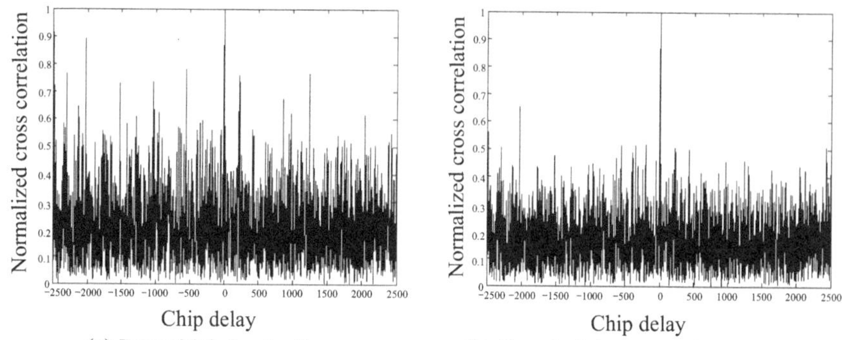

(a) Power minimization algorithm

(b) First order derivative constraint power minimization algorithm

Fig. 3.10 Comparisons of acquisition results

3.4 Null-Widen Spatial Domain Adaptive Filtering Based on Covariance Matrix Taper

3.4.1 Mailloux Algorithm

The Mailloux method [3] constructs a new covariance matrix using a cluster of artificial jamming sources to broaden nulls, thereby reducing the impacts on array performance by the changes of jamming direction.

1. Mailloux Algorithm Applied on Uniform Linear Array

Considering a uniform linear array with a half-wavelength array interval, there are $K - 1$ equal-strength incoherent artificial sources around each original interference source. ($K > M$ and K is an odd number), as shown in Fig. 3.11.

Fig. 3.11 Distribution of artificial jamming sources for Mailloux method

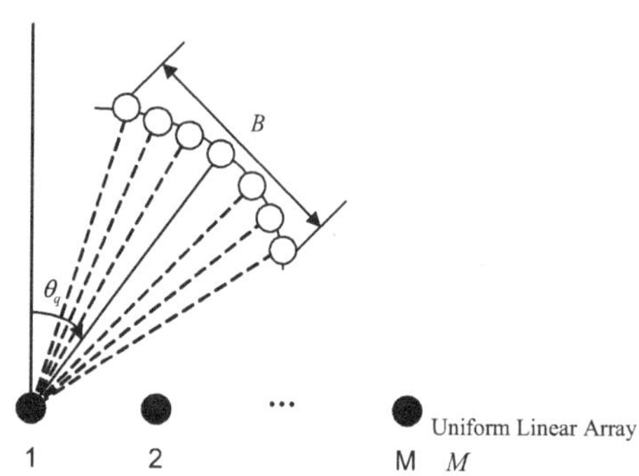

Then, the virtual signal received by the array is

$$\tilde{\mathbf{x}}(t) = \sum_{l=1}^{L} \mathbf{a}(\theta_l)\, s_l(t) + \sum_{q=1}^{Q} \sum_{k=-\frac{K-1}{2}}^{\frac{K-1}{2}} \mathbf{a}(\theta_q + k\beta)\, j_{qk}(t) + \mathbf{e}(t) \tag{3.31}$$

where θ_q is the initial DOA of the qth jamming signal; $\beta = B/(K-1)$ is the angle interval between the virtual jamming sources, which is usually very small; B is the angle distribution width for assumed jamming sources. Based on (2.12), mth element of the steering vector for the kth virtual jamming source near the qth jamming signal is

$$\mathbf{a}_m(\theta_q + k\beta) = \mathrm{e}^{-\mathrm{j}(m-1)\pi \sin(\theta_q + k\beta)} \tag{3.32}$$

By performing Taylor expansion on the $\sin(\theta_q + k\beta)$ in (3.32) at point θ_q, and by extracting the first two terms, we have

$$\sin(\theta_q + k\beta) \approx \sin(\theta_q) + \cos(\theta_q) k\beta \tag{3.33}$$

By substituting (3.33) into (3.32), we have

$$\mathbf{a}_m(\theta_q + k\beta) \approx \mathbf{a}_m(\theta_q) \mathrm{e}^{-\mathrm{j}(m-1)\pi k\beta \cos(\theta_q)} \tag{3.34}$$

where

$$\mathbf{a}_m(\theta_q) = \mathrm{e}^{-\mathrm{j}(m-1)\pi \sin(\theta_q)} \tag{3.35}$$

represents the mth element of steering vector for the qth jamming signal, and it can also be seen from (2.12). Equation (3.34) represents the angle expansion caused by the assumed jamming sources.

By substituting (3.34) into (3.31), it is not hard to derive that the element located at the mth row and the nth column of new covariance matrix $\overline{\mathbf{R}}_M$ corresponding to $\tilde{\mathbf{x}}(t)$ is

$$\overline{\mathbf{R}}_M(m,n) = \mathbf{T}_M(m,n) \sum_{q=1}^{Q} \sigma_q^2 \mathrm{e}^{-\mathrm{j}\pi(m-n)\sin\theta_q} + \sigma_e^2 \delta_{mn} \tag{3.36}$$

where

$$\mathbf{T}_M(m,n) = \frac{\sin(K\Delta_{mn})}{\sin(\Delta_{mn})} \tag{3.37}$$

represents the mth row and nth column element of the matrix $\mathbf{T}_M$, $\mathbf{T}_M$ is named as the expansion matrix, and it is related to the jamming DOA. And in addition,

$$\Delta_{mn} = (m-n)\pi\beta\cos\left(\theta_q\right)/2 \tag{3.38}$$

Based on Fig. 3.11 and (3.36), since there are far more jamming sources than array elements, when the data model given by (3.31) is used for beamforming, the adaptive array cannot form a null zone towards every direction of jamming, but it can form a broad null near every actual jamming source. Thereby in a high-dynamic scenario, the array covariance matrix in (3.36) can substitute the one in (3.2) to generate the wide null. However, the covariance matrix computed by the original array snapshot data is the matrix expressed in (3.2). Equation (3.36) represents the virtual covariance matrix, so it cannot be directly derived using snapshot data. Therefore we need to establish the connection between (3.2) and (3.36). For a uniform linear array with a half-wavelength interval between array elements, (3.2) becomes

$$\mathbf{R}_x(m,n) = \sum_{q=1}^{Q} \sigma_q^2 \mathrm{e}^{-\mathrm{j}\pi(m-n)\sin\theta_q} + \sigma_e^2\delta_{mn} \tag{3.39}$$

From this, we can obtain

$$\sum_{q=1}^{Q} \sigma_q^2 \mathrm{e}^{-\mathrm{j}\pi(m-n)\sin\theta_q(t)} = \mathbf{R}_x(m,n) - \sigma_e^2\delta_{mn} \tag{3.40}$$

By substituting (3.40) into (3.36), we can have

$$\overline{\mathbf{R}}_M(m,n) = \mathbf{T}_M(m,n)\mathbf{R}_x(m,n) - (\mathbf{T}_M(m,n)-1)\sigma_e^2\delta_{mn} \tag{3.41}$$

When $m \neq n$ and $\delta_{mn} = 0$, partial diagonal element of $\overline{\mathbf{R}}_M$ is

$$\overline{\mathbf{R}}_M(m,n) = \mathbf{T}_M(m,n)\mathbf{R}_x(m,n) \tag{3.42}$$

When $m = n$, $\delta_{mn} = 1$, $\mathbf{T}_M(m,m) = K$, then main diagonal element of $\overline{\mathbf{R}}_M$ is

$$\overline{\mathbf{R}}_M(m,n) = \mathbf{T}_M(m,n)\mathbf{R}_x(m,n) - (K-1)\sigma_e^2 \tag{3.43}$$

Thereby we have

$$\overline{\mathbf{R}}_M = \mathbf{T}_M \odot \mathbf{R}_x - (K-1)\sigma_e^2\mathbf{I} \tag{3.44}$$

Let

$$\widetilde{\mathbf{R}}_M = \mathbf{T}_M \odot \mathbf{R}_x \tag{3.45}$$

where $\odot$ denotes Hadamard product, i.e. multiplying the corresponding terms of the matrices. It can be seen by comparing (3.44) and (3.45) that, by using the $\widetilde{\mathbf{R}}_M$ in (3.45) to replace $\overline{\mathbf{R}}_M$ is equivalent to performing diagonal loading on $\overline{\mathbf{R}}_M$ (the loading amount is $(K-1)\sigma_e^2$), and as a result it can still form broad nulls on the DOAs of jamming. It can be known from the high-dynamic motion model that, when $\left|\cos\theta_q\right| = 1$, jamming direction changes the fastest, and as a result the null width that needs to be expanded is also the largest. Therefore, we can select the parameter $\beta_{\max}$ that is needed to generate the maximum null width to replace $\beta\cos(\theta_q)$, so a new expansion matrix can be formed, i.e.

$$\overline{\mathbf{T}}_M(m,n) = \frac{\sin(K\pi(m-n)\beta_{\max}/2)}{\sin(\pi(m-n)\beta_{\max}/2)} \tag{3.46}$$

Forming a new covariance matrix $\overline{\mathbf{R}}_M$ using (3.46) not only satisfies the maximum angle change, it also does not need the DOA information of jamming signals.

2. Mailloux Algorithm Applied on Uniform Circular Array

When an array is a circular array, based on Mailloux method, we assume that near every jamming source's pitch and azimuth, there are $K-1$ points distributed at equal intervals, and there exist $(K-1)\times(K-1)$ equi-power jamming signals that are independent from each other. After adding the virtual jamming source, the array's received signal becomes

$$\mathbf{x}(t) = \sum_{l=1}^{L}\mathbf{a}(\theta_l,\phi_l)\,s_l(t) + \sum_{q=1}^{Q}\sum_{k=-\frac{K-1}{2}}^{\frac{K-1}{2}}\sum_{k=-\frac{K-1}{2}}^{\frac{K-1}{2}}\mathbf{a}\left(\theta_q+k\beta,\phi_q+k\gamma\right)j_{q_{k,k}}(t) + \mathbf{e}(t) \tag{3.47}$$

In (3.47), $\beta = B_1/(K-1)$; $\gamma = B_2/(K-1)$ represent the intervals between virtual jamming sources' pitches and azimuths respectively. B_1 and B_2 represent angle distribution widths between the virtual jamming sources' pitches and azimuths angles respectively. Based on (3.18) we have

$$\mathbf{a}_m\left(\theta_q+k\beta,\phi_q+k\gamma\right) = \exp\left(-\mathrm{j}\frac{2\pi}{\lambda}\begin{bmatrix}\sin\left(\theta_q+k\beta\right)\cos\left(\phi_q+k\gamma\right)\\ \sin\left(\theta_q+k\beta\right)\sin\left(\phi_q+k\gamma\right)\end{bmatrix}^{\mathrm{T}}\mathbf{p}_m\right) \tag{3.48}$$

where $\mathbf{p}_m$ is the position vector of the mth array element. To perform Taylor series expansion on various terms of trigonometric functions in (3.48) at points θ_q and ϕ_q, and since β and γ values are very small, we take the first two terms and ignore the $\beta\gamma$ term, then we have

$$\begin{aligned}&\mathbf{a}_m(\theta_q+k\beta,\phi_q+k\gamma)\\&=\mathbf{a}_m(\theta_q,\phi_q)\exp\left(-\mathrm{j}\frac{2\pi}{\lambda}\begin{bmatrix}-k\gamma\sin\theta_q\sin\phi_q+k\beta\cos\theta_q\cos\phi_q\\k\gamma\sin\theta_q\cos\phi_q+k\beta\cos\theta_q\sin\phi_q\end{bmatrix}^{\mathrm{T}}\mathbf{p}_m\right)\end{aligned} \tag{3.49}$$

Thereby, it is not hard to derive that, by adding the element of row m and column n for new virtual jamming source's covariance matrix $\overline{\mathbf{R}}_M$

$$\overline{\mathbf{R}}_M(m,n)=\sum_{q=1}^{Q}\sigma_q^2\exp(-\mathrm{j}\mathbf{u}^T(\theta_q,\phi_q)(\mathbf{p}_m-\mathbf{p}_n))\frac{\sin(K\Delta_{mn})}{\sin(\Delta_{mn})}\frac{\sin(K\Lambda_{mn})}{\sin(\Lambda_{mn})}+\sigma_e^2\delta_{mn} \tag{3.50}$$

where

$$\Delta_{mn}=\frac{\pi\beta}{\lambda}[\mathbf{p}_m-\mathbf{p}_n]^{\mathrm{T}}\begin{bmatrix}\cos\theta_q\cos\phi_q\\\cos\theta_q\sin\phi_q\end{bmatrix} \tag{3.51}$$

$$\Lambda_{mn}=\frac{\pi\gamma}{\lambda}[\mathbf{p}_m-\mathbf{p}_n]^{\mathrm{T}}\begin{bmatrix}-\sin\theta_q\sin\phi_q\\\sin\theta_q\cos\phi_q\end{bmatrix} \tag{3.52}$$

Then we have

$$\mathbf{T}_M(m,n)=\frac{\sin(K\Delta_{mn})}{\sin(\Delta_{mn})}\frac{\sin(K\Lambda_{mn})}{\sin(\Lambda_{mn})} \tag{3.53}$$

By substituting Δ_{mn} and Λ_{mn} into (3.53) and rearrange, we can obtain

$$\begin{aligned}\mathbf{T}_M(m,n)=&\frac{\sin\{K\pi\beta d/\lambda[(\cos r_m-\cos r_n)\cos\theta_q\cos\phi_q+(\sin r_m-\sin r_n)\cos\theta_q\sin\phi_q]\}}{\sin\{\pi\beta d/\lambda[(\cos r_m-\cos r_n)\cos\theta_q\cos\phi_q+(\sin r_m-\sin r_n)\cos\theta_q\sin\phi_q]\}}\\&\frac{\sin\{K\pi\gamma d/\lambda[-(\cos r_m-\cos r_n)\sin\theta_q\sin\phi_q+(\sin r_m-\sin r_n)\sin\theta_q\cos\phi_q]\}}{\sin\{\pi\gamma d/\lambda[-(\cos r_m-\cos r_n)\sin\theta_q\sin\phi_q+(\sin r_m-\sin r_n)\sin\theta_q\cos\phi_q]\}}\end{aligned} \tag{3.54}$$

It can be found by observing (3.54) that, even if the maximum angle expansion is taken into consideration, the Mailloux method can be adopted to perform null-widen, but we need the information on the DOAs of the jamming signals.

3.4.2 Zatman Algorithm

1. Zatman Algorithm Applied on Uniform Linear Array

Zatman method [4] broadens nulls by assuming that the narrowband jamming signal actually received by an array has a certain bandwidth. Given that the virtual bandwidth for a jamming signal is B_w, and that it has a flat power spectrum inside the whole bandwidth, for a uniform linear array with array element interval d equaling to a half-wavelength, the element of mth row and nth column for the covariance matrix is

$$\overline{\mathbf{R}}_Z(m,n)=\sum_{q=1}^{Q} r_{mn}\left[\tau_q(m,n)\right]\sigma_q^2 e^{-\mathrm{j}\pi(m,n)\sin\theta_q}+\sigma_e^2\delta_{mn} \tag{3.55}$$

where $r_{mn}\left(\tau_q(m,n)\right)$ is the cross-correlation function between the qth jamming signal received by the mth array element and the qth jamming signal received by the nth array element; and $\tau_q(m,n)$ is the propagation delay of the qth jamming signal between these two array elements

$$\tau_q(m,n)=\frac{(n-m)\,d\sin\theta_q}{c} \tag{3.56}$$

Based on the jamming signal's virtual power spectrum, the jamming signal's auto-correlation function is

$$r(t)=\frac{\sin(\pi B_w t)}{\pi B_w t} \tag{3.57}$$

By substituting (3.57) into (3.55), and adopting derivation process similar to those of (3.40)–(3.45), we then have

$$\overline{\mathbf{R}}_Z(m,n)=\frac{\sin\left[\pi B_w\tau_q(m,n)\right]}{\pi B_w\tau_q(m,n)}\odot\mathbf{R}_x(m,n) \tag{3.58}$$

Thereby we can have the expansion matrix

$$\mathbf{T}_z(m,n)=\frac{\sin\left[\pi B_w\tau_q(m,n)\right]}{\pi B_w\tau_q(m,n)} \tag{3.59}$$

Unlike (3.45), both sides of the Eq. (3.58) are strictly equal to each other, instead of approximately equal to each other. The Zatman method does not change the contribution of the noise term in the covariance matrix, but (3.59) still contains information on DOAs of jamming signals. Considering the need for maximum expansion angle, we choose to make the sinusoid term in (3.56) as 1, then for a uniform linear array, the Zatman method does not need information on a jamming signal's DOA. In addition, even if the expansion matrix given by (3.59) is derived using a uniform linear array example, but this expression is proper for any array structure. The difference is that the values of $\tau_q(m,n)$ are different.

2. Zatman Algorithm Applied on Uniform Circle Array

For a uniform circular array composed of M array elements, when

$$\tau_q(m,n)=\frac{d}{c}\left[(\cos r_m-\cos r_n)\sin(\theta_q)\cos(\phi_q)+(\sin r_m-\sin r_n)\sin(\theta_q)\sin(\phi_q)\right] \tag{3.60}$$

By substituting (3.60) into (3.59), we have

$$\mathbf{T}_z(m,n)=\frac{\sin\left\{\frac{\pi dB_w}{c}\left[(\cos r_m-\cos r_n)\sin(\theta_q)\cos(\phi_q)+(\sin r_m-\sin r_n)\sin(\theta_q)\sin(\phi_q)\right]\right\}}{\frac{\pi dB_w}{c}\left[(\cos r_m-\cos r_n)\sin(\theta_q)\cos(\phi_q)+(\sin r_m-\sin r_n)\sin(\theta_q)\sin(\phi_q)\right]} \tag{3.61}$$

Obviously (3.61) and (3.54) are similar to each other. For circular arrays, to use the Zatman method for null-widen still requires information on the DOAs of jamming signals.

3.4.3 Laplace Algorithm

1. Laplace Null-Widen Algorithm Applied on Uniform Linear Array

As described above, both Mailloux algorithm and Zatman algorithm need information on jamming signals' DOAs when applied on circular arrays. In this section, we use statistical models for jamming DOA changes to discuss the null-widen algorithms that do not need jamming DOA information for circular and linear arrays.

Existing jamming DOA change statistics models mainly include Gaussian distribution, uniform distribution and two-point distribution. Among them, the Gaussian model is used in Ref. [14]. These models cannot simulate jamming change in a high-dynamic environment. Based on high-dynamic motion models discussed in Sect. 3.2, the plot for probability distribution of jamming DOA is shown in Fig. 3.12.

It can be seen in Fig. 3.12 that the change of jamming DOA mainly concentrates in the range of small angles, and approximately follows a Laplace distribution model. That is because within a very short time interval the jamming DOA has very small change, and has a very small probability of having a big angle change. Therefore relative to the Gaussian distribution, the Laplace distribution is more suitable to be used as the statistical model for jamming DOA changes under

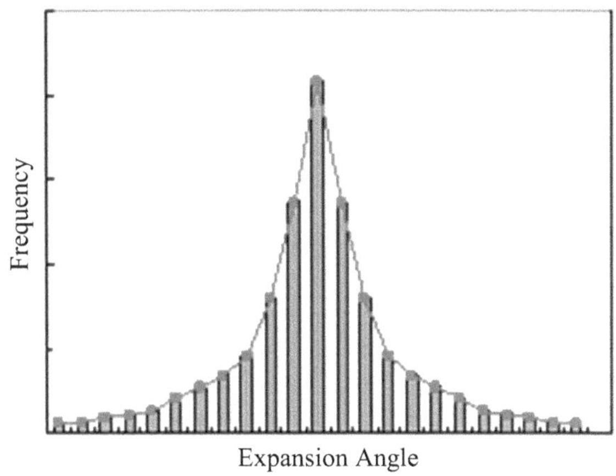

Fig. 3.12 Statistical probability distribution of jamming DOA change

high-dynamic conditions. Next, we derive Laplace null-widen algorithm [8] from the perspectives of a uniform linear array and a uniform circular array respectively.

From the perspective of jamming signal DOA angle change, the DOA of the qth jamming signal can be depicted as

$$\bar{\theta}_q = \theta_q + \Delta\theta_q \tag{3.62}$$

where $\Delta\theta_q$ denotes the change of jamming signal DOA, which follows a Laplace distribution of zero mean and $2\xi_q^2$ variance, i.e.

$$f(\Delta\theta_q) = \frac{1}{2\xi_q} \mathrm{e}^{-\frac{|\Delta\theta_q|}{\xi_q}} \tag{3.63}$$

Then, the average covariance matrix of the constructed array received signal is

$$\overline{\mathbf{R}}_L = \sum_{q=1}^{Q} \sigma_q^2 \int f(\Delta\theta_q)\mathbf{a}\left[\theta_q(t)\right]\mathbf{a}^{\mathrm{H}}\left[\theta_q(t)\right]\mathrm{d}\Delta\theta_q + \sigma_e^2\mathbf{I} \tag{3.64}$$

where the element at the mth row and the nth column of $\overline{\mathbf{R}}_L$ can be written as

$$\begin{aligned}\overline{\mathbf{R}}_L(m,n) &= \sum_{q=1}^{Q} \sigma_q^2 \int \frac{1}{2\xi_q} \mathrm{e}^{-\frac{|\Delta\theta_q|}{\xi_q}} \mathrm{e}^{-\mathrm{j}\pi(m-n)\sin\theta_q(t)} \mathrm{d}\Delta\theta_q + \sigma_e^2\delta_{mn} \\ &= \sum_{q=1}^{Q} \sigma_q^2 \mathrm{e}^{-\mathrm{j}\pi(m-n)\sin\theta_q} \sum_{q=1}^{Q} \int \frac{1}{2\xi_q} \mathrm{e}^{-\frac{|\Delta\theta_q|}{\xi_q}} \mathrm{e}^{-\mathrm{j}\pi(m-n)\cos\theta_q\Delta\theta_q} \mathrm{d}\Delta\theta_q + \sigma_e^2\delta_{mn}\end{aligned} \tag{3.65}$$

Let

$$\mathrm{A} = -\mathrm{j}\pi(m-n)\cos\theta_q \tag{3.66}$$

Then we have

$$\begin{aligned}\overline{\mathbf{R}}_L(m,n) &= \sum_{q=1}^{Q}\sigma_q^2 \mathrm{e}^{-\mathrm{j}\pi(m-n)\sin\theta_q}\sum_{q=1}^{Q}\int_{-\infty}^{0}\frac{1}{2\xi_q}\mathrm{e}^{\frac{\Delta\theta_q}{\xi_q}}\mathrm{e}^{\mathrm{A}\Delta\theta_q}\mathrm{d}\Delta\theta_q \\ &+ \int_{0}^{+\infty}\frac{1}{2\xi_q}\mathrm{e}^{-\frac{\Delta\theta_q}{\xi_q}}\mathrm{e}^{\mathrm{A}\Delta\theta_q}\mathrm{d}\Delta\theta_q + \sigma_e^2\delta_{mn} \\ &= \frac{1}{1-\mathrm{A}^2\xi_q^2}\sum_{q=1}^{Q}\sigma_q^2\mathrm{e}^{-\mathrm{j}\pi(m-n)\sin\theta_q} + \sigma_e^2\delta_{mn}\end{aligned} \tag{3.67}$$

After re-sorting, we can obtain the element at the mth row and the nth column of $\overline{\mathbf{R}}_L$ as

$$\overline{\mathbf{R}}_L(m,n) = \mathbf{T}_L(m,n)\sum_{q=1}^{Q}\sigma_q^2\mathrm{e}^{-\mathrm{j}\pi(n-m)\sin\theta_q} + \sigma_e^2\delta_{mn} \tag{3.68}$$

where the expansion matrix is

$$\mathbf{T}_L(m,n) = \frac{1}{1+\left(\Lambda_{mn}\xi_q\cos\theta_q\right)^2} \tag{3.69}$$

and in the above formula

$$\Lambda_{mn} = (m-n)\pi \tag{3.70}$$

By adopting the similar derivation process as used for (3.40)–(3.45), we have

$$\overline{\mathbf{R}}_L = \mathbf{R}_x \odot \mathbf{T}_L \tag{3.71}$$

We need to point out that, both sides of the equation in (3.71) equal to each other, instead of being just approximately equal. The Laplace method, similarly, does not change the contribution of the noise term in the covariance matrix. In the expression matrix, θ_q and ξ_q denote the DOAs for all jamming sources and their disturbance parameters. Based on the high-dynamic motion model, when $|\cos\theta_q| = 1$, the change of jamming DOA is at its peak, and consequently the needed null width expansion is the maximum as well. Therefore, we choose the parameter $\xi_{\max}$ that is needed for maximum null width to replace $\xi_q \cos\theta_q$ in order to form the new expansion matrix, i.e.

$$\overline{\mathbf{T}}_L(m,n) = \frac{1}{1+(\Lambda_{mn}\xi_{\max})^2} \tag{3.72}$$

It can be seen from (3.72), the new expansion matrix $\overline{\mathbf{T}}_L$ not only can satisfy the maximum angle change, it also does not need the information on the DOA of jamming signals. The detailed steps of the Laplace algorithm used for the uniform linear array are as below:

(1) Compute the received data's covariance matrix $\widehat{\mathbf{R}}_x$.
(2) Compute the expansion matrix $\overline{\mathbf{T}}_L$ based on (3.72).
(3) Re-modify (3.71) using $\overline{\mathbf{T}}_L$ to obtain the new covariance matrix $\overline{\mathbf{R}}_L$.
(4) Replace $\mathbf{R}_x$ in (3.3) using $\overline{\mathbf{R}}_L$, and compute the weight vector $\mathbf{w}$ for the antenna array.

We need to point out that, even though in this section we use a uniform linear array as an example to derive the Mailloux algorithm, the Zatman algorithm and the Laplace algorithm, the derivation process can apply to any linear array.

2. Laplace Null-Widen Algorithm Applied on Uniform Circular Array

For a uniform circular array, under the high-dynamic condition, the DOA of the qth jamming signal can be depicted as

$$\begin{cases} \bar{\theta}_q = \theta_q + \Delta\theta_q \\ \bar{\phi}_q = \phi_q + \Delta\phi_q \end{cases} \tag{3.73}$$

where $\Delta\theta_q$ follows a Laplace distribution of zero mean and $2\xi_{q1}^2$ variance, and $\Delta\phi_q$ follows a Laplace distribution of zero mean and $2\xi_{q2}^2$ variance. Then, to construct the average covariance matrix, we have

$$\overline{\mathbf{R}}_L(m,n) = \sum_{q=1}^{Q} \sigma_q^2 \iint f(\Delta\theta_q,\Delta\phi_q) \mathrm{e}^{-\mathrm{j}[\mathbf{p}_m-\mathbf{p}_n]^T \mathbf{u}(\theta_q+\Delta\theta_q,\phi_q+\Delta\phi_q)} d(\Delta\theta_q) d(\Delta\phi_q) + \sigma_e^2\delta_{mn} \tag{3.74}$$

where $f(\Delta\theta_q,\Delta\phi_q)$ denotes the joint probability density function of $(\Delta\theta_q,\Delta\phi_q)$. Since the expansion angles $\Delta\theta_q$ and $\Delta\phi_q$ are independent from each other, we can simplify (3.74) by performing a Taylor series expansion on the wave number vector $\mathbf{u}(\theta_q+\Delta\theta_q,\phi_q+\Delta\phi_q)$

$$\overline{\mathbf{R}}_L(m,n) \approx \sum_{q=1}^{Q} \sigma_q^2 e^{-j\frac{2\pi}{\lambda}[\mathbf{p}_m-\mathbf{p}_n]^{\mathrm{T}}\mathbf{u}(\theta_q,\phi_q)} \int f(\Delta\theta_q)\, e^{-j\frac{2\pi}{\lambda}[\mathbf{p}_m-\mathbf{p}_n]^{\mathrm{T}}\begin{bmatrix}\cos\theta_q\cos\phi_q\\ \cos\theta_q\sin\phi_q\end{bmatrix}\Delta\theta_q} d(\Delta\theta_q)$$
$$\int f(\Delta\phi_q) e^{-j\frac{2\pi}{\lambda}[\mathbf{p}_m-\mathbf{p}_n]^{T}\begin{bmatrix}-\sin\theta_q\sin\phi_q\\ \sin\theta_q\cos\phi_q\end{bmatrix}\Delta\phi_q} d(\Delta\phi_q) + \sigma_e^2\delta_{mn} \tag{3.75}$$

Same as the derivation process in (3.67), the first integration term in (3.75) can be simplified as

$$\int f(\Delta\theta_q)\, e^{-j\frac{2\pi}{\lambda}[\mathbf{p}_m-\mathbf{p}_n]^{T}\begin{bmatrix}\cos\theta_q\cos\phi_q\\ \cos\theta_q\sin\phi_q\end{bmatrix}\Delta\theta_q} d(\Delta\theta_q) = \frac{1}{1-D_{mn}^2\xi_{q1}^2} \tag{3.76}$$

where

$$D_{mn} = -j\frac{2\pi}{\lambda}[\mathbf{p}_m-\mathbf{p}_n]^{\mathrm{T}}\begin{bmatrix}\cos\theta_q\cos\phi_q\\ \cos\theta_q\sin\phi_q\end{bmatrix} \tag{3.77}$$

For the same reason, the second integration term in (3.76) can be simplified as

$$\int f(\Delta\phi_q) e^{-j\frac{2\pi}{\lambda}[\mathbf{p}_m-\mathbf{p}_n]^{\mathrm{T}}\begin{bmatrix}-\sin\theta_q\sin\phi_q\\ \sin\theta_q\cos\phi_q\end{bmatrix}\Delta\phi_q} d(\Delta\phi_q) = \frac{1}{1-F_{mn}^2\xi_{q2}^2} \tag{3.78}$$

where

$$F_{mn} = -j\frac{2\pi}{\lambda}[\mathbf{p}_m-\mathbf{p}_n]^{\mathrm{T}}\begin{bmatrix}-\sin\theta_q\sin\phi_q\\ \sin\theta_q\cos\phi_q\end{bmatrix} \tag{3.79}$$

By substituting (3.78) and (3.76) into (3.75), we can have

$$\overline{\mathbf{R}}_L(m,n) \approx \sum_{q=1}^{Q} \sigma_q^2 e^{-j\frac{2\pi}{\lambda}[\mathbf{p}_m-\mathbf{p}_n]^{\mathrm{T}}\begin{bmatrix}\sin\theta_q\cos\phi_q\\ \sin\theta_q\sin\phi_q\end{bmatrix}} \frac{1}{1-D_{mn}^2\xi_{q1}^2}\frac{1}{1-F_{mn}^2\xi_{q2}^2} + \sigma_e^2\delta_{mn} \tag{3.80}$$

Thereby we can have the element at the *m*th row and the *n*th column for the expansion matrix $\mathbf{T}_L$ as

$$\mathbf{T}_L(m,n) = \frac{1}{1 - D_{mn}^2 \xi_{q1}^2} \frac{1}{1 - F_{mn}^2 \xi_{q2}^2} \tag{3.81}$$

By substituting D_{mn} and F_{mn} into (3.81) and re-sorting, we can have

$$\begin{aligned} \mathbf{T}_L(m,n) = & \frac{1}{1 + \left[\xi_{q1}\left(G_{mn}\cos\theta_q\cos\phi_q + H_{mn}\cos\theta_q\sin\phi_q\right)\right]^2} \\ & \times \frac{1}{1 + \left[\xi_{q2}\left(-G_{mn}\sin\theta_q\sin\phi_q + H_{mn}\sin\theta_q\cos\phi_q\right)\right]^2} \end{aligned} \tag{3.82}$$

where

$$G_{mn} = \frac{2\pi}{\lambda} d(\cos r_m - \cos r_n) \tag{3.83}$$

$$H_{mn} = \frac{2\pi}{\lambda} d(\sin r_m - \sin r_n) \tag{3.84}$$

The expansion matrix in (3.82) still needs to know the information on the DOAs of jamming signals. Therefore, we need to satisfy the maximum angle needed for the motion, and when $\xi_{q1} = \xi_{q2} = \xi_q$, based on trigonometric expansions, (3.82) can be simplified as

$$\overline{\mathbf{T}}_L(m,n) = \frac{1}{1 + \left(\xi_{\max}\sqrt{G_{mn}^2 + H_{mn}^2}\right)^2} \tag{3.85}$$

By substituting (3.85) into (3.80), and adopting the derivation process similar to what is used for (3.40) and (3.45), we can have

$$\overline{\mathbf{R}}_L = \mathbf{R}_x \odot \overline{\mathbf{T}}_L \tag{3.86}$$

Obviously, by using the $\overline{\mathbf{R}}_L$ computed by (3.86) to replace the $\mathbf{R}_x$ in (3.3), we can form broad nulls towards the jamming DOAs, and there is no need to know the information on the DOAs of the jamming signals.

We need to highlight that, the Mailloux method, the Zatman method and the Laplace method discussed in this section can all be incorporated into a unified framework of covariance matrix taper, i.e. these methods can modify array sample covariance matrix using an expansion matrix to achieve the goal of null width expansion. Table 3.1 compares the adaptabilities of these three methods for the uniform linear array, the arbitrary linear array, and the uniform circular array (mainly considering whether or not the jamming DOA information is needed or not).

Table 3.1 Adaptability comparisons among the three methods for different array structures

Array structure	Algorithm		
	Mailloux method	Zatman method	Laplace method
Uniform linear array	Not required	Not required	Not required
Arbitrary linear array	Not required	Not required	Not required
Uniform circular array	Required	Required	Not required

3.4.4 Simulation Results

1. Uniform Linear Array Simulation Results

Our simulation adopts a 7-element half-wavelength equidistant linear array. Seven GNSS signals PRN4, PRN7, PRN9, PRN11, PRN20, PRN24, PRN28, and a jamming signal incident on the array, the initial DOA of the jamming is $-20°$, and the jamming DOA for the simulation changes gradually from $-20°$ to $-17°$. The jamming-to-noise ratio is 40 dB, and the signal-to-noise ratio is –20 dB. The received signal by the array has a 4.309 MHz digital IF frequency, and a 5.714 MHz sampling rate.

Figure 3.13 shows the comparison results on beam patterns between the power minimization algorithm and the three null-widen algorithms based on power minimization. To make it clear, Fig. 3.13b shows the partially enlarged graph for the null, where the dotted line corresponds to the power minimization algorithm, the dash-dot line corresponds to the Mailloux algorithm result, the thick solid line corresponds to the Zatman algorithm result, and the thin solid line corresponds to the Laplace algorithm result. It is clear from Fig. 3.13b that the three null-widen algorithms can effectively make the jamming nulls wider. In addition, when the widths of the null zones are the same, the beam patterns of the Mailloux algorithm and the Zatman algorithm almost coincident with each other, but the nulls formed by the Laplace algorithm are deeper than those formed by the other algorithms.

Figure 3.14 compares output signal-to-jamming-plus-noise ratios between the power minimization algorithm and the three null-widen algorithms. The dotted line corresponds to the results of the power minimization algorithm, and the dot-dash-line corresponds to the Mailloux algorithm result, the think solid line corresponds to the Zatman algorithm results, and the thick solid line corresponds to the Laplace algorithm results. It can be seen from Fig. 3.14a that the null-widen algorithm can significantly improve the output signal-jamming-plus-noise ratio. Figure 3.14b shows the amplified comparison, and from it we can see that the Zatman algorithm's output signal-jamming-plus-noise ratio is slightly higher than that of the Mailloux algorithm, but the output signal-to-jamming-plus-noise ratio obtained by the Laplace algorithm is the highest.

Figure 3.15 compares the receivers' acquisition results using the power minimization algorithm and the Laplace null-widen algorithm for jamming mitigation. In Fig. 3.15a, b, it can be seen that the jamming mitigation of the power

Fig. 3.13 Beam pattern comparisons between the power minimization algorithm and the null-widen algorithm based on the power minimization

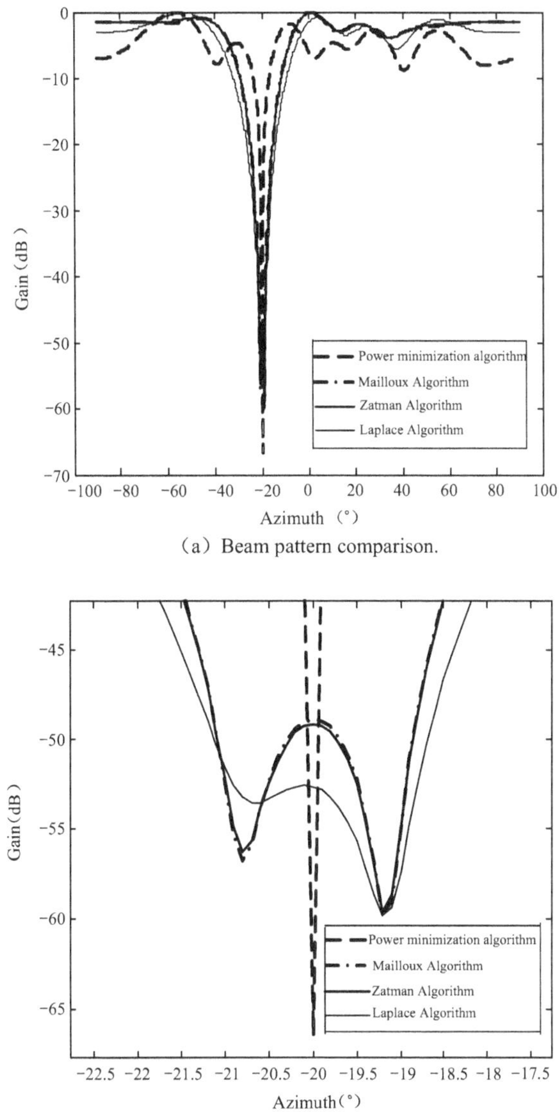

(a) Beam pattern comparison.

(b) Partial enlarged graph for the beam pattern's null zone.

minimization algorithm in a high-dynamic environment can not ensure the receiver' correct acquisition, but when using the Laplace null-widen algorithm for jamming mitigation, the receiver can acquire all seven GNSS signals.

2. Uniform Circular Array Simulation Results

This simulation uses a 7-element uniform circular array, four GNSS signal PRN1, PRN6, PRN7 and PRN14, which incident on the array from

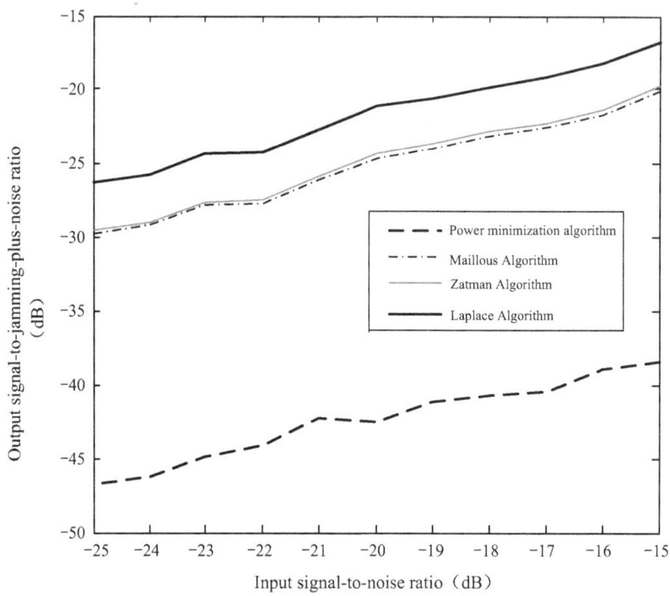

(a) Comparisons of output signal-jamming-plus-noise ratios.

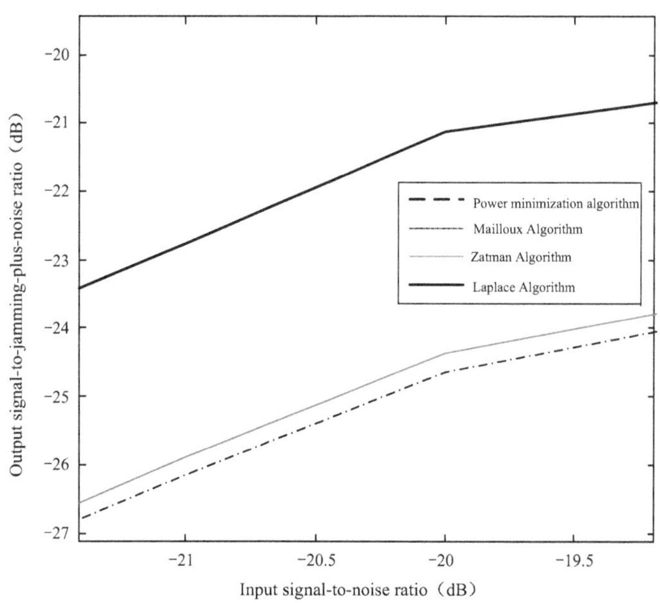

(b) Partially enlarged view of null-widen algorithms.

Fig. 3.14 Comparisons of output signal-to-jamming-plus-noise ratios between the power minimization algorithm and the three null-widen algorithms based on the power minimization

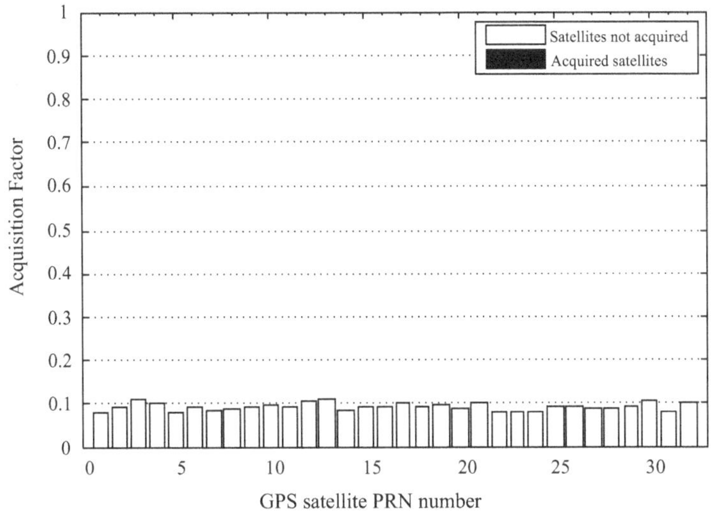

(a) Power minimization algorithm

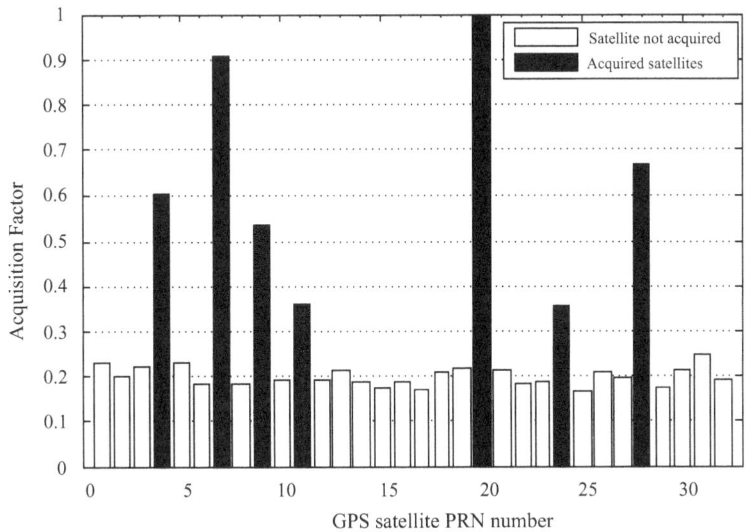

(b) Laplace algorithm

Fig. 3.15 Comparisons of acquisition results

$(-48°, 60°)$, $(10°, 120°)$, $(25°, 200°)$ and $(70°, 90°)$ respectively, and a jamming signal with an initial DOA of $(50°, 155°)$ which incidents on the array. During the simulation, the DOA direction gradually changes from $(50°, 155°)$ to $(54°, 159°)$. The jamming-to-noise ratio is 40 dB, and the signal-to-noise ratio is –20 dB. The

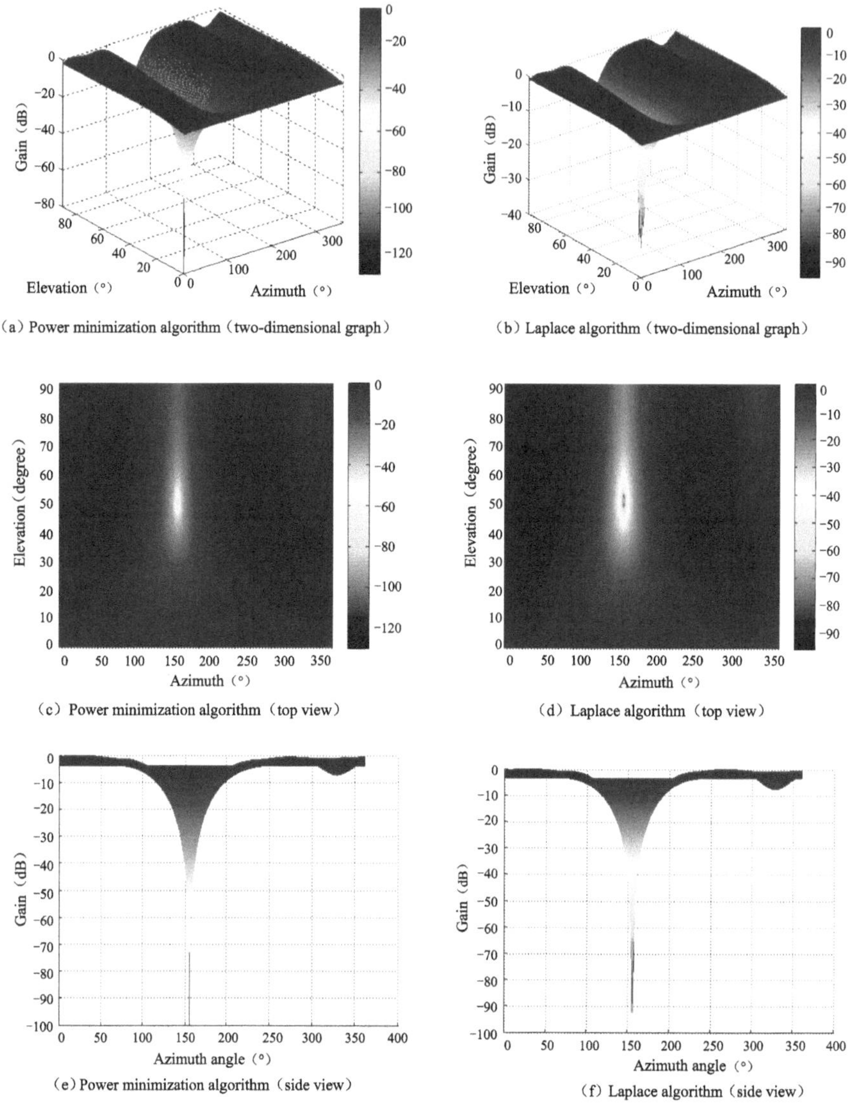

Fig. 3.16 Comparisons of beam pattern based on circular arrays

signal received by the array has a 4.309 MHz digital IF frequency and a 5.714 MHz sampling rate.

Figure 3.16 compares adaptive array beam patterns for the power minimization algorithm and the Laplace null-widen algorithm. In Fig. 3.16, subfigures (a), (c) and (e) are the beam pattern, top view and side view of the beam pattern for the power minimization algorithm respectively; subfigures (b), (d), and (f) are the beam pattern, top view and side view of the beam pattern for the Laplace null-widen

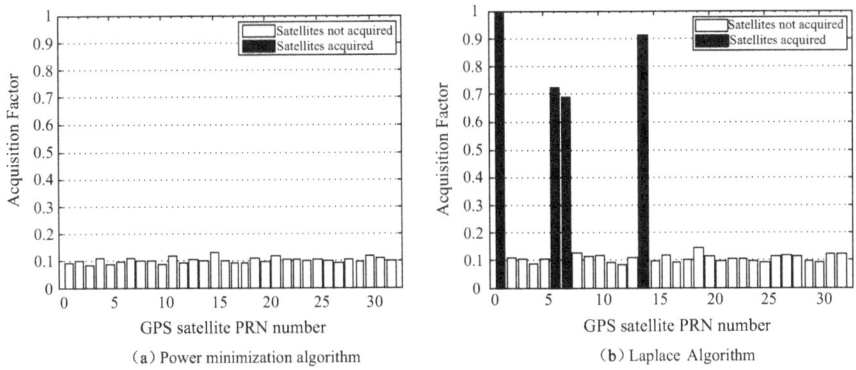

Fig. 3.17 Comparisons on acquisition results

algorithm respectively. It can be seen by comparing subfigures (a) and (b), (c) and (d) and (e) and (f) in Fig. 3.16, the null zone formed by the Laplace null-widen algorithm is wider but shallower compared with the null zone formed by the power minimization algorithm.

Figure 3.17 shows the receiver acquisition results after using the power minimization algorithm and the Laplace null-widen algorithm. By comparing Fig. 3.17a, b, it can be seen that, in a high-dynamic environment, the suppression effect of the power minimization algorithm cannot ensure correct acquisition by the receiver, while the Laplace null-widen algorithm can accurately acquire four GPS signals: PRN1, PRN6, PRN7 and PRN14.

3.5 Null-Widen Space-Time Adaptive Filtering

As described in Sect. 2.6, the space-time adaptive processing (STAP) [15, 16] has become a hot research topic because it can save spatial domain degrees of freedom, effectively suppressing narrowband jamming, wideband jamming, and dispersion multipath jamming. In this section, we expand the Laplace null-widen algorithm (introduced in Sect. 3.4.3), which has stable performance and wide applicability, into space-time domain. As a result the new technique can achieve high-dynamic jamming suppression using space-time processing.

3.5.1 Space-Time Laplace Null-Widen Algorithm

Considering a space-time uniform linear array with tap delay T, and for the sake of convenience, the array received signal model given in Sect. 2.6 can be rewritten as a high-dynamic space-time received data model, i.e.

$$\mathbf{x}(t) = \sum_{l=1}^{L} \mathbf{a}(\theta_l) \otimes \mathbf{s}_l(t) + \sum_{q=1}^{Q} \mathbf{a}\left[\theta_q(t)\right] \otimes \mathbf{j}_q(t) + \mathbf{e}(t) \tag{3.87}$$

where $\mathbf{x}(t)$ is the space-time data vector received by the array; $\mathbf{s}_l(t)$ denotes the temporal domain tap vector for the lth satellite; $\mathbf{a}(\theta_l)$ is the corresponding steering vector; $\mathbf{j}_q(t)$ denotes the temporal domain tap vector for the qth jamming; $\mathbf{a}\left[\theta_q(t)\right]$ is the corresponding time-varying steering vector, for a uniform linear array with a half-wavelength interval, we now have

$$\mathbf{a}\left[\theta_q(t)\right] = [1, \quad \mathrm{e}^{-\mathrm{j}\pi \sin\theta_q(t)}, \quad \ldots, \quad \mathrm{e}^{-\mathrm{j}(M-1)\pi \sin\theta_q(t)}]^{\mathrm{T}} \tag{3.88}$$

where $\mathbf{e}(t)$ denotes the thermal noise vector for the space-time receiver. Here we assume that GNSS signal, jamming signal and noise are not correlated with each other.

Based on what is described in Sect. 3.4.3, the jamming DOA $\theta_q(t)$ that changes over time can be described as the angle expansion of the initial jamming DOA θ_q, i.e.

$$\theta_q(t) = \theta_q + \Delta\theta_q \tag{3.89}$$

Since the value of $\Delta\theta_q$ is very small, based on the function's first-order Taylor expansion, we can have

$$\sin\theta_q(t) = \sin\left(\theta_q + \Delta\theta_q\right) \approx \sin\theta_q + \Delta\theta_q \cos\theta_q \tag{3.90}$$

By substituting (3.90) into (3.88), we can have

$$\mathbf{a}(\theta_q(t)) = \mathbf{a}(\theta_q) \odot \mathbf{b} \tag{3.91}$$

where

$$\mathbf{b} = \left[1, \ \mathrm{e}^{-\mathrm{j}\pi\Delta\theta_q \cos\theta_q}, \quad \ldots, \mathrm{e}^{-\mathrm{j}(M-1)\pi\Delta\theta_q \cos\theta_q}\right]^{\mathrm{T}} \tag{3.92}$$

For convenience, in this section $\mathbf{a}(\theta_q(t))$ is shorthanded as $\bar{\mathbf{a}}_q$ and $\mathbf{a}(\theta_q)$ is shorthanded as $\mathbf{a}_q$.

As described in Sect. 3.4.3, the expansion angle $\Delta\theta_q$ approximately follows a Laplace distribution, since the received signal by the array has an average space-time covariance matrix of

$$
\begin{aligned}
\overline{\mathbf{R}}_L &= \sum_{q=1}^{Q} \int f(\Delta\theta_q)\left(\bar{\mathbf{a}}_q \otimes \bar{\mathbf{a}}_q^{\mathrm{H}}\right) \otimes \mathbf{R}_q \mathrm{d}\Delta\theta_q + \sigma_e^2 \mathbf{I} \\
&= \sum_{q=1}^{Q} \int f(\Delta\theta_q)\left[\left(\mathbf{a}_q \otimes \mathbf{a}_q^{\mathrm{H}}\right) \odot \left(\mathrm{b} \otimes \mathrm{b}^{\mathrm{H}}\right)\right] \otimes \mathbf{R}_q \mathrm{d}\Delta\theta_q + \sigma_e^2 \mathbf{I}
\end{aligned}
\tag{3.93}
$$

where $\mathbf{R}_q$ denotes the time-domain correlation matrix of the jamming signal, i.e.

$$
\mathbf{R}_q = E\left\{\mathbf{j}_q(t)\mathbf{j}_q^{\mathrm{H}}(t)\right\} \tag{3.94}
$$

$f(\Delta\theta_q)$ is $\Delta\theta_q$'s probability density function, i.e.

$$
f(\Delta\theta_q) = \frac{1}{2\xi_q}\exp\left(-\frac{|\Delta\theta_q|}{\xi_q}\right) \tag{3.95}
$$

where $2\xi_q^2$ is the variance.

Using the similar derivation process as used in Sect. 3.4.3, (3.93) can be further simplified as

$$
\begin{aligned}
\overline{\mathbf{R}}_L &= \sum_{q=1}^{Q} (\mathbf{T}_L \otimes \mathbf{I}_{K\times K}) \odot \left[\left(\mathbf{a}_q \otimes \mathbf{a}_q^{H}\right) \otimes \mathbf{R}_q\right] + \sigma_e^2 \mathbf{I} \\
&= (\mathbf{T}_L \otimes \mathbf{I}_{K\times K}) \odot \mathbf{R}_x \triangleq \mathbf{T}_{STL} \odot \mathbf{R}_x
\end{aligned}
\tag{3.96}
$$

where the element in matrix $\mathbf{T}_L$ is

$$
\mathbf{T}_L(m,n) = \frac{1}{1 + \left(\Lambda_{mn}\xi_q \cos\theta_q\right)^2} \quad 1 \leqslant m,n \leqslant M \tag{3.97}
$$

and

$$
\Lambda_{mn} = (m-n)\pi \tag{3.98}
$$

Obviously, matrix $\mathbf{T}_{STL}$ is the expansion matrix for the space-time linear array Laplace null-widen algorithm. By observing (3.97), we know the matrix contains the jamming signals' DOA information. Similar to the spatial domain's uniform linear array algorithm, we can select parameter $\xi_{\max}$ to replace $\xi_q \cos\theta_q$ and form a new expansion matrix, consequently we do not need information on jamming signals anymore.

The computation steps for the space-time Laplace null-widen algorithm are listed below:

(1) Solve the space-time covariance matrix for received signals by the array $\widehat{\mathbf{R}}_x$.
(2) Based on (3.97) and (3.96), compute the expansion matrix that does not contain DOA information $\mathbf{T}_{STL}$.
(3) Compute $\overline{\mathbf{R}}_L$ based on (3.96), then replace $\mathbf{R}_x$.
(4) Based on $\overline{\mathbf{R}}_L$, compute antenna array's weight vector $\mathbf{w}$ using the space-time power minimization algorithm.

3.5.2 Reduced Rank Space-Time Null-Widen Algorithm

As described in Sect. 2.6.3, STAP processing involves the inverses of two-dimensional high-order matrices, consequently the computation complexity increases significantly, and has very high demands on the number of snapshots. This is even harder to achieve in a high-dynamic environment (because the number of snapshots is very low), so a reduced rank processing is needed [17–19]. In this section, we combine the CSA-MWF in Sect. 2.6.4 and the space-time null widen algorithm to solve this problem.

CSA-MWF is one of the multi-stage Wiener filters (described in Sect. 2.6.4). It does not need to compute the block matrix explicitly, and has smaller computation complexity compared with the GRS-MWF, and can achieve even better performance using fewer snapshots. It can be seen from Table 2.3, for the whole iterative process of the CSA-MWF, there is no need to calculate the input signal's covariance matrix. Additionally, the space-time null-widen algorithm achieves the objective of broadening nulls by reconstructing the covariance matrix. In this section, we perform equivalent processing on the CSA-MWF algorithm, to establish the relation between the weight vector obtained from the algorithm and the covariance matrix, and thereby applying directly onto the null-widen algorithm [19].

For the power minimization algorithm, the CSA-MWF weight given by (2.192) can be re-written as

$$\mathbf{w}_{CSA-MWF} = \mathbf{h}_0 - \mathbf{T}_D \mathbf{w}_d \tag{3.99}$$

where

$$\mathbf{h}_0 = \boldsymbol{\delta}_{MK} \tag{3.100}$$

$$\mathbf{w}_d = \mathbf{R}_d^{-1} \boldsymbol{r}_{dd_0} = \left(\mathbf{T}_D^{\mathrm{H}} \mathbf{R}_{x_0} \mathbf{T}_D\right)^{-1} \boldsymbol{r}_{dd_0} = \left(\mathbf{T}_D^{\mathrm{H}} \mathbf{R}_{x_0} \mathbf{T}_D\right)^{-1} \mathbf{T}_D^{\mathrm{H}} \boldsymbol{r}_{x_0 d_0} \tag{3.101}$$

$$\mathbf{T}_D = [\mathbf{h}_1, \quad \mathbf{h}_2, \quad \ldots, \quad \mathbf{h}_D] \tag{3.102}$$

is the reduced rank matrix, and

$$\mathbf{h}_i = \frac{\boldsymbol{r}_{x_{i-1}d_{i-1}}}{\|\boldsymbol{r}_{x_{i-1}d_{i-1}}\|}, \quad i = 1, 2, \ldots D \tag{3.103}$$

Since $\mathbf{x}_0(n) = (\mathbf{I} - \mathbf{h}_0\mathbf{h}_0^{\mathrm{H}})\mathbf{x}(n)$, $d_0(n) = \mathbf{h}_0^{\mathrm{H}}\mathbf{x}(n)$ (see Table 2.3), then

$$\mathbf{r}_{x_0d_0} = E(\mathbf{x}_0(n)d_0^*(n)) = (\mathbf{I} - \mathbf{h}_0\mathbf{h}_0^{\mathrm{H}})\mathbf{R}_x\mathbf{h}_0 \tag{3.104}$$

$$\mathbf{R}_{x_0} = (\mathbf{I} - \mathbf{h}_0\mathbf{h}_0^{\mathrm{H}})\mathbf{R}_x(\mathbf{I} - \mathbf{h}_0\mathbf{h}_0^{\mathrm{H}})^{\mathrm{H}} \tag{3.105}$$

Also because

$$\mathbf{x}_i(n) = (\mathbf{I} - \mathbf{h}_i\mathbf{h}_i^{\mathrm{H}})^{\mathrm{H}}\mathbf{x}_{i-1}(n) = (\mathbf{I} - \mathbf{h}_0\mathbf{h}_0^{\mathrm{H}} - \cdots - \mathbf{h}_i\mathbf{h}_i^{\mathrm{H}})\mathbf{x}(n) \tag{3.106}$$

$$d_i(n) = \mathbf{h}_i^{\mathrm{H}}\mathbf{x}_{i-1}(n) = \mathbf{h}_i^{\mathrm{H}}\mathbf{x}(n) \tag{3.107}$$

then

$$\mathbf{r}_{x_{i-1}d_{i-1}} = \left(\mathbf{I} - \sum_{j=0}^{i-1}\mathbf{h}_j\mathbf{h}_j^{\mathrm{H}}\right)\mathbf{R}_x\mathbf{h}_{i-1} \tag{3.108}$$

$$\mathbf{h}_i = \frac{\left(\mathbf{I} - \sum\limits_{j=0}^{i}\mathbf{h}_j\mathbf{h}_j^{\mathrm{H}}\right)\mathbf{R}_x\mathbf{h}_{i-1}}{\left\|\left(\mathbf{I} - \sum\limits_{j=0}^{i}\mathbf{h}_j\mathbf{h}_j^{\mathrm{H}}\right)\mathbf{R}_x\mathbf{h}_{i-1}\right\|} \tag{3.109}$$

It can be seen in (3.109) that $\mathbf{T}_D$ can be determined by $\mathbf{R}_x$, and by substituting (3.104) and (3.105) into (3.101), we can have

$$\mathbf{w}_d = \left(\mathbf{T}_D^{\mathrm{H}}(\mathbf{I} - \mathbf{h}_0\mathbf{h}_0^{\mathrm{H}})\mathbf{R}_x(\mathbf{I} - \mathbf{h}_0\mathbf{h}_0^{\mathrm{H}})^{\mathrm{H}}\mathbf{T}_D\right)^{-1}\mathbf{T}_D^{\mathrm{H}}(\mathbf{I} - \mathbf{h}_0\mathbf{h}_0^{\mathrm{H}})\mathbf{R}_x\mathbf{h}_0 \tag{3.110}$$

Also because

$$\mathbf{h}_i^{\mathrm{H}}\mathbf{h}_j = \begin{cases} 0, i \neq j \\ 1, i = j \end{cases} \tag{3.111}$$

Table 3.2 Comparisons of computation complexities

Algorithm	Number of multiplications and additions
Space-time null-widen algorithm based on CSA-MWF	$(8D+2N)(MK)^2+O(D^3)+(2D^2+4D)MK+D^3-D^2-D$
Space-time Laplace null-widen algorithm	$(2N+4)(MK)^2+O\left((MK)^3\right)+MK-1$

then

$$\mathbf{T}_D^{\mathrm{H}}\left(\mathbf{I}-\mathbf{h}_0\mathbf{h}_0^{\mathrm{H}}\right)=\mathbf{T}_D^{\mathrm{H}} \tag{3.112}$$

thereby

$$\mathbf{w}_d=\left(\mathbf{T}_D^{\mathrm{H}}\mathbf{R}_x\mathbf{T}_D\right)^{-1}\mathbf{T}_D^{\mathrm{H}}\mathbf{R}_x\mathbf{h}_0 \tag{3.113}$$

So that the CSA-MWF's weight can be written as

$$\mathbf{w}_{CSA-MWF}=\mathbf{h}_0-\mathbf{T}_D\left(\mathbf{T}_D^{\mathrm{H}}\mathbf{R}_x\mathbf{T}_D\right)^{-1}\mathbf{T}_D^{\mathrm{H}}\mathbf{R}_x\mathbf{h}_0 \tag{3.114}$$

It can be seen from (3.114) that the equivalent weight solution formula is only related to covariance matrix, we can combine the CSA-MWF and the space-time Laplace null-widen algorithm. For the space-time Laplace algorithm, we just need to use the $\overline{\mathbf{R}}_L$ in (3.96) to replace the $\mathbf{R}_x$ in (3.114).

Assuming that the number of array elements is M, the number of time delay is K, the reduced number of dimension is D, and the number of samples is N, the comparisons of computation complexities between the space-time Laplace null-widen algorithm and the reduced rank space-time null-widen algorithm based on CSA-MWF is shown in Table 3.2. It can be seen from the Table 3.2 that the reduced rank space-time null-widen algorithm can reduce the algorithm computation complexity.

3.5.3 Simulation Results

This simulation uses a 5-element half-wavelength equidistant linear array, and every element has 5 delay taps, with a tap delay $T=50\,\text{ns}$. Four GNSS signals PRN2, PRN4, PRN5 and PRN9 incident onto the array from the angles of $-48°$, $10°$, $25°$ and $50°$, and a jamming signal with an initial DOA of $-20°$ incidents on the array. For the simulation, the jamming signal's DOA gradually changes from $-20°$ to $-17°$. The jamming-to-noise ratio is 40 dB, and the signal-to-noise ratio is $-20\,\text{dB}$.

Figure 3.18 compares beam patterns for the space-time power minimization algorithm and the space-time Laplace null-widen algorithm. In the figure, the subfigures (a), (c), and (e) are the beam pattern, the top view and side view of the beam patterns for the space-time power minimization algorithm respectively. The

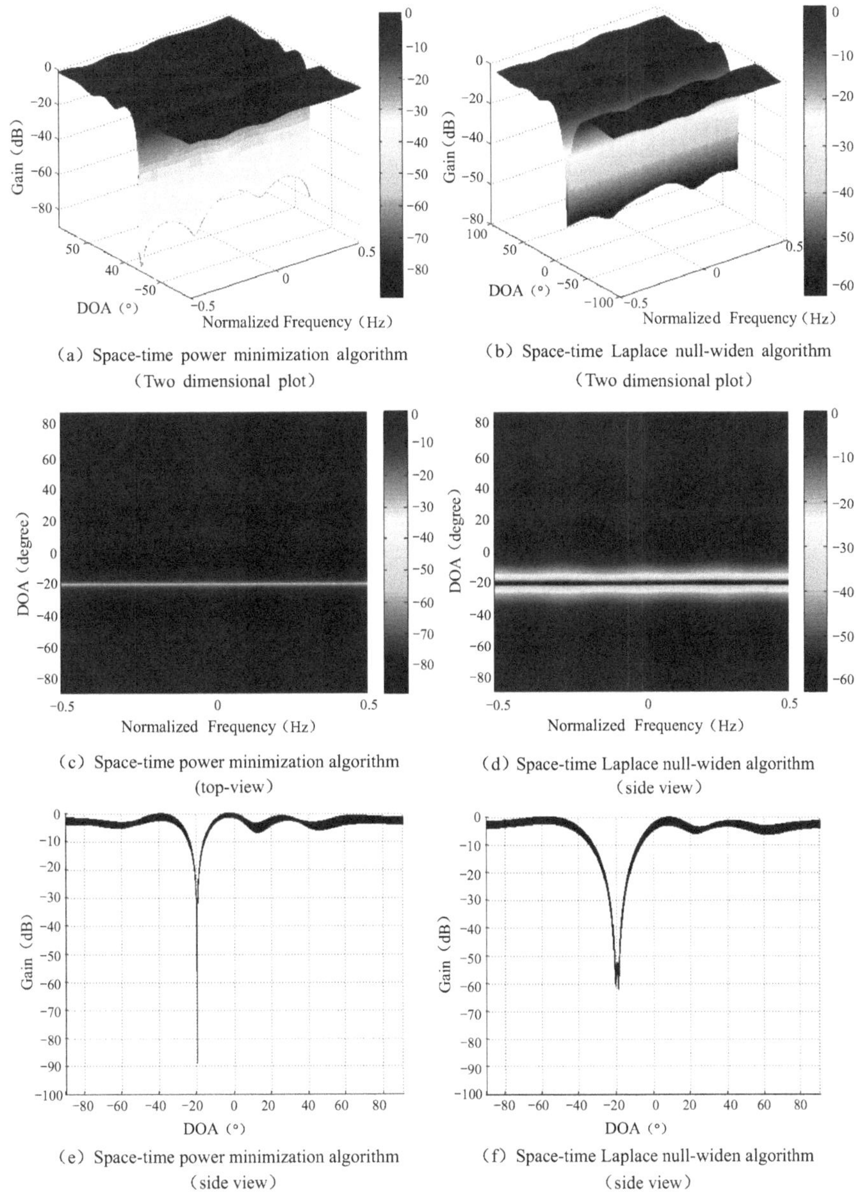

(a) Space-time power minimization algorithm (Two dimensional plot)

(b) Space-time Laplace null-widen algorithm (Two dimensional plot)

(c) Space-time power minimization algorithm (top-view)

(d) Space-time Laplace null-widen algorithm (side view)

(e) Space-time power minimization algorithm (side view)

(f) Space-time Laplace null-widen algorithm (side view)

Fig. 3.18 Comparisons of adaptive array beam patterns

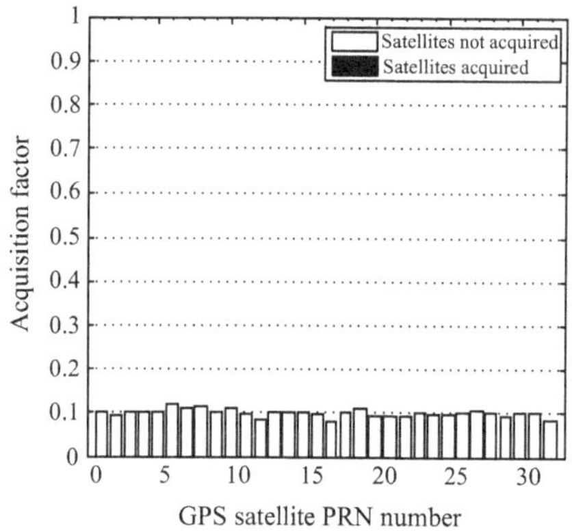

(a) Space-time power minimization algorithm

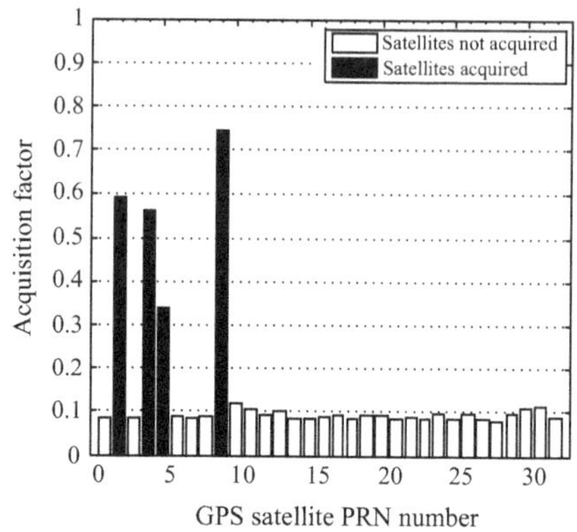

(b) Space-time Laplace null-widen algorithm

Fig. 3.19 Comparisons of acquisition results

subfigures (b), (d) and (f) are the beam pattern, the top view, and side view of the beam patterns for the space-time Laplace null-widen algorithm. By comparing the subfigures (a), (b), (c), (d), (e) and (f) in Fig. 3.18, it can be seen that compared with the space-time power minimization algorithm, the beam pattern for the space-time Laplace algorithm can form a broad null towards the DOA of the jamming.

Figure 3.19 compares the acquisition results. By comparing the subfigures (a) and (b) in Fig. 3.19, the Laplace null-widen algorithm can adapt to the time-varying jamming signal DOA in a high-dynamic environment, and effectively suppress the jamming signal associated with DOA changes, thus ensuring the correct signal acquisition by GPS receivers. On the other hand, the nulls formed by the space-time power minimization method are too narrow, so it cannot effectively suppress the jamming in a high-dynamic environment, and as a result it cannot ensure the correct signal acquisition by GPS receivers.

Figure 3.20 shows the comparison results of beam patterns for the space-time null-widen algorithm and the reduced space-time null-widen algorithm based on 50 Monte Carlo simulations. To make it easier for readers, every curve in Fig. 3.20's subfigures (a) and (b) represents the cross-section results of the space-time tow-dimensional beam patterns obtained from one simulation run. It can be seen from Fig. 3.20a, b that the beam patterns formed by the reduced rank space-time null-widen algorithm is more stable and more robust, so it is suitable for high-dynamic environments.

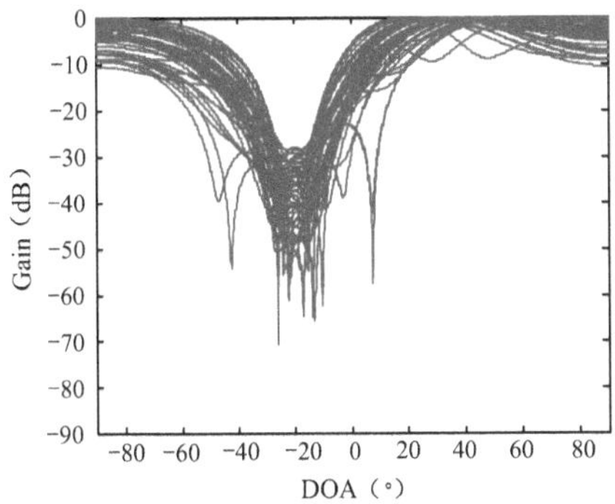

(a) Space-time Laplace null-widening algorithm

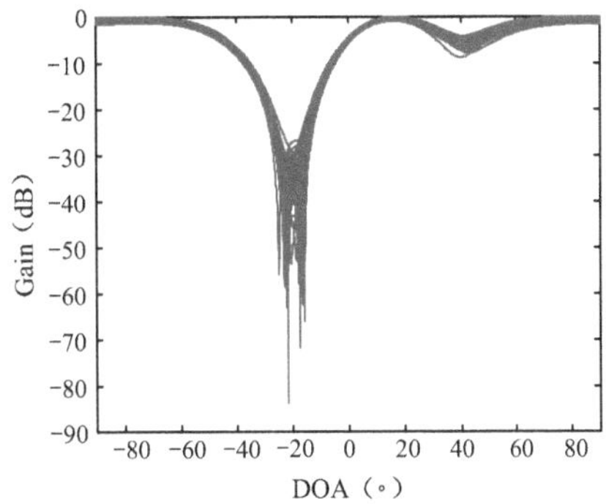

(b) Reduced rank space-time Laplace null-widen algorithm

Fig. 3.20 Beam pattern comparisons

3.6 Sparsely Represented Jamming DOA Estimation Algorithm Based on Small Number of Snapshots

For the jamming suppression algorithms in a high-dynamic environment described before, some of them need to know the jamming DOA information beforehand; consequently they must perform jamming DOA estimation in high-dynamic environments. Even though there exist many matured DOA estimation algorithms such as the MUSIC (MUltiple SIgnal Classification) algorithm based on sub-space decomposition [20] and the ESPRIT (Estimation of Signal Parameters via Rotational Invariance Technique) algorithm [21], these algorithms need to accurately estimate covariance matrices, which requires a constant jamming DOA within the processing time interval. A fast change of the jamming DOA can result in an insufficient number of snapshots and reduced estimation accuracy on a covariance matrix. Under those conditions, the performances of the regular DOA estimation algorithms decrease significantly, or become completely ineffective.

In recent years, a new type of algorithm for high precision DOA estimation has been formulated by taking advantage of the sparse characteristics of the signal DOA space spectrum and performing DOA estimation based on compressed sensing [22]. Especially, this type of algorithm can still be effective for a single snapshot. Thereby, in this section, we study on how to take advantage of the space sparse characteristics of the jamming DOA, so that we can have an algorithm that only need one single snapshot to estimate DOA. Therefore we can solve the problem of knowing the jamming DOA in a high-dynamic environment, and then verify the effectiveness of the algorithm using simulations.

3.6.1 Convex Optimization Algorithm

Considering an M-element uniform equidistant linear array, if we assume that Q jamming signals incident onto the array, then the snapshot data vector of the array antenna's received signal $\mathbf{x}(t)$ can be represented as

$$\mathbf{x}(t) = \sum_{q=1}^{Q} \mathbf{a}\left(\theta_q j_q(t) + \mathbf{e}(t)\right) \tag{3.115}$$

where $j_q(t)$ is the qth jamming signal; $\mathbf{a}\left(\theta_q\right)$ is the corresponding steering vector; $\mathbf{e}(t)$ is the thermal noise vector of the receiver. For the sake of convenience, we already include the GNSS signal in the noise vector.

Since the number of jamming sources is limited in the space, and is usually smaller than the number of array elements, the scenario satisfies the condition of space "sparseness" [22]. Thereby it is feasible to apply a compressed sensing algorithm for DOA estimation. Figure 3.21 shows the space sparse characteristics for jamming DOA. It can be seen in the figure that, the whole possible jamming DOA space is $[-90°, 90°]$, but only a small limited number of angles can have the incoming jamming. This represents the space sparseness characteristics of the jamming signal DOAs.

By converting the assumed array signal model to a sparse model, we perform discrete sampling on the complete possible jamming DOA space. This method converts the original received signal to a spare representation using a mathematic transformation, so that there is no need to make any change on the GPS receiver's front end. The DOA space for the whole jamming signal is $[-90°, 90°]$. Assuming the discrete sampling is performed using 1° space interval, then the space under consideration can be divided into

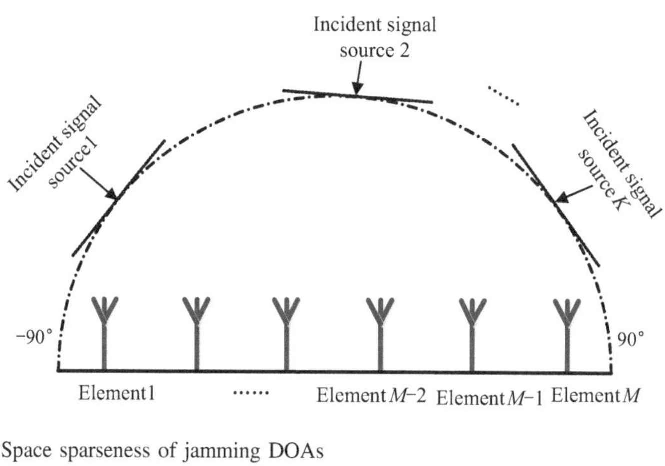

Fig. 3.21 Space sparseness of jamming DOAs

$$\Psi = \{-90^{\circ}, -89^{\circ}, \ldots, 90^{\circ}\} \tag{3.116}$$

Ψ is the discrete angle set, and the Ψ_i is used to represent the ith element. Usually we can assume the jamming signal's DOA belongs to the discrete angle set. In reality, when this assumption is not satisfied, we can further reduce the space sampling interval, in order to reduce jamming DOA estimation error. We can let $\theta_q = \Psi$, and then take

$$\bar{j}_i(t) = \begin{cases} j_q(t), & \theta_q = \Psi_i \\ 0, & \text{Others} \end{cases} \tag{3.117}$$

Then an $L \times 1$ dimension sparse vector $\bar{\mathbf{j}}(t) = [\bar{j}_1(t), \bar{j}_2(t), \ldots, \bar{j}_L(t)]^{\mathrm{T}}$ $(L = 181)$ can be constructed. That means, if the corresponding DOA has real jamming, the element corresponding to the sparse vector is the jamming signal, otherwise it is zero. Then for $\bar{\mathbf{j}}(t)$ only Q elements are not zero. Under general conditions, Q is way smaller than L, which reflects the jamming space's sparseness characteristics.

At the same time, we can take

$$\widetilde{\mathbf{A}} = [\mathbf{a}(-90^{\circ}), \mathbf{a}(-89^{\circ}), \ldots, \mathbf{a}(90^{\circ})] \tag{3.118}$$

$\widetilde{\mathbf{A}}$ includes all steering vectors corresponding to the discrete angle set. In the sparse representation theory, this is known as the super complete dictionary. Then the signal received by the GNSS receiver can be represented using a sparse format

$$\mathbf{x}(t) = \widetilde{\mathbf{A}}\,\bar{\mathbf{j}}(t) + \mathbf{e}(t) \tag{3.119}$$

We can estimate and obtain $\bar{\mathbf{j}}(t)$ based on the signal reconstruction algorithm, where the angles corresponding to the non-zero elements are the jamming DOAs. The most direct way to solve it is to solve for the l_0 norm optimization problem in (3.120)

$$\begin{aligned} &\min \left\|\bar{\mathbf{j}}(t)\right\|_0 \\ &\text{s.t.} \left\|\mathbf{x}(t) - \widetilde{\mathbf{A}}\,\bar{\mathbf{j}}(t)\right\|_2 \leq \sigma_e \end{aligned} \tag{3.120}$$

where l_0 norm represents the number of non-zero elements.

Since the solution for (3.120) is a NP-hard problem, a solution is hard to obtain. Usually an l_1 norm is used to substitute an l_0 norm [23, 24]. Therefore, The optimization problem described in (3.120) can be converted to the following optimization problem

$$\begin{aligned} &\min \left\|\bar{\mathbf{j}}(t)\right\|_1 \\ &\text{s.t.} \left\|\mathbf{x}(t) - \widetilde{\mathbf{A}}\,\bar{\mathbf{j}}(t)\right\|_2 \le \sigma_e \end{aligned} \tag{3.121}$$

The problem described in (3.121) is a convex optimization problem, so that the convex optimization algorithm can be adopted to recover the sparse vector $\bar{\mathbf{j}}(t)$.

Convex optimization is a nonlinear optimization method used to solve for the extreme values on a convex set of a convex objective function. Its main characteristic is that when the objective function is a convex function, the stagnation point of the objective function is also its minimum point, and in addition it is the global minimum value as well. When the objective function is a strict convex function, the number of global minimum is 1. For the compressed sensing recovery problem, the l_1 norm is a function that is separable and has sparse feature, thereby it can be used as the objective function for a convex optimization problem. The frequently used convex optimization methods are basis pursuit method [25], interior-point method [24], sparse gradient projection method [26], iterative shrinkage threshold method [27] etc. Below we briefly introduce the basis pursuit method and interior-point method.

Basis Pursuit (BP) method looks to obtain the sparsest representation of the signal from the complete function set, i.e. use as few bases as possible to accurately represent the original signal, in order to obtain the signal's intrinsic essential characteristics. The basis pursuit method defines the signal sparse representation problem as a type of constrained extreme value problem by minimizing l_1 norm, and then it can be converted to and solved as a linear programming problem.

Interior-point method is a classic method to solve constrained optimization problems. It is one of the penalty function methods. The basic idea is to convert the constraints into certain functions which can be added into the objective function based on the characteristics of constrained conditions. The objective is to convert a constrained optimization problem to an unconstrained optimization problem for solution.

In addition, mathematics software can be used to directly solve (3.121), e.g. MATLAB CVX toolbox. Once the estimated value $\hat{\bar{\mathbf{j}}}(t)$ of the spare vector $\bar{\mathbf{j}}(t)$ is obtained, the peak value positions can be searched and the corresponding steering vector is the steering vector of the jamming, then the estimated jamming DOA $\hat{\theta}_q$ $(q = 1, 2, \ldots, Q)$ can be obtained.

3.6.2 Greeding Algorithm

To reduce the complexity of using the convex optimization algorithm to directly solve the l_1 norm optimization problem, many greedy algorithms have been proposed. Compared with the convex optimization algorithm, the complexity has been reduced greatly, but the performance is slightly worse than that of the convex optimization algorithm. The main idea of the greedy algorithm is to treat the observed signal as a linear combination of certain atoms in the super complete dictionary, and use certain iterations to obtain an approximate value of the original signal. Typical greedy algorithms are the Marching Pursuit (MP) algorithm [20] and the Orthogonal Matching Pursuit (OMP) algorithm [28]. Some improved greedy algorithms have also been proposed, and their main differences are to adopt different criteria of selecting atoms during the signal reconstruction process. The most representative algorithms in this category are the backtracking greedy algorithm and the staged greedy algorithm. A representative algorithm of the backtracking greedy algorithm is the Compressing Sampling Matching Pursuit (CoSaMP) algorithm [29]. The CoSaMP algorithm introduces a rejection mechanism so that it is very robust. Examples of the segmented greedy algorithms are the Regularized OMP (ROMP) algorithm [30] and the Stage wise OMP (StOMP) algorithm [31]. The staged greedy algorithm is different from the original greedy algorithm because it selects multiple atoms for different iterations, so the signal reconstruction can speed up.

Based on the advantages of the OMP algorithm, in this section we adopt the OMP algorithm, and only use a single snapshot of data to estimate the jamming signal DOA. For time t, the corresponding jamming DOA estimation steps are as below:

(1) Initialization: The residue amount $\mathbf{r}_0 = \mathbf{x}(t)$, the super complete dictionary is $\widetilde{\mathbf{A}}$, the index set Γ_0 is a null set $\Gamma_0 = \varnothing$, and the number of iterations $n = 1$.
(2) Look for the maximum correlation between the residue amount $\mathbf{r}_{n-1}$ and all the steering vectors $\mathbf{a}(\Psi_i)$ (each row of $\widetilde{\mathbf{A}}$'s), and the corresponding index value is $\lambda_n = \arg\max_{i=1,2,\ldots,L}|\langle \mathbf{r}_{n-1}, \mathbf{a}(\Psi_i)\rangle|$.
(3) Update index set $\Gamma_n = \Gamma_{n-1} \cup \{\lambda_n\}$.
(4) Use a least squares method to re-estimate the sparse vector $\hat{\bar{\mathbf{j}}}_n(t) = \arg\min_{\mathbf{z}:supp(z)\subseteq\Gamma_n}\left\|\mathbf{x}(t) - \widetilde{\mathbf{A}}\mathbf{z}\right\|_2$, where $supp\,(z)\subseteq\Gamma_n$ represents $z_j \neq 0 (j \in \Gamma_n)$ and $z_j = 0 (j \notin \Gamma_n)$.
(5) Update the residue amount $\mathbf{r}_n = \mathbf{r}_{n-1} - \widetilde{\mathbf{A}}\,\hat{\bar{\mathbf{j}}}_n(t)$, and let $h = h + 1$.
(6) Repeat step (2) to step (5) iteratively, the number of iteration is Q (i.e. the number of jamming sources).

Similarly, after the OMP algorithm iteration, we can obtain the estimated value $\hat{\bar{\mathbf{j}}}_n(t)$ of the sparse vector $\bar{\mathbf{j}}_n(t)$, and $\hat{\bar{\mathbf{j}}}_n(t)$ only has Q non-zero elements, where every non-zero element corresponds to one row in $\widetilde{\mathbf{A}}$, and the corresponding steering vector is the jamming steering vector, and thus we obtain the estimated jamming DOA $\hat{\theta}_q$ $(q = 1, 2, \ldots, Q)$.

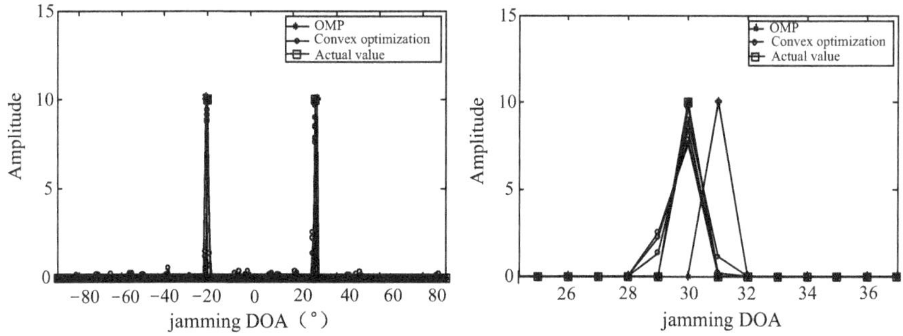

(a) Comparison of DOA Estimations (b) Partially enlarged view

Fig. 3.22 Performance comparisons of jamming DOA estimations between the OMP algorithm and the convex optimization algorithm

3.6.3 Simulation Results

This simulation only considers a 10-element half-wavelength equidistant linear array. Two jamming sources are added with $INR = 20$ dB, and the jamming DOAs are $-20°$ and $30°$ respectively. Only single snapshot is used in the simulation for jamming DOA estimation.

Figure 3.22 shows the results of using the OMP algorithm and convex optimization algorithm for jamming DOA estimation. To have a clear display, the plot only shows 5 Monte Carlo simulation results. Among all the subfigures, the Fig. 3.22b is the Fig. 3.21a's partially enlarged view. It can be seen on the figure that the convex optimization algorithm has better accuracy than the OMP algorithm, but the OMP algorithm's complexity is lower than that of the convex optimization algorithm.

It is noticeable that the methods adopted by this section avoid estimating the receiver signal's covariance matrix. In addition, since single snapshot processing is used, the correlations among jamming signals have no impact on the algorithm.

3.7 Summary

In this chapter, we mainly discuss high-dynamic GNSS jamming mitigation techniques. For linear arrays, those algorithms based on power minimization algorithm, such as the derivative constraint method, the Mailloux method, the Zatman method and the Laplace method, do not need information on jamming signals' DOAs. For uniform circular arrays, except for the Laplace method, all the other methods all need information on jamming signals' DOAs. By combining the reduced rank technique and the space-time null-widen technique, the space-time high-dynamic jamming suppression algorithm can be implemented better. We can solve the jamming DOA estimation problem in high-dynamic environments by using a sparse representation model.

References

1. Gershman AB, Serebryakov GV, Bohme JF. Constrained Hung-Turner adaptive beam-forming algorithm with additional robustness to wideband and moving jammers. IEEE Trans Antennas Propag. 1996;44(3):361–7.
2. Gershman AB, Nickel U, Bohme JF. Adaptive beamforming algorithms with robustness against jammer motion. IEEE Trans Signal Process. 1997;45(7):1878–85.
3. Mailloux RJ. Covariance matrix augmentation to produce adaptive array pattern troughs. Electron Lett. 1995;31(10):771–2.
4. Zatman M. Production of adaptive array troughs by dispersion synthesis. Electron Lett. 1996;31(25):2141–2.
5. Guerci JR. Theory and application of covariance matrix tapers for robust adaptive beamforming. IEEE Trans Signal Process. 1999;47(4):977–85.
6. Wu R, Li C, Lu D. Power minimization with derivative constraints for high-dynamic GPS interference suppression. Sci China: Inf Sci. 2012;55(4):857–66.
7. Lu D, Li S, Wu R. Simultaneous null and main lobe widening for jamming suppression in attitude high-dynamic GNSS receiver. Tempa, USA: ION GNSS+; 2015.
8. Ma Y, Lu D, Wang W, et al. A high-dynamic null-widen GPS anti-jamming algorithm based on statistical model of the changing interference DOA. In: China satellite navigation conference. 2014. p. 694–702.
9. Ge L, Lu D, Wang W, Wang L, et al. A high-dynamic null-widen GNSS anti-jamming algorithm based on reduced-dimension space-time adaptive processing. In: Proceedings of 2015 China satellite navigation conference. Berlin, Heidelberg: Springer; 2015.
10. Hinedi S, Statman JI. High-dynamic GPS tracking final report. JPL Publication; 1988. p. 35–88.
11. Gecan A, Zoltowski M. Power minimization technique for GPS null steering antennas. In: Institute of Navigation conference; 1995. vol 12, no 1, p. 13–5.
12. Hatke GF. Multipath effects on F-15 and F-16 multi-channel GPS antenna performance. In: Thirty-third Asilomar conference on signals, systems, and computers, CA. 1999. p. 922–6.
13. Mohamed EA, Tan ZZ. Adaptive antenna utilizing power inversion and linearly constrained minimum variance algorithms. Chin J Aeronaut. 2005;18(2):153–60.
14. Zetterberg P, Ottersten B. The spectrum efficiency of a base station antenna array system for spatially selective transmission. IEEE Conf Veh Technol. 1995;44(3):1517–21.
15. Wang H, Cai L. On adaptive spatial-temporal processing for airborne surveillance radar systems. IEEE Trans Aerosp Electron Syst. 1994;30(3):660–9.
16. Wu R, Bao Z. Array pattern distortion and remedies in space-time adaptive processing for airborne radar. IEEE Trans Antennas Propag. 1998;46(7):963–70.
17. Myrick WL, Goldstein JS, Zoltowski MD. Anti-jam space-time preprocessor for GPS based on multistage nested wiener filter. IEEE Mil Commun Conf (Atlantic NJ). 1999;1:675–81.
18. Joham M, Zoltowski MD. Interpretation of the multi-stage nested wiener filter in the Krylov subspace framework. Technical Report, TR-ECE-00-51, Purdue University; 2000.
19. Lu D, Ge L, Wu R, et al. High-dynamic wideband interference suppression in GNSS via reduced-rank STAP. Tempa, USA: ION GNSS+2015; 2015.
20. Mallat SG, Zhang Z. Matching pursuits with time-frequency dictionaries. IEEE Trans Signal Process. 1993;41(12):3397–415.

21. Roy R, Kailath T. ESPRIT—estimation of signal parameters via rotational invariance techniques. IEEE Trans Acoust Speech Signal Process. 1989;37(7):984–95.
22. Malioutov D, Cetin M, Willsky AS. A sparse signal reconstruction perspective for source localization with sensor arrays. IEEE Trans Signal Process. 2005;53(8):3010–22.
23. Chen SS, Donoho DL, Saunders MA. Atomic decomposition by basis pursuit. SIAM J Sci Comput. 1998;20(1):33–61.
24. Kim SJ, Koh K, Lustig M, et al. An interior-point method for large-scale l1-regularized least squares. IEEE J Sel Top Signal Process. 2007;1(4):606–17.
25. Schmidt RO. Multiple emitter location and signal parameter estimation. IEEE Trans Antennas Propag. 1986;34(3):276–80.
26. Figueiredo MAT, Nowak RD, Wright SJ. Gradient projection for sparse reconstruction: application to compressed sensing and other inverse problems. IEEE J Sel Top Signal Process. 2007;1(4):586–97.
27. Wright SJ, Nowak RD, Figueiredo MAT. Sparse reconstruction by separable approximation. IEEE Trans Signal Process. 2009;57(7):2479–93.
28. Tropp JA, Gilbert AC. Signal recovery from random measurements via orthogonal matching pursuit. IEEE Trans Inf Theory. 2007;53(12):4655–66.
29. Needell D, Tropp JA. CoSaMP: iterative signal recovery from incomplete and inaccurate samples. Appl Comput Harmon Anal. 2009;26(3):301–21.
30. Needell D, Vershynin R. Uniform uncertainty principle and signal recovery via regularized orthogonal matching pursuit. Found Comput Math. 2009;9(3):317–34.
31. Donoho DL, Tsaig Y, Drori I, et al. Sparse solution of underdetermined systems of linear equations by stagewise orthogonal matching pursuit. IEEE Trans Inf Theory. 2012;58 (2):1094–112.

Chapter 4
Spoofing Countermeasure Techniques

4.1 Introduction

As described in Chap. 1, common types of intentional interference mainly include jamming and spoofing. Chapters 2 and 3 mainly discuss jamming. In this chapter we discuss techniques to suppress the interference emitted by spoofers. Different from jamming, the power level, signal format and frequency spectrum structure of interference emitted by spoofers are similar to the authentic satellite signals. The intent of spoofing is to trick a receiver locking onto interference without awareness, so a navigational positioning result, which seems to be reliable but actually misleading, can be produced. In the worst case, a receiver may be controlled by spoofers. Under the impacts of spoofing, receivers usually are not aware of the spoofers, so consequently spoofing is more harmful than jamming [1–5]. In 2012, the radio navigation research group from University of Texas successfully took control of an unmanned aerial vehicle (UAV) and made it deviate away from the planned route. The research group also analyzed the impacts from spoofing on the intelligent grid on which the GPS system timing was used. In 2013, the same research group implemented spoofing on a private yacht that was sailing the ocean, and consequently led it off course [6]. All these real examples have fully verified the harms brought by spoofing.

Present antispoofing techniques can be categorized into two types: spoofing detection and spoofing countermeasure. Spoofing detection technology detects spoofing by monitoring abnormal changes in signal characteristics [7–28], including detection techniques based on signal power [7–11], space characteristics [12–16], navigation messages [17], Signal Quality Monitoring (SQM) [18–21], integrated navigation [22, 23], and encryption authentication [24–28]. Spoofing detection techniques are the main methods used for antispoofing, but they cannot help the receivers restore positioning and navigation capabilities, which is the main objective of spoofing countermeasure techniques. Spoofing countermeasure techniques mainly include the method that can remove the measured abnormal values

R. Wu et al., *Adaptive Interference Mitigation in GNSS*, Navigation: Science and Technology, https://doi.org/10.1007/978-981-10-5571-3_4

using the Receiver Autonomous Integrity Monitoring (RAIM) technique [29, 30], and the method that can suppress the spoofing in the space domain using array signal processing techniques [31–36]. Between them, the latter takes advantage of the space domain characteristics of spoofing: it contains multiple pseudo random codes, which are transmitted by the same antenna. Therefore the latter approach is full of potential.

In this chapter, we give definitions and mathematical models of spoofing, and analyze their impacts on GNSS receiver acquisition, tracking and positioning. Then we review various spoofing detection methods, and study multiple space domain spoofing countermeasure methods in different environments. Finally we will verify the antispoofing capabilities of these methods by performing simulations.

4.2 Spoofing and the Impacts of Spoofing

Spoofing tricks a GNSS receiver into having wrong navigational and positioning results by using a spoofer to transmit an interference that is the same as or similar to the authentic GNSS signal. In terms of signal generation, spoofing can be categorized into two types: receiver based spoofing and generator based spoofing. Figure 4.1 shows a possible categorization of spoofing on GNSS receivers.

4.2.1 Receiver Based Spoofing

Figure 4.2 shows the mechanism of receiver based spoofing. Receiver based spoofing receives an authentic GNSS signal first, and then after certain time delay, amplifies the signal and transmits out. It uses this approach to trick the target GNSS receiver. This retransmitted GNSS signal is known as the spurious PRN signal. Compared with authentic GNSS signals, spurious PRN signals are different in terms of time delay and power level, but the navigation messages are the same. To allow receiver based spoofing to successfully enter the acquisition stage of a GNSS

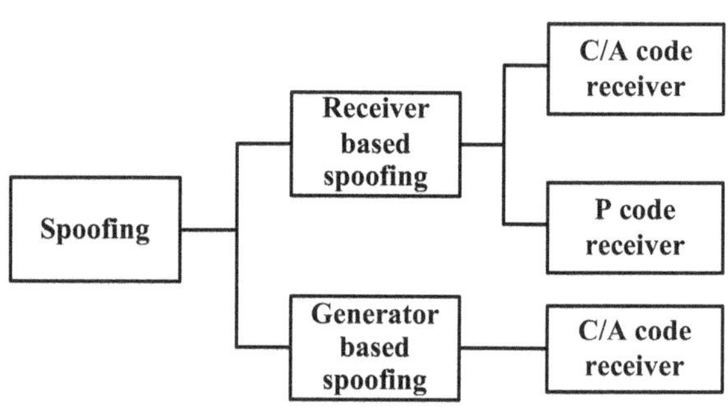

Fig. 4.1 Spoofing on GNSS receivers

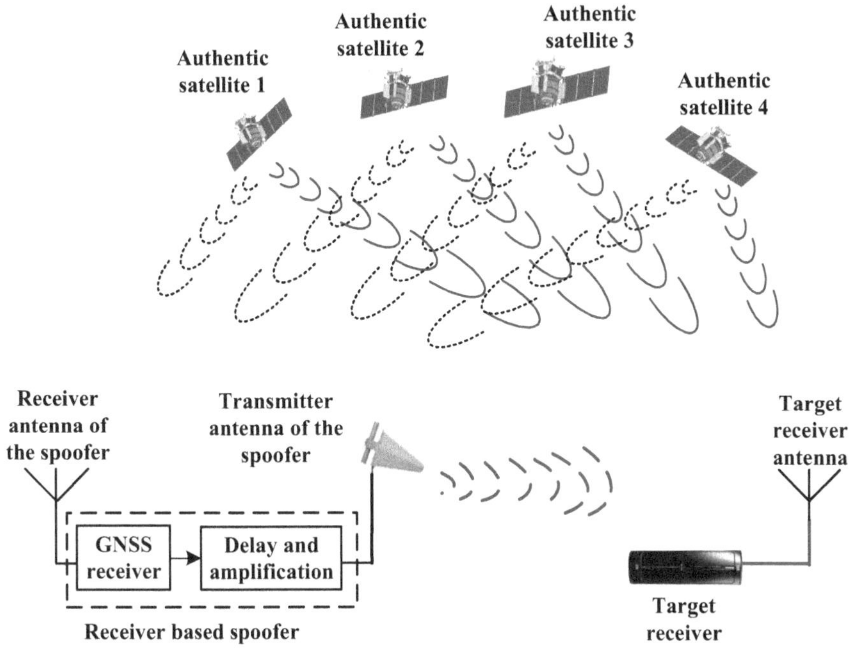

Fig. 4.2 Receiver based spoofing mechanism

receiver, usually the spurious PRN signal's power is enhanced so it is higher than the power of the authentic GNSS signal by around 2 dB [2]. Thereby, GNSS receivers without anti-spoofing measures are vulnerable to receiver based spoofing. A target GNSS receiver, after acquiring the receiver based spoofing, calculates a wrong pseudorange, thereby it cannot solve for the proper positions.

Receiver based spoofing can be categorized based on different implementation complexities as well: Type one uses a single receiver antenna to receive all authentic satellite signals, and delay, amplify, and retransmit them [3]. Under this scenario, if the delay is sufficiently short, we can assume that the interference is synchronized with the authentic GNSS signal which it pretends to be. This type of receiver based spoofing can force GNSS receivers to acquire and track the interference. Type two uses one array antenna to form high gain narrow beams for every authentic GNSS signal respectively. So it can perform separate receptions for various authentic GNSS signals, and consequently add different delays on different GNSS signals before retransmissions. This type of receiver based spoofing can achieve the objective of tricking the receiver into a set position [4].

Receiver based spoofing technique is the most important means of spoofing in present days. The main topic of this chapter is to discuss this type of spoofing.

4.2.2 *Generator Based Spoofing*

Figure 4.3 shows the mechanism for generator based spoofing. For generator based spoofing, a spoofer generates interference consistent with the authentic GNSS signal format, and then uses a transmitter to send out the interference. Based on different implementation complexities, there could be different implementations: For the first type, a simulator is used directly to transmit the interference, but this implementation does not consider the synchronization with the authentic GNSS signal. As a result when a receiver is in a state of tracking, the generator based spoofing's signal is equivalent to a weak interference that only reduces output carrier-to-noise ratio. Nevertheless, when a receiver is under the impact of jamming or in a cold-start state in which the receiver needs to re-acquire the GNSS signal, it becomes vulnerable to such a deception [4]. For the second type, a generator based spoofer first receives the authentic GNSS signal and demodulates the navigation message. Then the spoofer modifies the received navigation message, and uses the authentic GNSS signal's parameters such as chip delay and carrier phase as a reference, and uses a simulator to re-spread the counterfeit message, and then retransmits. This type of generator based spoofing synchronizes with the authentic

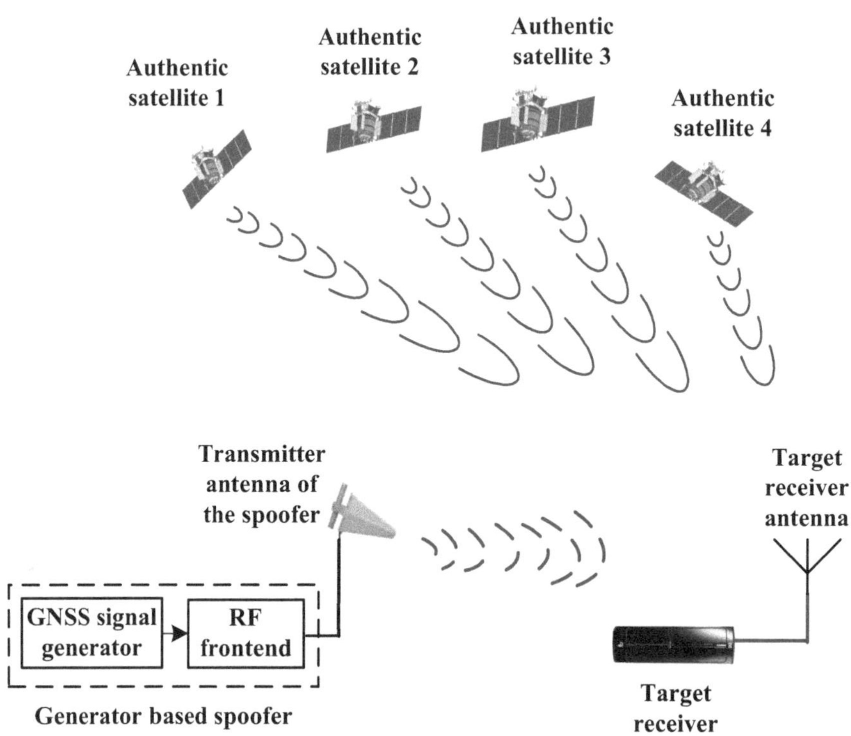

Fig. 4.3 The mechanism of generator based spoofing

GNSS signal, therefore it can force GNSS receivers to acquire and track the interference, leading to erroneous calculation results [3]. For the third type, the generator based spoofer first receives the authentic GNSS signal, and demodulates the navigation message, and uses the authentic GNSS signal's parameters such as chip delay and carrier phase as the reference, and under the scenario of accurately knowing the target receiver's position, it can produce an interference exactly the same as the authentic GNSS signal with the reverse phase. This generator based spoofing can use the interference to cancel out the authentic GNSS signal, so the GNSS receiver cannot position itself [5].

At present day, most structures of GNSS civil signals are publically known, so it is completely feasible to design a generator based spoofing scheme towards civil signals, but the techniques are relatively complex, and they are very hard to be implemented [3]. Since the military GNSS signals have longer sequences and their structures are kept secret, it is technically very hard to decode GNSS military code and thereby produces high fidelity spoofing received by GNSS receivers. Consequently, it is very hard to implement generator based spoofing towards GNSS military signals. For GNSS military signals, usually receiver based spoofing is used.

4.2.3 Spoofing Data Model

Based on the previous analyses on spoofing, this type of interference has the same signal structure as the authentic GNSS signal and the signal model is

$$j^s(t) = \sum_{l=1}^{L} \sqrt{p_l^s} D_l^s\left(t - \tau_l^s\right) c_l^s\left(t - \tau_l^s\right) e^{j\left(\omega_l^s t + \phi_l^s\right)} \tag{4.1}$$

where $D_l^s(t)$ and $c_l^s(t)$ represent the lth spurious PRN signal's navigation message and the spread spectrum code within the spoofer; ϕ_l^s, ω_l^s, p_l^s and τ_l^s represent the corresponding phase, Doppler angular frequency, power and chip delay respectively; and the superscript s represents spoofing. The interference usually contains multiple spurious PRN signals, and the L in (4.1) represents the number of spurious PRN signals within the spoofer. Unless otherwise specified, the interference signal in this chapter refers to spoofer-emitted interference.

When signals from $\bar{M}$ satellites and $\bar{K}$ spoofing sources arrive at a GNSS receiver at the same time, the intermediate frequency (IF) signal in the GNSS receiver can be represented as

$$x(t) = \sum_{m=1}^{\bar{M}} s_m^a(t) + \sum_{k=1}^{\bar{K}} j_k^s(t) + e(t) \tag{4.2}$$

where $s_m^a(t)$ represents the mth authentic satellite signal; $j_k^s(t)$ represents the interference transmitted by the kth spoofer; the superscript a and s denote the authentic satellite signal and interference; $e(t)$ is the additive Gaussian white noise.

Assuming the authentic satellite signal and multiple interferences incident on the array simultaneously, in this chapter a uniform linear array (ULA) with M elements is used as an example, then the intermediate frequency (IF) signal received by the GNSS receiver antenna array is

$$\mathbf{x}(t) = \sum_{m=1}^{\bar{M}} \mathbf{a}(\theta_m^a) s_m^a(t) + \sum_{k=1}^{\bar{K}} \mathbf{a}(\theta_k^s) j_k^s(t) + \mathbf{e}(t) \tag{4.3}$$

where $\mathbf{a}(\theta_m^a)$ and $\mathbf{a}(\theta_k^s)$ represent the steering vectors for the mth authentic satellite signal and the kth spoofing source; θ_m^a and θ_k^s represent the DOAs for the mth authentic satellite signal and the kth spoofing source; $\mathbf{e}(t)$ is the additive Gaussian white noise vector.

4.2.4 Analyses on the Impacts on GNSS Receivers by Spoofing

As described by the spoofing data model in Sect. 4.2.3, the signal structure of the interference is the same as that of the authentic GNSS signal. As long as correlation peaks formed by the interferences satisfy the acquisition decision condition, it is possible for the interference to be acquired, and thus has impact on the subsequent signal processing on the receiver. In this section, we analyze the impacts on receivers from the perspectives of acquisition, tracking, and positioning.

1. Impacts on Acquisition

During the acquisition process, when the authentic GNSS signal has the same carrier frequency and chip delay as that of the local signal, the correlation reaches the maximum. Similarly, if the interference has the same carrier frequency and chip delay as the local signal, the correlation with the interference also reaches the maximum. Eventually the correlation peak satisfying the receiver acquisition decision condition (which can correspond to the authentic GNSS signal, or to the interference) is acquired by the receiver.

For receiver based spoofing, since the interference and the authentic satellite signal both contain the same PRN signal, two obvious correlation peaks exist within the two-dimensional search results of the acquired PRN, one for the authentic satellite signal and the other for the interference. Figure 4.4 shows the two-dimensional search results for acquisition on a PRN signal when the GNSS receiver is spoofed, where x axis represents the chip delay, y axis represents the Doppler frequency, and z axis represents the correlated accumulation value. In Fig. 4.4, since the interference's power is slightly higher than the power of the

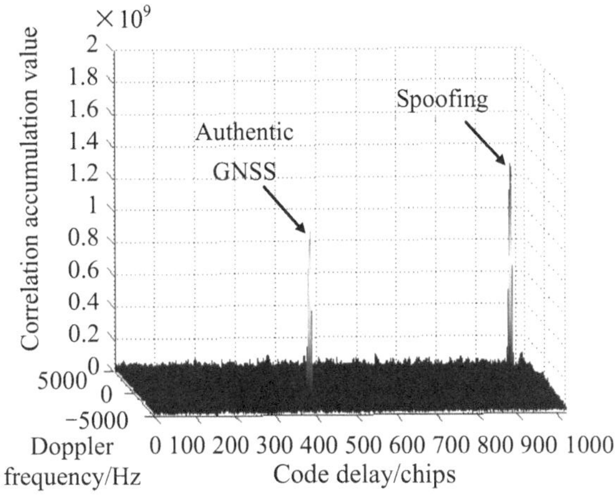

Fig. 4.4 Two-dimensional search results for acquisition when GNSS receiver is spoofed

authentic satellite signal, the correlation peak caused by the interference would be significantly higher than the correlation peak caused by the authentic satellite signal. As a result, the receiver only acquires the interference. The two correlation peaks have different chip delays, and the difference between the chip delays is the time delay between the interference and the authentic GNSS signal.

For generator based spoofing, when the interference PRN being generated is contained within the authentic GNSS signal, the impact on the acquisition process is similar to the impact of the receiver based spoofing as described above; when the interference PRN being generated isn't contained within the authentic GNSS signal, the receiver is not be aware of the interference, and acquires it as the authentic GNSS signal.

2. Impacts on Tracking

In the subsequent tracking process, for receiver based spoofing, since the interference and the authentic satellite signal both contain the same PRN, the impact on a receiver's tracking processing is similar to the multipath effect on the authentic satellite signal (more details can be found in Chap. 5). The effects mainly reflect in the form of tracking errors on the phase locked loop (PLL) and delayed locked loop (DLL), the errors are correlated with the spread spectrum code's autocorrelation function, the gain factor of the interference power towards the GNSS signal power, and the time delay and phase bias of the interference relative to the GNSS signal. Then, if the delay between the interference and the GNSS signal is within 1.5 chips, the interference can cause discriminant function distortion on the Phase Locked Loop (PLL) and Delay Locked Loop (DLL), leading to larger additional tracking errors. On the other hand if the delay between the interference and the GNSS signal is more than 1.5 chips, the interference would not generate additional tracking errors, consequently the receiver is not aware of the interference and would track the interference as the authentic GNSS signal.

For generator based spoofing, when the spurious PRN is contained in the authentic GNSS signal, the impact on the tracking process is similar to the impact of receiver based spoofing described above. If the generated spurious PRN is not contained in the authentic GNSS signal, the receiver would not be aware of the interference and would track the interference as the authentic GNSS signal.

3. Impacts on Positioning

Spoofing makes its impact on positioning results by introducing range measurement error. Receiver based spoofing changes the GNSS positioning calculation results by controlling retransmitting delay. To use the same retransmitting delay is a relative simple implementation under which, when the PRNs of the spurious signal and the authentic GNSS signal are consistent with each other, the receiver uses the interference PRN signal to perform positioning calculation. In that case, since a fixed retransmitting delay can introduce the same pseudorange error, the position calculated by the receiver is the location of the spoofer, thus the receiver is very easy to be controlled by the spoofer. When the PRNs in the interference and the authentic GNSS signal are not consistent, the receiver would only have positioning errors, but it would not be controlled by the spoofer. A more sophisticated intelligent receiver based spoofing can add different retransmitting delays in the interference PRN signal, in order to trick the receiver to a set position.

Generator based spoofing changes the GNSS receiver positioning calculation by adding fake parameters in the navigation message, which mainly include two perspectives: firstly, it achieves the objectives of generating satellite ephemeris information error and satellite clock adjustment error by changing the parameters of navigation messages, leading to the miscalculation of satellite position by the receiver because the wrong parameters are used in the calculation process. Secondly, it generates an incorrect pseudorange value after peusdorange correction by directly modifying the pseudorange correction parameters in the navigation message (i.e. modifying the correction parameters such as Ionospheric delay error etc.). And as a final result, it achieves the objective of deceiving the target receiver.

4.3 Spoofing Detection

Present day spoofing techniques can be classified into two categories: spoofing detection and spoofing countermeasure. The spoofing detection techniques' main approach is to recognize whether a GNSS receiver is under attack from a spoofer. On the other hand, the approach of spoofing countermeasure technique is to help a victim receiver to recover its capabilities of positioning and navigation. In this section we briefly discuss some existing spoofing detection techniques.

1. Detection Techniques Based on Signal Power

Signal power detection means that a GNSS receiver continuously monitor the received signals on parameters related to the power, and determines whether the receiver is attacked by a spoofer by checking for abnormal values. These parameters related to the power mainly include Carrier-to-Noise ratio (C/N_0), absolute power and the signal power received by the receiver when it moves relative to the GNSS satellites.

Carrier-to-Noise Ratio Monitoring: C/N_0 is an important parameter for measuring GNSS signal quality. Only the GNSS satellite movement and ionospheric variation can generate smooth variation on the GNSS signal power. When a receiver is spoofed, the C/N_0 of the received GNSS signal has a sudden change. Therefore, a receiver can detect spoofing attack by continuously monitoring signal C/N_0 and looking for abnormal variations [7]. The GNSS receiver can also be installed on a moving platform, since the distance between the spoofer's transmitting antenna and the target receiver is closer, the relative motion between a receiver and spoofer's antenna can lead to significant variation on the spoofing signal's C/N_0, so the detection of spoofing can be achieved by monitoring C/N_0 [10, 11].

Absolute Power Monitoring: The power of the authentic GNSS signal received by the ground receiver is very low. Using the GPS L1 signal as an example, its maximum power is approximately −153 dBw [9]. Since the path between the spoofer's antenna and the receiver is complex and changes frequently, the spoofer is hard to transmit an interference that can deceive the receiver while still not having a level that surpasses the power level of the authentic GNSS signal. Therefore, when the signal power received by a receiver is significantly stronger than the expected authentic GNSS signal's power, it can be interpreted that spoofing is detected. The absolute power monitoring technique requires that the receiver has higher precision on measuring a received signal's amplitude, so the receiver's hardware complexity increases accordingly.

2. Detection Techniques Based on Space Characteristics

The detection techniques based on space characteristics take advantage of the fact that a spoofer uses the same antenna to transmit several interference PRN signals, but GNSS satellites transmitting GNSS signals come from different directions. As a result, interference PRN signals are spatially correlated, so the spoofing can be recognized by determining the received signals' spatial correlation using the space domain processing technique.

The detection technique based on space characteristics can be implemented using dual antennas [12], so the phase difference between two fixed antennas can be continuously observed, and at the same time the motion trajectories of the antenna array and the GNSS satellite can be used to calculate the phase difference between the two antennas. By comparing the observed phase difference with the computed phase difference, if the distinction is larger, a spoofing threat can be assumed. The main drawback of this method is that a fairly long period of time is needed to make observations. When there are multiple antennas [15], we can recognize whether signals come from the same angular region by comparing different PRN signal correlators' output phase measurements, so the spoofing detection can be achieved.

3. Detection Techniques Based on Navigation Message

For detection techniques based on navigation messages, we have the relative delay detections between L1 and L2 signals, and the detections on signal's code rate and phase rate.

Relative delay between L1/L2 Signals: The P(Y) codes after encryption can be modulated onto L1 and L2 frequency bands. Since the ionosphere has different responses towards different frequencies, as a result there is a relative time delay between the L1 and L2 signals received by the receiver. By using a dual frequency receiver to obtain correlation between L1 and L2 signals, the correlated results only have one peak value [11], and the time delay corresponding to that peak value is the relative time delay between the signals located at the L1 and L2 frequencies. Since the relative time delay is approximately known, we can use it to recognize spoofing. This recognition technique can only be applied on generator based spoofing, but it is not effective for receiver based spoofing.

Detection Techniques Based on Code Rate and Phase Rate: For authentic GNSS signals, Doppler frequency and code rate are determined by the relative motion between the GNSS satellite and the receiver, therefore the two have consistency [17]. Equation (4.4) expresses the consistency

$$f_l^a = -f_{RF}\dot{\tau}_l^a \tag{4.4}$$

where f_{RF} represents the RF frequency of the GNSS signal; $\dot{\tau}_l^a$ represents the chip delay rate of the lth authentic signal; f_l^a represents the Doppler frequency of the lth authentic signal. A simple interference cannot maintain such a consistency between its Doppler frequency and code rate. Therefore when the two are not consistent, the spoofing can be detected.

4. Detection Techniques Based on Signal Quality Monitoring (SQM)

When there is no spoofing, due to the autocorrelation characteristics of the ranging code, the output of the GNSS correlator is a symmetric triangle. SQM technique takes advantage of this characteristic to monitor the quality of a GNSS correlator's output peak [18]. When the relative delay between the interference and the authentic signal is less than 1.5 chips, the impact of the spoofing on the correlator output is similar to the results of a multipath interference [19], because both can lead to correlator output distortion. Triangulation detection technique based on SQM can be used to detect whether a GNSS's correlator output peak has abnormal asymmetry caused by spoofing, and consequently it can recognize whether a receiver is being spoofed or not [20]. SQM technique can detect the existence of spoofing very well in an environment without multipath signals. But in an environment with multipath interference, present SQM technique cannot differentiate spoofing and multipath interference. In addition, when the delay between the interference and the authentic signal is more than 1.5 chips, this method is also ineffective.

5. Detection Techniques Based on Integrated Navigation

Integrated navigation combines one or more systems such as GNSS, radio navigation, astronomy navigation etc. with the inertial navigation system, to form an integrated navigation system. Combining a GNSS navigation system with the data from the other navigation systems (e.g. inertial navigation system) can help receivers to recognize spoofing effectively [22]. By comparing the positioning result calculated by the GNSS receiver with the position results obtained by the other positioning systems, if the confidence intervals of different systems have no overlap, then it is highly probable that the receiver is under spoofing attack. However, the integrated navigation detection techniques can increase implementation complexities of a GNSS receiver's hardware and software.

6. Detection Techniques Based on Encryption Authentication

For civil and military equipments, encryption authentication technologies can be used to detect spoofing [24]. Most encryption authentication technologies need to make some changes on the structure of GNSS signals. Therefore, the encryption authentication technologies still cannot be applied at least in the near future.

4.4 Spoofing Countermeasure

In this section, we list several spoofing countermeasure algorithms using array signal processing in different spoofing environments, including algorithms based on DOA estimation, signal separation estimation theory, de-spread re-spread technique, and diagonal loading technique. These algorithms can solve the spoofing countermeasure problems in the environments of a single receiver based spoofing source, multiple receiver based spoofing sources and coexistence of jamming and spoofing.

4.4.1 Single Spoofer Suppression

A single receiver based spoofing source uses one spoofing device to transmit multiple interference signals. In a scenario of a single receiver based spoofing source, the antenna array's received signal after sampling based on the model established in Sect. 4.2.3 is

$$\mathbf{x}(n) = \sum_{m=1}^{\bar{M}} \mathbf{a}(\theta_m^a) s_m^a(n) + \mathbf{a}(\theta^s) j^s(n) + \mathbf{e}(n) \tag{4.5}$$

which is equivalent to a case described in (4.3) where the number of receiver based spoofing source was $\bar{K} = 1$, and the definitions on the symbol notations can be found in (4.3). Authentic GNSS signals are transmitted from different directions by different satellites, but the interference PRN signals are all transmitted from one direction. At the same time, to deceive GNSS receivers to acquire the interference PRN signals with a higher probability, the powers of the interferences are usually much higher than the powers of the authentic GNSS signals. The power received from the direction of the receiver based spoofing source is the summation of all interference PRN signals' powers. As a result, the energy radiation on that direction is much larger than that on the directions of authentic GNSS satellites.

In an environment of single receiver based spoofing source, we propose the spoofing countermeasure algorithm based on DOA estimation [36]. The method takes advantage of the feature that the radiation is at its strongest towards the direction of the single receiver based spoofing source, and under the condition of a known array manifold, this method estimates the directions of the interferences first, then uses the estimated directions to construct a interference orthogonal projection matrix, and then projects the received signals onto that space, in order to implement spoofing countermeasure. Among these steps, the DOA estimation can be implemented using the regular beamforming technique. Beamforming is a nonparametric estimation method that is equivalent to the maximum likelihood estimation method in the case of single signal source. Below we discuss detailed implementation process of the spoofing countermeasure algorithm based on DOA estimation.

First, beamforming can be used to estimate the DOA of a single spoofer, which can be summarized as: By searching for the highest peak of the signal's spatial spectral function $\mathbf{a}^{\mathrm{H}}(\theta)\mathbf{R}_x\mathbf{a}(\theta)$, the DOA corresponding to the highest peak is regarded as the estimated DOA of the spoofer $\hat{\theta}$, i.e.

$$\hat{\theta} = \arg \max_{\theta} \mathbf{a}^{\mathrm{H}}(\theta)\mathbf{R}_x\mathbf{a}(\theta) \tag{4.6}$$

where the steering vector of the array is $\mathbf{a}(\theta)$; $\mathbf{R}_x$ is the covariance matrix of the received signals, and

$$\begin{aligned}\mathbf{R}_x &= E\{\mathbf{x}(n)\mathbf{x}^{\mathrm{H}}(n)\} \\ &= \sum_{m=1}^{M} p_m^a \mathbf{a}(\theta_m^a)\left[\mathbf{a}(\theta_m^a)\right]^{\mathrm{H}} + p_{\Sigma}^s \mathbf{a}(\theta^s)[\mathbf{a}(\theta^s)]^{\mathrm{H}} + \sigma_e^2\mathbf{I}\end{aligned} \tag{4.7}$$

where p_m^a, p_{Σ}^s, and σ_e^2 denote the signal power of the authentic GNSS satellite, interference power of the receiver based spoofing source, and noise power respectively. In (4.6), $\mathbf{R}_x$ usually can be substituted by the sample covariance matrix $\hat{\mathbf{R}}_x = \frac{1}{N}\sum_{n=1}^{N}\mathbf{x}(n)\mathbf{x}^{\mathrm{H}}(n)$, where N is the number of samples.

By substituting (4.7) into (4.6), we can find that since $p_{\Sigma}^{s} \gg p_{m}^{a}$, the position of the highest peak for the spatial spectrum function aims towards the direction of spoofer θ^s, i.e. $\hat{\theta} \approx \theta^s$. After estimating the DOA of spoofer, an orthogonal projection matrix of the interference subspace can be constructed

$$\mathbf{P}_{\perp} = \mathbf{I}_M - \frac{1}{M}\mathbf{a}(\hat{\theta})\mathbf{a}^{\mathrm{H}}(\hat{\theta}) \tag{4.8}$$

Finally, the received signal by the array can be projected onto the interference orthogonal subspace to eliminate spoofing effects.

Considering that a single spoofer only needs to consume one system degree of freedom and based on the signal correlations between antennas, the PLAN lab of the University of Calgary, Canada proposed a low-complexity spoofing countermeasure algorithm based on dual antennas spoofing cancellation [31]. First, the space phase difference and gain difference between two antennas are estimated

$$\hat{\theta}_{\Delta} = \measuredangle \sum_{n=1}^{N}\left(x_1^*(n)x_2(n)\right) \tag{4.9}$$

$$\hat{u} \approx \left\{\frac{\left|\sum_{n=1}^{N} x_2(n)x_2^*(n-T)\right|}{\left|\sum_{n=1}^{N} x_1(n)x_1^*(n-T)\right|}\right\}^{\frac{1}{2}} \tag{4.10}$$

where $x_1(n)$ and $x_2(n)$ represent the signals received by the two antennas; T represents a C/A code period. Then, the estimated spoofing signal is subtracted from the signal received by the antenna No. 2, in order to achieve the goal of removing the spoofing effect.

$$y(n) = x_2(n) - \hat{u}e^{j\hat{\theta}_{\Delta}}x_1(n) \tag{4.11}$$

4.4.2 Multiple Spoofer Suppression

In reality, GNSS might be under the impacts of spoofing signals retransmitted by receiver based spoofing devices coming from different DOAs. In this case, the model of the signals received by the antenna array corresponds to (4.3) in Sect. 4.2.3, and the spoofing countermeasure algorithm introduced in the

Sect. 4.4.1 does not apply. In this section, we introduce two approaches of suppressing multiple spoofers.

1. Spoofing Countermeasure Based on Signal Separation Estimation Theory

In an environment of multiple receiver based spoofing sources, we propose a spoofing countermeasure algorithm based on signal separation estimation theory [36]. First, the proposed method estimates the amplitude and DOA of every spoofer one by one, and recognizes the number of spoofers using the observation that the radiation energy from the direction of any spoofer is larger than that of the authentic GNSS signals. Then an interference orthogonal projection matrix using all estimated spoofing DOAs can be constructed, and the received signals by the array antennas are projected onto this matrix, to achieve the goal of spoofing suppression. One channel of the data after the spoofing countermeasure can be directed into a common receiver for acquisition, tracking and positioning, which is equivalent to a single beam. Or we use the signal separation estimation theory on the data after the spoofing suppression to estimate the DOA of every authentic GNSS signal one by one, then we use the estimated DOAs to perform multiple beamforming, in order to achieve the array signal processing gain.

The DOAs of multiple spoofers and authentic GNSS signals can be estimated using the CLEAN algorithm and the RELAX algorithm [36]. As described in Sect. 2.4.3, compared with the CLEAN algorithm, the RELAX algorithm has better estimation performance, but its implementation complexity is higher. Considering that the null created by orthogonal projection is very sharp, to prevent the problem of having the spoofing DOA shifting out of the null zone due to DOA estimation error, the precision requirement on the DOA estimation on multiple spoofers is higher. But when beam direction errors exist due to GNSS signal DOA estimation errors, the only consequence is a slight reduction on the GNSS signal's array signal processing gain, i.e. the requirements on the GNSS signal's DOA estimation precision are not very strict. Consequently, the RELAX algorithm should be used to estimate DOAs of multiple spoofers, but both the RELAX algorithm and the CLEAN algorithm can be used to estimate DOAs of GNSS signals. Below we introduce the detailed implementation of the spoofing countermeasure method based on the signal separation estimation theory.

First, the amplitude and DOA of every spoofer can be estimated using the RELAX algorithm [37, 38]. We rewrite the array received signal model (4.3) after sampling in a multiple spoofer environment as below

$$\begin{aligned}\mathbf{x}(n) &= \sum_{m=1}^{\bar{M}} \mathbf{a}(\theta_m^a) s_m^a(n) + \sum_{k=1}^{\bar{K}} \mathbf{a}(\theta_k^s) j_k^s(n) + \mathbf{e}(n) \\ &= \mathbf{A}(\boldsymbol{\theta}) \boldsymbol{\beta}(n) + \mathbf{e}(n)\end{aligned} \tag{4.12}$$

where

$$\begin{aligned}\mathbf{A}(\boldsymbol{\theta}) &= \left[\mathbf{a}(\theta_1^s), \mathbf{a}(\theta_2^s), \ldots, \mathbf{a}(\theta_{\bar{K}}^s), \mathbf{a}(\theta_1^a), \mathbf{a}(\theta_2^a), \ldots, \mathbf{a}(\theta_{\bar{M}}^a)\right] \\ \boldsymbol{\theta} &= \left[\theta_1^s, \theta_2^s, \ldots, \theta_{\bar{K}}^s, \theta_1^a, \theta_2^a, \ldots, \theta_{\bar{M}}^a\right]^{\mathrm{T}} \\ \boldsymbol{\beta}(n) &= \left[j_1^s(n), j_2^s(n), \ldots, j_{\bar{K}}^s(n), s_1^a(n), s_2^a(n), \ldots, s_{\bar{M}}^a(n)\right]^{\mathrm{T}}\end{aligned} \tag{4.13}$$

As a result $\boldsymbol{\beta}$ and $\boldsymbol{\theta}$ can be estimated by minimizing the cost function F below

$$F[\boldsymbol{\theta}, \boldsymbol{\beta}(1), \boldsymbol{\beta}(2), \ldots, \boldsymbol{\beta}(N)] = \sum_{n=1}^{N} [\mathbf{x}(n) - \mathbf{A}(\boldsymbol{\theta})\boldsymbol{\beta}(n)]^H [\mathbf{x}(n) - \mathbf{A}(\boldsymbol{\theta})\boldsymbol{\beta}(n)] \tag{4.14}$$

The nonlinear least squares problem in (4.14) can be solved using the RELAX algorithm. First, we can do some preparatory works.

Given L signals that need to be estimated, and we have $\left\{\hat{\theta}_l, \hat{\beta}_l(1), \hat{\beta}_l(2), \ldots, \hat{\beta}_l(N)\right\}_{l=1, l\neq k}^{L}$, what has already been estimated, and we can define

$$\hat{\mathbf{x}}_k(n) = \mathbf{x}(n) - \sum_{l=1, l\neq k}^{L} \mathbf{a}(\hat{\theta}_l)\hat{\beta}_l(n), \quad n = 1, 2, \ldots, N \tag{4.15}$$

Then the cost function for the kth signal's parameters can be written as

$$F_k\left[\hat{\theta}_k, \hat{\beta}_k(1), \hat{\beta}_k(2), \ldots, \hat{\beta}_k(N)\right] = \sum_{n=1}^{N} \left|\hat{\mathbf{x}}_k(n) - \mathbf{a}(\theta_k)\hat{\beta}_k(n)\right|^2 \tag{4.16}$$

By minimizing the (4.16), we can obtain

$$\hat{\theta}_k = \arg\max_{\theta_k} \mathbf{a}^{\mathrm{H}}(\theta_k)\left[\sum_{n=1}^{N} \hat{\mathbf{x}}_k(n)\hat{\mathbf{x}}_k^{\mathrm{H}}(n)\right]\mathbf{a}(\theta_k) \tag{4.17}$$

$$\hat{\beta}_k(n) = \left.\frac{\mathbf{a}^{\mathrm{H}}(\hat{\theta}_k)\hat{\mathbf{x}}_k(n)}{M}\right|_{\theta_k = \hat{\theta}_k}, \quad n = 1, 2, \ldots, N \tag{4.18}$$

The above formulas are similar to the (4.6) which is used for the single spoofer. Below we give the steps for the RELAX algorithm:

(1) Given $L = 1$, let $\hat{\mathbf{x}}_1(n) = \mathbf{x}(n)$, we can use (4.17) and (4.18) to estimate $\hat{\theta}_1$ and $\hat{\beta}_1(n)$, $n = 1, 2, \ldots, N$.
(2) Given $L = 2$, we can use $\left\{\hat{\theta}_1, \hat{\beta}_1(n)\right\}$ from (4.15) to solve for $\hat{\mathbf{x}}_2(n)$, and then use $\hat{\mathbf{x}}_2(n)$ from (4.17) and (4.18) to estimate $\left\{\hat{\theta}_2, \hat{\beta}_2(n)\right\}$.

Then the $\left\{\hat{\theta}_2, \hat{\beta}_2(n)\right\}$ estimated at step 2 can be used to solve for the new $\hat{\mathbf{x}}_1(n)$ from (4.15), and this $\hat{\mathbf{x}}_1(n)$ can be used to re-estimate the $\left\{\hat{\theta}_1, \hat{\beta}_1(n)\right\}$ from (4.17) and (4.18). We can iterate the above two steps until convergence.

(3) Given $L = 3$, we can use previously estimated $\left\{\hat{\theta}_l, \hat{\beta}_l(n)\right\}_{l=1}^{2}$ to derive $\hat{\mathbf{x}}_3(n)$ from (4.15), then use the $\hat{\mathbf{x}}_3(n)$ from (4.17) and (4.18) to estimate $\left\{\hat{\theta}_3, \hat{\beta}_3(n)\right\}$.

Next, we can use $\left\{\hat{\theta}_l, \hat{\beta}_l(n)\right\}_{l=2}^{3}$ to derive $\hat{\mathbf{x}}_1(n)$ from (4.15), re-estimate the $\left\{\hat{\theta}_1, \hat{\beta}_1(n)\right\}$ using (4.17) and (4.18), use $\left\{\hat{\theta}_l, \hat{\beta}_l(n)\right\}_{l=1,l\neq 2}^{3}$ to derive $\hat{\mathbf{x}}_2(n)$ from (4.15), and estimate $\left\{\hat{\theta}_2, \hat{\beta}_2(n)\right\}$ based on (4.17) and (4.18); and so on and so forth, until we can eventually obtain a set of $\left\{\hat{\theta}_l, \hat{\beta}_l(n)\right\}_{l=1}^{3}$.

We can repeat the above steps until L equals the number of spoofers. Then we can obtain estimated parameters for all spoofers $\left\{\hat{\theta}_l, \hat{\beta}_l(n)\right\}_{l=1}^{L}$.

Since the power of the spoofing signal is higher than that of the authentic GNSS signal, the RELAX algorithm can estimate the amplitude and the DOA of the spoofer firstly, and then estimate the amplitude and the DOA of the authentic GNSS signal. After estimating a signal, the power of the signal can be computed. If the power of the $(P+1)$th signal is significantly smaller than the power of the Pth signal, then the first P signals are the interferences, and the number of spoofers is $L = P$.

The interference subspace can be constructed using the estimated interference DOA $\left\{\hat{\theta}_1, \hat{\theta}_2, \ldots, \hat{\theta}_P\right\}$ by the RELAX algorithm, and it is denoted as

$$\mathbf{U} = span\left\{\mathbf{a}(\hat{\theta}_1), \mathrm{a}(\hat{\theta}_2), \ldots, \mathbf{a}(\hat{\theta}_P)\right\} \tag{4.19}$$

Then the interference subspace's orthogonal complement space projection matrix is

$$\mathbf{P}_{\perp} = \mathbf{I} - \mathbf{U}\mathbf{U}^{\mathrm{H}}/M \tag{4.20}$$

And the data after projecting the array received signal is

$$\mathbf{y}(n) = \mathbf{P}_{\perp}\mathbf{x}(n) \tag{4.21}$$

We can take one channel of the projected data to directly feed into a regular receiver to perform acquisition, tracking, and position computation.

To achieve a good processing gain on the array signal, we hope that the beam formed can aim toward the authentic GNSS satellites. Then, for the projected data $\mathbf{y}(n)$, we can use the CLEAN algorithm discussed in Sect. 2.4.3 or the RELAX algorithm discussed in this section to estimate the DOAs of all authentic GNSS signals, and use the estimated DOAs to perform multiple beamforming on the spoofing mitigated data. A multi-beam method needs to calculate a multi-beam weighted vector for every GNSS satellite, and then uses a multi-beam receiver to receive every beam output. For this implementation, since we need to estimate the DOAs of all authentic GNSS satellites, the complexity is higher, but the array signal processing gain can be achieved.

The spoofing countermeasure method based on the signal separation estimation theory does not need any feedback from the receiver, so it can be embedded into a navigation receiver as an independent spoofing countermeasure module, but it needs to know the array manifold information. On the other hand, even though the RELAX algorithm can separate out various space signal sources, the task to determine whether the signal is a spoofing signal or an authentic GNSS signal can be implemented by comparing the power levels for various signals. As a result, it needs to use the power relationship between spoofing signals and authentic GNSS signals.

2. Spoofing Countermeasure Based on De-spread and Re-spread

Taking into consideration that the condition of array manifold information is unknown, we propose a spoofing countermeasure method based on the de-spread re-spread algorithm [39]. This method is based on the de-spread re-spread jamming countermeasure techniques introduced in Sect. 2.5 of this book [40, 41], which cleverly takes advantage of the fact that the de-spread re-spread weight vector is proportional to the signal steering vector. Since the authentic GNSS signals are transmitted by different GNSS satellites, and all retransmitted spurious PRN signals come from the same direction, the de-spread re-spread weight vectors corresponding to the spurious PRN signals are highly spatially correlated. However the de-spread re-spread weight vectors between the spurious PRN signals and authentic GNSS signals, and among the authentic GNSS signals, are not highly spatially correlated. The spoofing countermeasure method based on the de-spread re-spread algorithm first derives the de-spread re-spread weight vectors for every acquired satellite signal (these signals could be authentic GNSS signals, and some of them might be interference signals), then it distinguishes the interferences by observing the correlations among these weight vectors, and then performs spoofing suppression by projecting onto the interference orthogonal space and establishing multiple beamformings. And consequently the authentic GNSS signals can go through without distortion. This method does not need to know array manifold, and not sensitive to array manifold errors. Below we introduce detailed implementation process of the spoofing countermeasure method based on de-spread re-spread algorithm.

First of all, the de-spread re-spread weight vector for the lth GNSS satellite can be solved based on the flow illustrated in Fig. 4.5 $\mathbf{w}_l$ $(l = 1, 2, \ldots, L_{acq})$, and L_{acq} represents the number of acquired GNSS satellites.

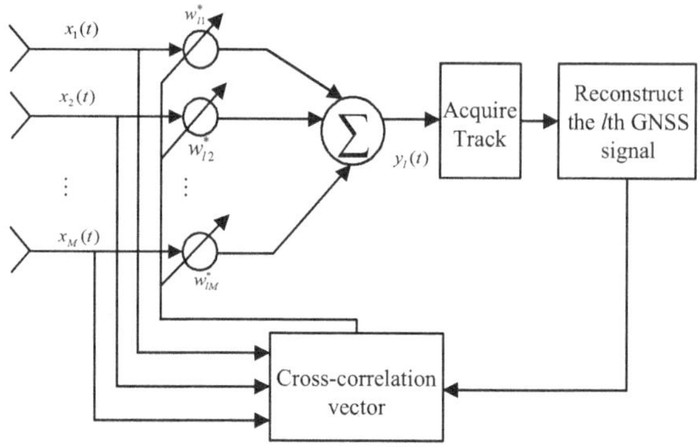

Fig. 4.5 Flow chart of solving for the de-spread re-spread weight vectors

By writing the de-spread re-spread weight vectors corresponding to all L_{acq} GNSS satellites in matrix format, it can be denoted that

$$\mathbf{W} = \left[\mathbf{w}_1, \mathbf{w}_2, \ldots, \mathbf{w}_{L_{acq}}\right] \tag{4.22}$$

Since the de-spread re-spread weight vectors are proportional to the steering vectors, we can determine whether spoofing exists or not by checking if **W** has full rank, i.e.

$$rank[\mathbf{W}] < \min\{M, L_{acq}\} \tag{4.23}$$

If (4.23) is not valid, that means the de-spread re-spread weight vectors $\mathbf{w}_1, \mathbf{w}_2, \ldots, \mathbf{w}_{Lacq}$ are not correlated to each other, so there are no signals coming from the same direction, therefore among the GNSS signals presently received and tracked by the receiver, there are no spoofing signals. On the other hand, if (4.23) is valid, that means spoofing exists among the acquired signals by the receiver. Under this condition, further processing is needed to recognize which de-spread re-spread weight vectors correspond to which spoofer, i.e. the following decisions need to be made:

Given the de-spread re-spread weight vectors for two GNSS signals s_1 and s_2 are $\mathbf{w}_1$ and $\mathbf{w}_2$, ρ denotes their normalization correlation coefficient. When the normalization correlation coefficient ρ is larger than a certain threshold D_Δ, i.e.

$$\rho = \left| \left(\frac{\mathbf{w}_1}{|\mathbf{w}_1|}\right)^{\mathrm{H}} \frac{\mathbf{w}_2}{|\mathbf{w}_2|} \right| \geq D_\Delta \tag{4.24}$$

These two signals can be regarded as highly correlated in space, i.e. they come from the same direction, and as a result the existence of spoofer can be detected. We can use (4.24) to determine the correlation coefficients among column vectors of **W**. In that case, the de-spread re-spread weight vectors corresponding to the interferences transmitted by the same spoofer (denoted as spoofer k) have very high spatial correlations and (4.24) is satisfied. Two or more de-spread re-spread weight vectors can be classified into a set, and proper methods can be used to average the de-spread re-spread in the same set, to obtain the averaged vector $\tilde{\mathbf{w}}_k^s$. The derived vector corresponds to the steering vector of the spoofer k. The estimated steering vectors for all $\bar{K}$ spoofers $\{\tilde{\mathbf{w}}_1, \tilde{\mathbf{w}}_2, \ldots, \tilde{\mathbf{w}}_{\bar{K}}\}$ can be derived in turn using this method, and these steering vectors can construct the interference subspace, which can be denoted as

$$\mathbf{U} = span\{\tilde{\mathbf{w}}_1, \tilde{\mathbf{w}}_2, \ldots, \tilde{\mathbf{w}}_{\bar{K}}\} \tag{4.25}$$

Then the orthogonal complement space projection matrix for the interference subspace is

$$\mathbf{P}_{\perp} = \mathbf{I} - \mathbf{U}(\mathbf{U}^{\mathrm{H}}\mathbf{U})^{-1}\mathbf{U}^{\mathrm{H}} \tag{4.26}$$

The data after projecting the array received signals is

$$\mathbf{y}(n) = \mathbf{P}_{\perp}\mathbf{x}(n) \tag{4.27}$$

A channel of the projected data can be directly fed into a regular receiver for acquisition, tracking and navigational positioning calculation. If we want to have signal processing gain, we would hope the formed beams have directional gains and point at authentic GNSS signals. In that case, de-spread re-spread weight vectors need to be calculated on the projected data, to obtain the de-spread re-spread weight vectors $\mathbf{w}_l^a$ $(l = 1, 2, \ldots, \bar{M})$ for the authentic GNSS signals. Multiple beamforming can be performed on the data $\mathbf{y}(n)$ after anti-spoofing using the authentic GNSS signals' de-spread re-spread weight vectors.

$$z_l(n) = \left(\mathbf{w}_l^a\right)^{\mathrm{H}}\mathbf{y}(n) \tag{4.28}$$

where, subscript l represents the output signal of the lth authentic GNSS signal, and a multi-beam receiver can be used to receive outputs of various beams.

4.4.3 Combined Suppression on Jamming and Spoofing

The spoofing signal suppression methods introduced in the previous sections are for environments in which only spoofing exists. In this section we will consider more complex environments. When jamming and spoofing co-exist, receivers cannot acquire GNSS signals due to jamming, and as a result they cannot function. If a receiver successfully suppresses the jamming, it could lock onto a spoofing signal with higher power when it starts to work again. Consequently the receiver provides navigational positioning information that seems to be reliable, but in reality it is incorrect. Thus, jamming and spoofing combine to form a joint threat to the receivers, and the methods designed against jamming-only or spoofing-only environments cannot work in such complex environments. In this case, receives need to use algorithms that have combined capabilities to suppress both jamming and spoofing.

1. Combined Suppression Based on Diagonal Loading

Before we introduce the spoofing suppression algorithm based on diagonal loading [42], we first analyze the characteristics of array received data's covariance matrix using eigen-decomposition. The eigen-decomposition can be performed on the covariance matrix $\mathbf{R}_x$ of the antenna array's received data:

$$\mathbf{R}_x = \sum_{i=1}^{M} \lambda_i \mathbf{u}_i \mathbf{u}_i^{\mathrm{H}} \tag{4.29}$$

where λ_i $(i = 1, 2, \ldots, M)$ denotes eigenvalues, which can be ranked in descending order $\lambda_1 \geq \lambda_2 \geq \cdots \geq \lambda_K \gg \lambda_{K+1} = \cdots = \lambda_M = \sigma^2$; K is the number of strong interference; σ^2 is the power of additive Gaussian white noise; The eigenvectors corresponding to the M eigenvalues are $\mathbf{u}_i (i = 1, 2, \ldots, M)$. As a result, the inverse matrix of the covariance matrix $\mathbf{R}_x^{-1}$ can be represented as

$$\begin{aligned}
\mathbf{R}_x^{-1} &= \sum_{i=1}^{K} \frac{1}{\lambda_i} \mathbf{u}_i \mathbf{u}_i^{\mathrm{H}} + \frac{1}{\sigma^2} \sum_{i=K+1}^{M} \mathbf{u}_i \mathbf{u}_i^{\mathrm{H}} \\
&= \frac{1}{\sigma^2} \left[\sum_{i=K+1}^{M} \mathbf{u}_i \mathbf{u}_i^{\mathrm{H}} + \sum_{i=1}^{K} \frac{\sigma^2}{\lambda_i} \mathbf{u}_i \mathbf{u}_i^{\mathrm{H}} \right] \\
&= \frac{1}{\sigma^2} \left[\mathbf{I} - \sum_{i=1}^{K} \left(1 - \frac{\sigma^2}{\lambda_i} \right) \mathbf{u}_i \mathbf{u}_i^{\mathrm{H}} \right]
\end{aligned} \tag{4.30}$$

If $\lambda_1 \geq \lambda_2 \geq \cdots \geq \lambda_K \gg \sigma^2$ (corresponding to strong jammings), (4.30) can be approximated as

$$\mathbf{R}_x^{-1} \approx \frac{1}{\sigma^2} \left[\mathbf{I} - \sum_{i=1}^{K} \mathbf{u}_i \mathbf{u}_i^{\mathrm{H}} \right] = \frac{1}{\sigma^2} \mathbf{P}_{\perp} \tag{4.31}$$

where $\mathbf{P}_{\perp}$ represents the orthogonal complement space projection matrix of the interference subspace. Since the eigenvectors corresponding to large eigenvalues can be regarded as the eigenvectors of those strong interference, when the interferences are strong, $\mathbf{R}_x^{-1}$ and $\mathbf{P}_{\perp}$ are equivalent.

However, $\mathbf{R}_x^{-1}$ does not have good suppression performance on weak interference. When only one weak interference exists, we can perform eigen-decomposition similar to (4.29) on the covariance matrix $\mathbf{R}_x$. By defining σ_j^2 as the power of the weak interference then $\lambda_1 = M\sigma_j^2 + \sigma^2$ is the maximum eigenvalue corresponding to the weak interference signals, $\lambda_2 = \cdots = \lambda_N = \sigma^2$. Based on (4.30), we can derive

$$\mathbf{R}_x^{-1} = \frac{1}{\sigma^2}\left[\mathbf{I} - \left(1 - \frac{\sigma^2}{\lambda_1}\right)\mathbf{u}_1\mathbf{u}_1^{\mathrm{H}}\right] = \frac{1}{\sigma^2}\left[\mathbf{I} - \frac{1}{1 + \frac{\sigma^2}{M\sigma_j^2}}\mathbf{u}_1\mathbf{u}_1^{\mathrm{H}}\right] \tag{4.32}$$

When $\frac{\sigma^2}{M\sigma_j^2} \gg 1$,

$$\mathbf{R}_x^{-1} \approx \frac{1}{\sigma^2}\mathbf{I} \tag{4.33}$$

$\mathbf{R}_x^{-1}$ does not have interference suppression capability.

Based on the analyses above, for strong jamming/spoofing, $\mathbf{R}_x^{-1}$ has stronger suppression capability. However, usually jamming is stronger, and spoofing is not very strong. Directly using $\mathbf{R}_x^{-1}$ cannot suppress spoofing very well, and instead we can use the diagonal loading factor technique [43–45] to improve interference-to-noise ratio.

Given a diagonal loading factor of $F\sigma^2$ $(0 \le F < 1)$, the covariance matrix after diagonal loading factor can be represented as

$$\tilde{\mathbf{R}}_x = \mathbf{R}_x - F\sigma^2\mathbf{I} \tag{4.34}$$

Since the diagonal loading factor does not change eigenvectors, it can only change the eigenvalues. Thereby, eigenvectors of $\tilde{\mathbf{R}}_x$ and $\mathbf{R}_x$ are the same, and the relation between the eigenvalue of $\tilde{\mathbf{R}}_x$ $\tilde{\lambda}_i$ and the eigenvalue of $\mathbf{R}_x$ is

$$\tilde{\lambda}_i = \lambda_i - F\sigma^2 \tag{4.35}$$

Using a single weak jamming/spoofer as an example, matrix inversion can be performed on the covariance matrix $\tilde{\mathbf{R}}_x$ after the diagonal loading factor, and (4.32) can be converted to

$$\begin{aligned}\tilde{\mathbf{R}}_x^{-1} &= \frac{1}{(1-F)\sigma^2}\left[\mathbf{I} - \left(1 - \frac{(1-F)\sigma^2}{M\sigma_j^2 + (1-F)\sigma^2}\right)\mathbf{u}_1\mathbf{u}_1^{\mathrm{H}}\right] \\ &= \frac{1}{(1-F)\sigma^2}\left[\mathbf{I} - \frac{1}{1 + \frac{(1-F)\sigma^2}{M\sigma_j^2}}\mathbf{u}_1\mathbf{u}_1^{\mathrm{H}}\right]\end{aligned} \tag{4.36}$$

To make $\frac{(1-F)\sigma^2}{M\sigma_j^2} \ll 1$ by adjusting F value, it is implied that the interference-to-noise ratio can be greatly improved using diagonal loading factor technique. In this case the single weak interference is equivalent to a single strong interference, then (4.36) can be approximated as

$$\tilde{\mathbf{R}}_x^{-1} \simeq \frac{1}{(1-F)\sigma^2}\left[\mathbf{I} - \mathbf{u}_1\mathbf{u}_1^{\mathrm{H}}\right] \tag{4.37}$$

It can be seen from (4.37), after diagonal loading factor, $\tilde{\mathbf{R}}_x^{-1}$ can successfully suppress the weak interference.

Diagonal loading factor is the same as lowering the noise level, which converts the spoofing signal from a weak interference to a strong interference. Meanwhile, jamming could become stronger. Consequently, a power minimization algorithm based on diagonal loading factor can suppress jamming and spoofing simultaneously. When this algorithm is used, the selection of the diagonal loading factor $F\sigma^2$ must satisfy certain criteria, i.e. we must ensure that we can suppress the spoofing as much as we could, while simultaneously ensuring that useful GNSS signals are not excessively attenuated.

2. Combined Suppression Algorithm Based on De-spread Re-spread

The combined suppression algorithm on jamming and spoofing based on diagonal loading factor's minimum power is easy to implement, but the selection of diagonal loading factor is impacted by the power relation between the spoofers and authentic GNSS signals, and the number of antenna arrays. Also, this method cannot provide array signal processing gain. Combined suppression algorithm based on the de-spread re-spread algorithm can solve the above problems. When jammings and spoofings coexist, since jammings are very strong, the powers from the spoofers are larger than those of GNSS signals, but still are much smaller than the noise power. As a result, the covariance matrix of an array antenna's received signals is mainly determined by the covariance matrices of the jamming and noise. i.e.

$$\mathbf{R}_x \approx \mathbf{R}_j + \mathbf{R}_v \tag{4.38}$$

In (4.38), $\mathbf{R}_x$ denotes the covariance matrix of the received signals; $\mathbf{R}_j$ denotes the covariance matrix of the jamming signals; $\mathbf{R}_v$ denotes the noise covariance matrix (containing statistical information of spoofing, noise, and GNSS signals). In this case, $\mathbf{R}_x^{-1}$ is equivalent to the orthogonal complement space projection matrix of the strong interference subspace. By projecting the array received signal towards $\mathbf{R}_x^{-1}$, we can obtain

$$\mathbf{y}(n) = \mathbf{R}_x^{-1}\mathbf{x}(n) \tag{4.39}$$

$\mathbf{R}_x^{-1}$ can effectively suppress jamming, and it also has some attenuation effects on spoofing. The spoofing countermeasure algorithm based on de-spread re-spread technique described in Sect. 4.4.2 can be cascaded to further suppress the residue spoofing. Detailed processing steps are the same as (4.22)–(4.28).

4.4.4 Simulation Results

In this section, we perform simulations in different interference environments such as a single receiver based spoofer, multiple receiver based spoofers and coexistence of jamming and spoofing. The goal is to compare the performances of the spoofing countermeasure algorithms discussed in this chapter. The data used for the simulations is a L1 signal generated by the GPS simulator, with a sampling frequency of 5.714 MHz, and an intermediate frequency of 4.309 MHz. A uniform linear array with 10 array elements spaced by a half-wavelength is used.

1. Simulation Results in the Environment of a Single Spoofer

Authentic GNSS signals are PRN 1, PRN 3, PRN 22 and PRN 25; the signal-to-noise ratio is -20 dB; the incident angles are 0°, -35°, 10° and -10° respectively. The spoofers amplify the received GNSS signals and retransmitted after a delay. We assume that every spoofing PRN signal has a signal-to-noise ratio of -18 dB, and the DOA of the spoofing is 35°.

Figure 4.6 shows the array antenna pattern derived using the spoofing countermeasure algorithm based on DOA estimation. The solid line in the figure represents the DOAs of authentic GNSS signals; the dotted line represents the DOAs of spoofing signals. It can be seen that the antenna array pattern has relatively flat gain on the directions of authentic GNSS signals, so that authentic GNSS signals can pass through, but at the same time very deep nulls are formed towards the DOAs of spoofing signals.

Figure 4.7 shows the acquisition results on GNSS signals before and after interference suppression. In the Fig. 4.7a, c, e, g are for results without interference

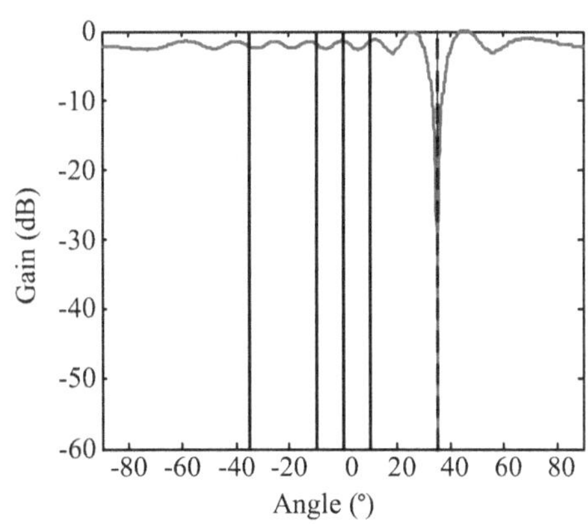

Fig. 4.6 Antenna array pattern

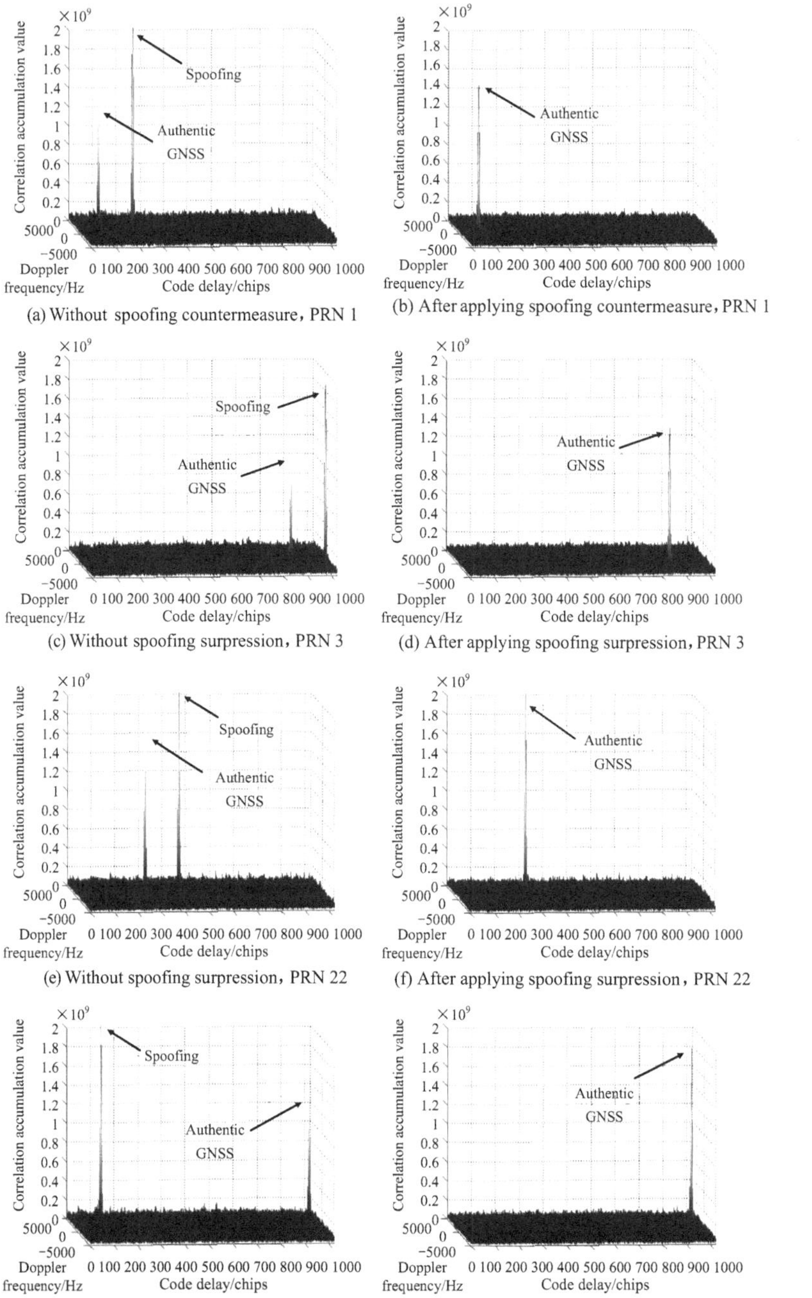

(a) Without spoofing countermeasure, PRN 1

(b) After applying spoofing countermeasure, PRN 1

(c) Without spoofing surpression, PRN 3

(d) After applying spoofing surpression, PRN 3

(e) Without spoofing surpression, PRN 22

(f) After applying spoofing surpression, PRN 22

(g) Without spoofing surpression, PRN 25

(h) After applying spoofing surpression, PRN 25

Fig. 4.7 Comparisons of GNSS signal acquisition results by the receiver before and after applying spoofing countermeasure suppression, and Fig. 4.7b, d, f, h are for results after applying interference suppression. It can be seen from the Fig. 4.7a, c, e, g that without interference suppression, the two-dimensional search result for the GNSS signals received by the receiver has two obvious correlation peaks: one correlation peak is for the authentic GNSS signal, and the other correlation peak is for the interference. Since the interference power is higher than the powers of the GNSS signals, the interference correlation peak value is higher than the authentic GNSS signals. Therefore, without interference suppression, the receiver will acquire interference correlation peak. After interference suppression, in Fig. 4.7b, d, f, h, only correlation peaks for authentic GNSS signals exist, which shows that the interference has already been effectively suppressed, so the receiver can robustly acquire the correlation peaks of the authentic GNSS signals.

2. Simulation in the Environment of Multiple Spoofers

Authentic GNSS signals are PRN1, PRN2, PRN3 and PRN22, and the signal-to-noise ratio is -20 dB, and the incident angles are $0°$, $10°$, $-10°$ and $-35°$ respectively. The spoofing signal transmitted by spoofer 1 incidents onto the array from the direction of $40°$, and the spoofing signal transmitted by the spoofer 2 incidents onto the array from the direction of $25°$. The signal-to-noise ratios for all interference PRN signals are all -18 dB.

Figure 4.8 shows the array antenna patterns derived using the spoofing countermeasure algorithm introduced in Sect. 4.4.2. Among the subfigures, subfigures (a) and (b) correspond to the receiver based spoofing suppression algorithm based on signal separation estimation theory (in our simulation, the estimations on interferences' DOAs and GNSS signals' DOAs all are implemented using the RELAX algorithm). Subfigure (c) corresponds to the spoofing countermeasure algorithm based on the de-spread re-spread technique. The solid lines in the figure represent the DOAs of the authentic GNSS signals; the dotted lines represent the DOAs of the interferences. It can be seen that both methods can form nulls towards the DOAs of the interferences. Towards the authentic GNSS signals, subfigure (a)'s antenna pattern has slight attenuation; and in subfigures (b) and (c), the beams formed by authentic GNSS satellites all point towards the DOAs of the respective GNSS satellites, i.e. beamforming gains are provided on the DOAs of authentic GNSS signals.

Figure 4.9 shows the two-dimensional search result of the GNSS PRN1 signal by the receiver before and after applying spoofing countermeasure. It can be seen that it contains one correlation peak corresponding to the authentic GNSS signals and two correlation peaks corresponding to the two interferences, which is consistent with the simulation conditions, and the two correlation peaks corresponding to the two interferences are both higher than the authentic GNSS signals. Thereby, the receiver acquires the highest peak which corresponds to one of the interferences. Figure 4.9b, c, d are the results after applying spoofing countermeasure. It can be seen that every one of them shows correlation peaks corresponding to the authentic GNSS signals. In addition, the correlation peak values in the Fig. 4.9c, d are

Fig. 4.8 Comparisons of antenna directions obtained by different spoofing countermeasure algorithms

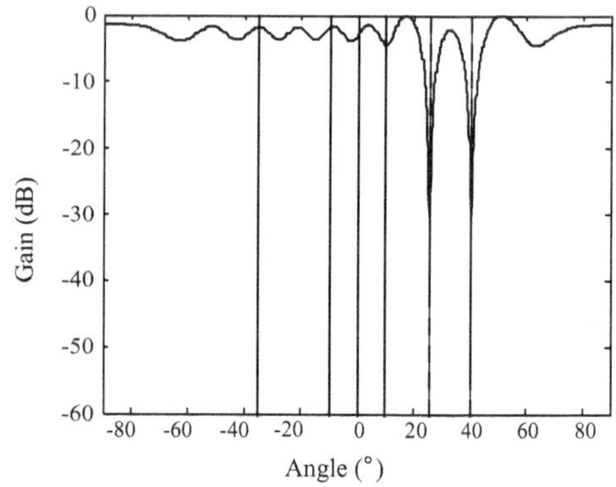

(a) The algorithm based on signal separation estimation theory (single beam).

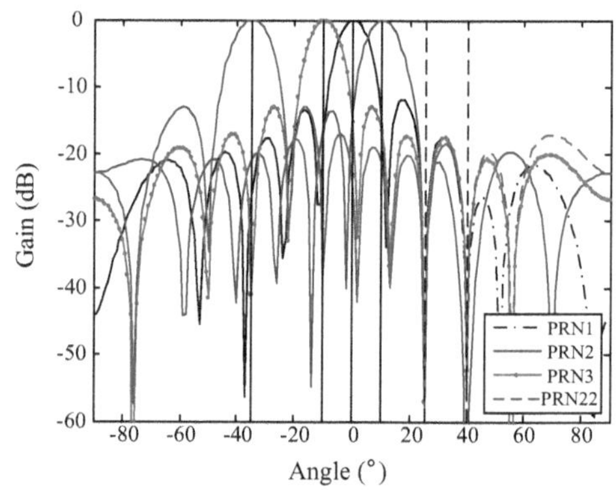

(b) The algorithm based on signal separation estimation theory (multiple beams).

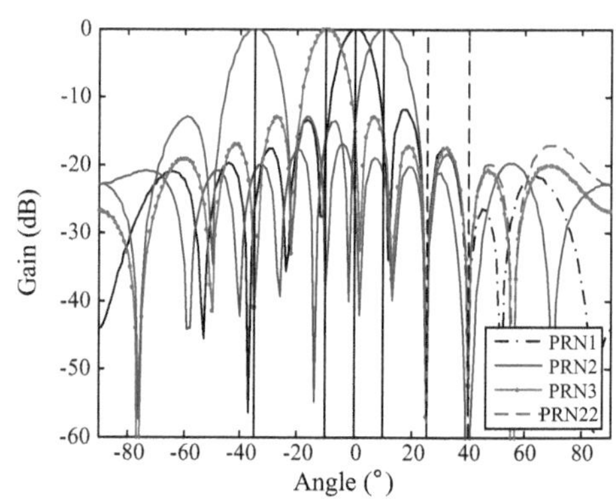

(c) The algorithm based on de-spread re-spread technqiue.

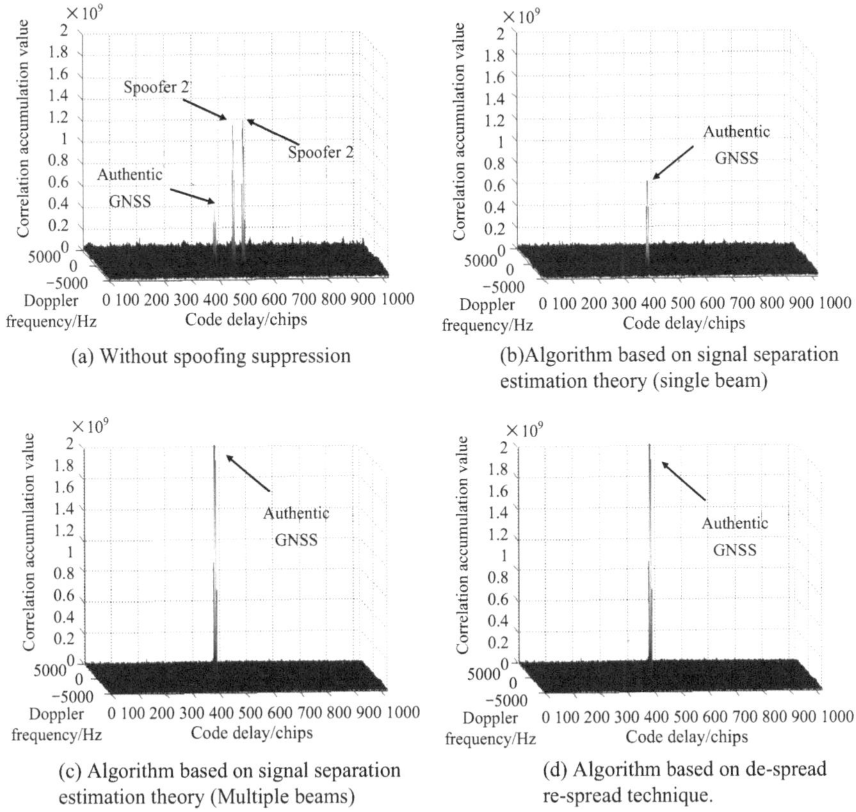

(a) Without spoofing suppression

(b)Algorithm based on signal separation estimation theory (single beam)

(c) Algorithm based on signal separation estimation theory (Multiple beams)

(d) Algorithm based on de-spread re-spread technique.

Fig. 4.9 Comparisons of acquisition results on PRN1 GNSS signal before and after suppressing the multiple receiver based spoofing interferences

approximately 10 times higher than the correlation peak values in the Fig. 4.9a, b, and have a low noise level.

3. Simulation Results when Jamming and Spoofing Coexist

Four GPS satellite signals PRN1, PRN2, PRN3 and PRN22 incident onto the array from the directions of $-5°$, $5°$, $-25°$ and $15°$, and the signal-to-noise ratio is -20 dB. The interferences incident onto the array from the directions of $25°$ and $-40°$. The signal-to-noise ratio of each interference PRN signal is -18 dB. The jamming signal incidents onto the array from the direction of $60°$, and its interference-to-noise ratio is 40 dB.

Figure 4.10 shows the antenna patterns of using power minimization algorithm and power minimization algorithm based on diagonal loading factor. In the figure, the solid line aims towards the DOAs of spoofer 1 and 2, and the jamming signals. The antenna patterns for these two methods both form very deep nulls towards the jamming DOA. In the directions of spoofers, it can be seen in Fig. 4.10 that the

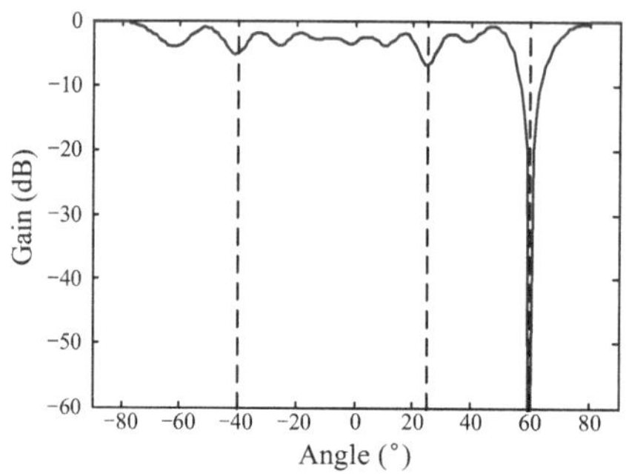

(a) Power minimization algorithm loading factor.

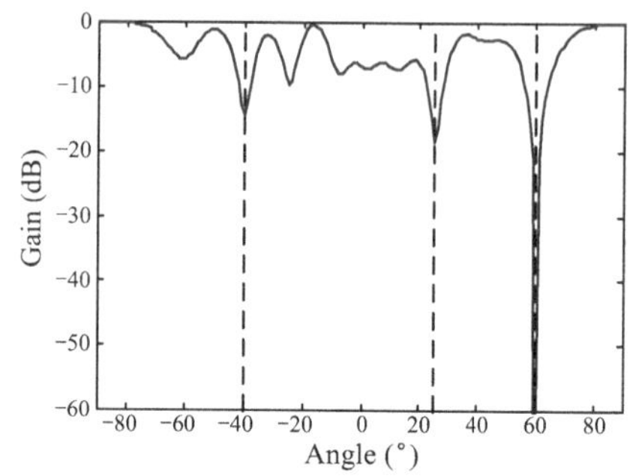

(b) Power minimization algorithm based on diagonal

Fig. 4.10 Antenna patterns for different methods

power minimization algorithm based on diagonal loading factor can form deeper nulls, and can suppress the jamming and spoofing simultaneously, but the power minimization algorithm cannot effectively suppress spoofing.

In Fig. 4.10b, the antenna pattern for the power minimization algorithm based on diagonal loading factor has fluctuation on the sidelobe, which is the impact of increased small eigenvalue divergence of covariance matrix caused by diagonal loading factor. The fluctuation on the sidelobe can lead to certain attenuation on authentic GNSS signals, which is a shortcoming of the method. A joint interference countermeasure method based on the de-spread re-spread algorithm that is introduced in Sect. 4.4.3 can overcome this shortcoming. Below we list simulation results of applying the de-spread re-spread algorithm under the same simulation conditions.

Figure 4.11 is the overall pattern for all authentic GNSS signals using the combined jamming and spoofing suppression algorithm based on de-spread re-spread technique. The solid lines in the figure represent the authentic GNSS signals' DOAs, and the dotted lines show the DOAs of spoofers 1, 2 and the jamming signal. It can be seen in Fig. 4.11, the de-spread re-spread weight vector of an authentic GNSS signal can be used to form a beam towards the satellite signal's DOA with an obvious beam gain, and at the same time form nulls towards the spoofing and jamming signals. More details can be seen in Fig. 4.12.

Next, we discuss the acquisition results of different algorithms. Figure 4.13 shows the comparisons of acquisition results before suppressing the spoofing and jamming signals. Figure 4.13a shows that without suppressing the spoofing and jamming signals, the receiver cannot acquire the GNSS signals under the influence of jamming. Figure 4.13b, c show the acquisition results after applying combined jamming and spoofing suppression based on diagonal loading factor, and combined jamming and spoofing suppression based on de-spread re-spread technique. It can be seen that the Fig. 4.13b, four GNSS signals are successfully acquired, but the noise level is relatively high. The de-spread re-spread method is a multi-beam

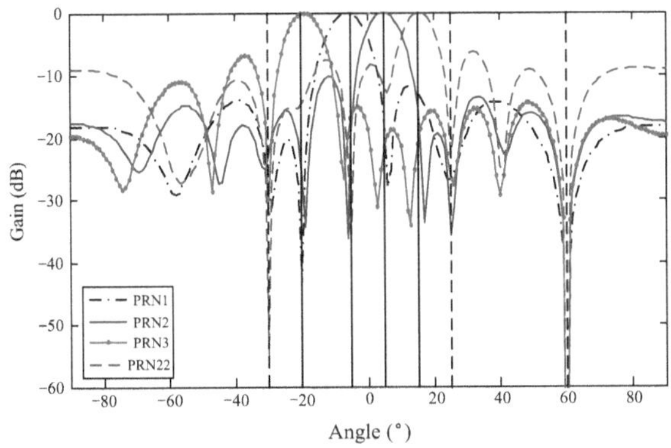

Fig. 4.11 Antenna patterns for all authentic GNSS signals after applying the combined jamming and spoofing suppression algorithm based on de-spread re-spread technique

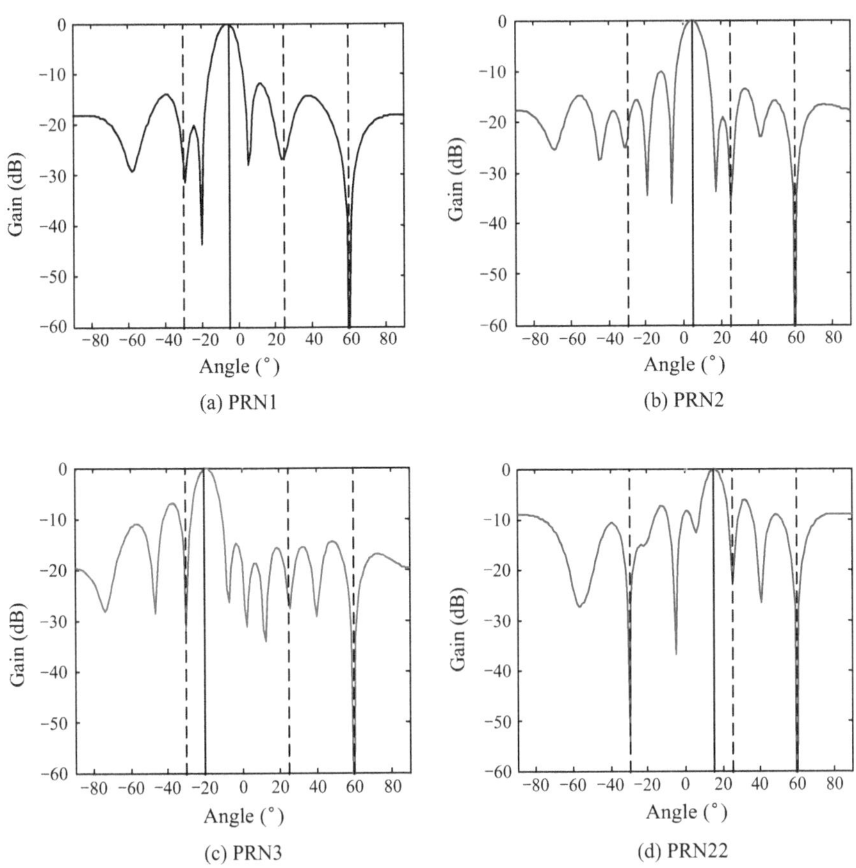

Fig. 4.12 Antenna patterns for various authentic GNSS signals after applying the combined jamming and spoofing suppression algorithm based on de-spread re-spread technique

Fig. 4.13 Comparisons of acquisition results before and after suppressing the jamming and spoofing

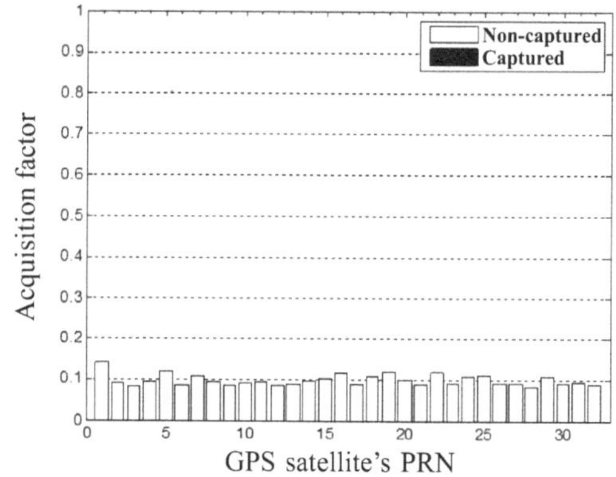

(a) Without jamming and spoofing suppression

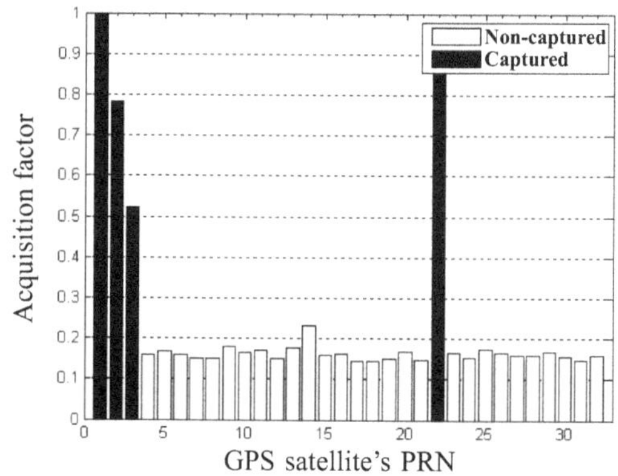

(b) After applying the power minization algorithm based on diagonal loading factor.

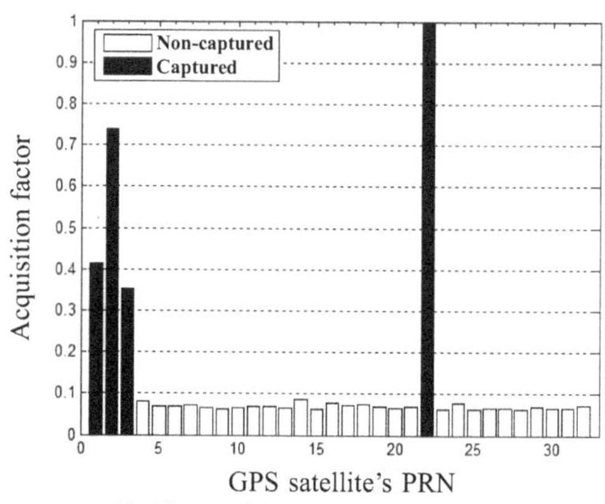

(c) After applying the suppresion algorithm based on de-spread re-spread technique.

method, where every formed beam aims towards one GNSS signal. Figure 4.13c only shows the beam output acquisition result corresponding to the PRN22 (the results for all the other GNSS signals are similar, so we choose not to show the results for them). It can be seen that PRN22 has a higher acquisition factor, and the noise level is much lower than those shown in Fig. 4.13a, b.

4.5 Summary

We mainly study GNSS spoofing countermeasure techniques in this chapter. Firstly, we introduce the spoofing generating methods, give data models, and analyze the impacts of spoofing on GNSS receivers. Then several common spoofing detection methods are introduced. Finally, we introduce different spoofing countermeasure algorithms aiming towards various spoofing environments such as single spoofer, multiple spoofers, and coexistence of jamming and spoofing, and analyze the characteristics and applicability of various algorithms.

References

1. John A. Vulnerability assessment of the transportation infrastructure relying on the global positioning system. Volpe National Transportation Systems Center; 2001.
2. Shepard DP, Humphreys TE. Characterization of receiver response to a spoofing attack. In: Proceedings of the 24th international technical meeting of the satellite division of the Institute of Navigation (ION GNSS 2011), September 20–23, Portland, OR. 2011. p. 2608–18.
3. Humphreys TE, Ledvina BM, Psiaki ML, et al. Assessing the spoofing threat: development of a portable GPS civilian spoofer. In: Proceedings of ION GNSS 21st international technical meeting of the satellite division, September 16–19, Savannah, GA. 2008. p. 2314–25.
4. Montgomery YP, Humphreys TE, Ledvina BM. Receiver-autonomous spoofing detection: experimental results of a multi-antenna receiver defense against a portable civil GPS spoofer. In: Proceedings of the Institute of Navigation, national technical meeting, January 26–28, Anaheim, CA. 2009. p. 124–30.
5. Humphreys TE. Robust and resilient navigation. ION GNSS + 2015 Tutorial, September 15, 2015, Tampa, USA. 2015.
6. http://www.engr.utexas.edu/features/superyacht-gps-spoofing.
7. Jafarnia JA, Broumandan A, Nielsen J, et al. GPS spoofer countermeasure effectiveness based on signal strength, noise power and C/N_0 observables. Int J Satell Commun Netw. 2012;30 (4):181–91.
8. Dehghanian V, Nielsen J, Lachapelle G. GNSS spoofing detection based on receiver C/N_0 estimates. In: Proceedings of the ION GNSS meeting, Nashville, Tennessee. Institute of Navigation; 2012.
9. Kaplan ED, Hegarty CJ. Understanding GPS principles and applications. 2nd ed. Boston, Mass, USA: Artech House; 2006.
10. Juang JC. GNSS spoofing analysis by VIAS. Coordinates Magazine. 2011.
11. Wen HQ, Huang PYR, Dyer J, et al. Countermeasures for GPS signal spoofing. In: Proceedings of the 18th international technical meeting of the satellite division of the Institute of Navigation (ION GNSS'05), September 2005, Long Beach, CA, USA. 2015. p. 1285–90.

12. Montgomery PY, Humphreys TE, Ledvina BM. Receiver-autonomous spoofing detection: experimental results of a multi-antenna receiver defense against a portable civil GPS spoofer. In: Proceedings of the Institute of Navigation—international technical meeting (ITM'09), January 2009, Anaheim, CA, USA. 2009. p. 124–30.
13. Psiaki ML, O'Hanlon BW, Powell, et al. GNSS spoofing detection using two-antenna differential carrier phase. In: Proceedings of the ION GNSS + Meeting. Tampa, FL: Institute of Navigation; 2014.
14. Castaneda MH, Stein M, Antreich F, et al. Joint space-time interference mitigation for embedded multi-antenna GNSS receivers. 2013.
15. McDowell CE. GPS spoofer and repeater mitigation system using digital spatial nulling—US Patent 7250903 B1. 2007.
16. Nielsen J, Broumandan A, Lachapelle G. Spoofing detection and mitigation with a moving handheld receiver. GPS World. 2010;21(9):27–33.
17. Moshavi S. Multi-user detection for DS-CDMA communications. IEEE Commun Mag. 1996;34(10):124–36.
18. Phelts RE. Multicorrelator techniques for robust mitigation of threats to GPS signal quality. Palo Alto, CA, USA: Standford University; 2001.
19. Shepard DP, Humphreys TE. Characterization of receiver response to a spoofing attack. In: Proceedings of the 24th international technical meeting of the satellite division of the Institute of Navigation (ION GNSS'11), September 2011, Portland, OR, USA. 2011. p. 2608–18.
20. Cavaleri A, Motella B, Pini M, et al. Detection of spoofed GPS signals at code and carrier tracking level. In: Proceedings of the 5th ESA workshop on satellite navigation technologies and European workshop on GNSS signals and signal processing (NAVITEC'10), December 2010. 2010. p. 1–6.
21. Pini M, Fantino M, Cavaleri A, et al. Signal quality monitoring applied to spoofing detection. In: Proceedings of the 24th international technical meeting of the satellite division of the Institute of Navigation (ION GNSS'11), September 2011, Portland, OR, USA. 2011. p. 1888–96.
22. Jafarnia-Jahromi A, Lin T, Broumandan A, et al. Detection and mitigation of spoofing attack on a vector based tracking GPS receiver. In: Proceedings of the international technical meeting of the Institute of Navigation, January 2012, Newport Beach, CA, USA. 2012. p. 790–800.
23. White NA, Maybeck PS, Devilbiss SL. Detection of interference/jamming and spoofing in a DGPS-aided inertial system. IEEE Trans Aerosp Electron Syst. 34(4):1208–17.
24. Cheng X, Xu J, Cao K, et al. An authenticity verification scheme based on hidden messages for current civilian GPS signals. In: Proceedings of the 4th international conference on computer sciences and convergence information technology, November 2009, Wuhan, China. 2009. p. 345–52.
25. Humphreys TE. Detection strategy for cryptographic GNSS anti-spoofing. IEEE Trans Aerosp Electron Syst. 2013;49(2):1073–90.
26. Wesson KD, Rothlisberger MP, Humphreys TE. Practical cryptographic civil GPS signal authentication, navigation. J Inst Navig. 2012;59(3):177–93.
27. Hernandez IF, Rijmen V, Granados GS, et al. Design drivers, solutions and robustness assessment of navigation message authentication for the Galileo open service. In: Proceedings of the ION IEEE+meeting. 2014.
28. Kerns AJ, Wesson KD, Humphreys TE. A blueprint for civil GPS navigation message authentication. In: Proceedings of the IEEE/ION PLANS meeting, May 2014.
29. Ledvina BM, Bencze WJ, Galusha B, et al. An in-line anti-spoofing device for legacy civil GPS receivers. In: Proceedings of the Institute of Navigation—international technical meeting. 2010. p. 698–712.
30. Khanafseh S, Roshan N, Langel S, et al. GPS spoofing detection using RAIM with INS coupling, in position, location and navigation symposium-PLANS 2014, 2014 IEEE/ION. IEEE; 2014. p. 1232–9.

31. Daneshmand S, Jafarnia-Jahromi A, Broumandan A, et al. A low complexity GNSS spoofing mitigation technique using a double antenna array. GPS World Mag. 2011;22(12):44–6.
32. Psiaki ML, O'Hanlon BW, Bhatti JA, et al. GPS spoofing detection via dual-receiver correlation of military signals. IEEE Trans Aerosp Electron Syst. 2013;49(4):2250–67.
33. Wesson KD, Shepard DP, Bhatti JA. An evaluation of the vestigial signal defense for civil GPS anti-spoofing. In: Proceedings of the 2011 ION GNSS conference, Portland, OR. 2011. p. 2646–56.
34. Le Liboux JC, Descomps N, Luneau G. Interference signal reduction method and receiver. 2010, No. 8345730 B2.
35. Daneshmand S, Jafarnia-Jahromi A, Broumandan A, et al. A low-complexity GPS anti-spoofing method using a multi-antenna array. In: ION GNSS12 conference, Session B3, Nashville, TN. 2012. p. 1233–43.
36. Zhang Y, Wang L, Wang W, et al. Spoofing jamming suppression techniques for GPS based on DOA estimating. In: Proceedings of 2014 China satellite navigation conference. Berlin, Heidelberg: Springer; 2014. p. 683–94.
37. Li J, Stoica P. Angle and waveform estimation via RELAX. IEEE Trans Aerosp Electron Syst. 1997;33(3):1077–87.
38. Li J, Wu R. An efficient algorithm for time delay estimation. IEEE Trans Signal Process. 1998;46(8):2231–5.
39. Zhang Y, Wang L, Wang W, et al. Spoofing interference suppression for GNSS via estimating steering vectors. In: Proceedings of 2015 China satellite navigation conference. Berlin, Heidelberg: Springer; 2015.
40. Wang L, Wu R, Zhang Y, et al. Multi-type interference suppression for GNSS based on despread-respread method. Tempa, USA: ION GNSS+2015; 2015.
41. Lu D, Wu R, Wang L, et al. A blind adaptive GPS interference method based on code word, China: ZL200910069091.3. 2009.
42. Bao L, Wu R, Lu D, et al. A novel adaptive algorithm for spoofing and jamming suppression in GNSS. J Commun Technol Electron. 2015.
43. Inder JG. SMI adaptive antenna arrays for weak interfering signals. IEEE Trans Antennas Propag. 1986;34(10):1237–42.
44. Wu R, Bao Z, Ma Y. Control of peak sidelobe level in adaptive arrays. IEEE Trans Antennas Propag. 1996;44(10):1341–7.
45. Wu R, Bao Z. Array pattern distortion and remedies in space-time adaptive processing for airborne radar. IEEE Trans Antennas Propag. 1998;46(7):963–70.

Chapter 5
Multipath Interference Suppression

5.1 Introduction

Interferences on GNSS can be categorized into two types: intentional interference and unintentional interference. Jamming and spoofing, mainly discussed in Chaps. 2 and 4 in this book, belong to the category of intentional interference. Multipath interference studied in this chapter, and pulse interference that will be studied in Chap. 6 both belong to unintentional interference.

Multipath interference refers to the interference signals formed by reflected GNSS signals due to objects around the receiver antenna. It is one of the common error sources of GNSS. Multipath interference can change the phase characteristics of the receiver tracking loop, leading to tracking and measurement errors. Studies have shown that the resulting pseudorange errors can reach several to hundreds of meters [1, 2], which are sufficient to degrade system reliability and positioning accuracy. Since the impacts of multipath interferences show significant differences for different observation time and station environment, presently there is not a general and accurate mathematical model to correct multipath errors. Consequently, multipath interference suppression is always a hot research topic in the field of GNSS receiver design [1–4].

Present-day multipath interference suppression technologies mainly expand along two approaches: receiver antenna design and signal processing. The antenna enhancement technologies, from the antenna design perspective, include measures such as drawing a multipath environment diagram around the antenna, using special types of antenna, and selecting proper sites to set up antenna [5–10]. Multipath interference suppression technologies, based on signal processing, include two major categories: temporal domain signal processing and spatial domain signal processing. Commonly used temporal domain processing methods include enhanced correlator techniques represented by the narrow correlator technique [11, 12] and parameter estimation methods represented by the Multipath Estimate Delay Lock Loop (MEDLL) [13, 14]. Spatial signal processing mainly takes advantage of

R. Wu et al., *Adaptive Interference Mitigation in GNSS*, Navigation: Science and Technology, https://doi.org/10.1007/978-981-10-5571-3_5

the fact that the line of sight (LOS) signal and multipath interference arrive at the receiver from different spatial directions, therefore the multipath interference suppression can be achieved using adaptive array signal processing with multiple antennas [15–19].

Focusing on the multipath interference suppression problem for GNSS, this chapter gives the received signal model when multipath interferences exist, and analyzes the impacts on GNSS receivers by multipath interferences. Then we introduce the MEDLL technique based on temporal domain processing, and introduce a new temporal domain multipath suppression method based on Weighted Fourier Transform and RELAXation (WRELAX) parameter estimation. Lastly, by taking advantage of the periodic repetitive characteristic of GNSS spreading codes, we introduce a multipath interference suppression technique based on spatial domain decoupled parameter estimation theory.

5.2 Multipath Signal Model and Impact Analyses

To study effective methods to counter multipath interferences, we must establish a received signal model where multipath interferences exist, and analyze the impacts of multipath interferences on the GNSS signal. To facilitate discussion, we use the GPS signal as an example, and the related conclusions can be generalized to other GNSS systems.

5.2.1 *Signal Model*

The transmitting GNSS signal can be represented as

$$s(t) = D(t)c(t)\cos(\omega_c t) \tag{5.1}$$

where $D(t)$ denotes the navigation message of the GNSS signal; $c(t)$ represents the C/A code; ω_c is the carrier frequency. Assuming that a GPS receiver receives a LOS signal combined with multipath interferences from P reflection paths, then the received signal can be represented as

$$\bar{\bar{x}}(t) = \sum_{p=0}^{P} \tilde{\alpha}_p D(t-\tau_p)c(t-\tau_p)\cos\left(\omega_c(t-\tau_p)+\tilde{\varphi}_p\right) \tag{5.2}$$

where $\tilde{\alpha}_p$, τ_p , $\tilde{\varphi}_p$ represent the amplitude, code delay and initial phase for the pth multipath signal; $p=0$ represents the LOS signal arrived through a direct path. Since the period of the navigation message is much longer than the period of the C/A code, and the data block length required by the subsequent code delay estimation is relatively short (only few C/A code periods needed). During the data duration, the navigation message $D(t - \tau_p) = \pm 1$, and assuming no data bit change, we can further merge the value of $D(t - \tau_p)$ in (5.2) within the amplitude $\tilde{\alpha}_p$, and the new variable can be denoted as α'_p. Then, (5.2) is converted to

$$\bar{\bar{x}}(t) = \sum_{p=0}^{P} \alpha'_p c(t - \tau_p) \cos(\omega_c (t - \tau_p) + \tilde{\varphi}_p) \tag{5.3}$$

To derive the Doppler frequencies for signals arrived through different paths, we need to consider the Doppler frequency of the LOS signal. We can observe that

$$\omega_c \tau_0 = \omega_c \frac{R_0(t)}{c} \tag{5.4}$$

where c is the light speed; $R_0(t) = r_0 + v_0 t$ represents the distance between the receiver and the satellite at time t; r_0 is the distance between the receiver and the satellite at the initial observational time; v_0 represents the redial relative speed between the receiver and the satellite. The above formula can be further expressed as

$$\omega_c \tau_0 = \omega_c \frac{r_0 + v_0 t}{c} = 2\pi \left(\frac{r_0}{\lambda} + \frac{v_0 t}{\lambda} \right) \tag{5.5}$$

When the redial relative speed v_0 is a constant, the Doppler frequency is

$$\omega_{d0} = 2\pi \frac{v_0}{\lambda} \tag{5.6}$$

The propagation distance of a multipath interference signal can be expressed as

$$R_p(t) = r_0 + v_0 t + \Delta R_p(t) \tag{5.7}$$

where $\Delta R_p(t)$ is the propagation distance difference between a multipath interference and the LOS signal. When the distance between the reflecting point and the receiver is far away so the code delay between the multipath interference and the LOS signal is more than 1.5 chips, the multipath interference's impacts on the receiver can be ignored. Since the satellite is far away from the receiver on the ground, by assuming that the multipath reflecting points are not too far from the receiver, we can assume that $\Delta R_p(t)$ does not change during a short duration. Then $\omega_c \tau_p$ can be further expressed as

$$\omega_c \tau_p = \omega_c \frac{r_0 + v_0 t + \Delta R_p}{c} = 2\pi \left(\frac{r_0}{\lambda} + \frac{v_0 t}{\lambda} + \frac{\Delta R_p}{\lambda} \right) \tag{5.8}$$

It can be seen that the Doppler frequency of the pth multipath interference $\omega_{dp} \approx 2\pi v_0/\lambda$, i.e. the LOS signal and the multipath interference approximately have the same Doppler frequency, and we can denote them using a unified representation ω_{d0}. Consequently (5.3) can be further simplified as

$$\bar{\bar{x}}(t) = \sum_{p=0}^{P} \alpha'_p c(t - \tau_p) \cos\left[(\omega_c + \omega_{d0})t + \varphi'_p\right] \tag{5.9}$$

where $\varphi'_0 = \tilde{\varphi}_0 + 2\pi r_0/\lambda$ is the phase of the LOS signal; $\varphi'_p = \varphi'_0 + \Delta\varphi_p$, $(p = 1, 2, \ldots, P)$ is the phase of the pth multipath signal; $\Delta\varphi_p = 2\pi\Delta R_p/\lambda$ is the additional phase difference between the pth multipath interference and the LOS signal.

5.2.2 *Impacts of Multipath Interferences*

Assuming that the carrier recovery is entirely accurate, the signal after the carrier loop demodulation can be represented as

$$\bar{x}(t) = \sum_{p=0}^{P} \alpha'_p c(t - \tau_p) \cos\left(\varphi'_p - \hat{\varphi}'_0\right) \tag{5.10}$$

where $\hat{\varphi}'_0$ represents the estimated carrier phase of the LOS signal. If the interval between the early code correlator and the late code correlator is d, then the locally generated early and late codes can be represented, respectively as

$$s_E(t) = c(t - \hat{\tau}_0 - d/2) \tag{5.11}$$

$$s_L(t) = c(t - \hat{\tau}_0 + d/2) \tag{5.12}$$

where $\hat{\tau}_0$ represents the estimated code delay of the LOS signal. After the demodulation, the correlations between the signal and the locally generated early/late codes $R_E(\varepsilon)$ and $R_L(\varepsilon)$ can be represented, respectively, as [4]:

$$R_E(\varepsilon) = \sum_{p=0}^{P} \alpha'_p R(\varepsilon + \Delta\tau_p - d/2) \cos\left(\varphi'_p - \hat{\varphi}'_0\right) \tag{5.13}$$

$$R_L(\varepsilon) = \sum_{p=0}^{P} \alpha'_p R(\varepsilon + \Delta\tau_p + d/2) \cos\left(\varphi'_p - \hat{\varphi}'_0\right) \tag{5.14}$$

where $R(\cdot)$ represents the autocorrelation function of the C/A code; $\varepsilon = \tau_0 - \hat{\tau}_0$ is the code delay estimation error of the LOS signal; $\Delta\tau_p = \tau_p - \tau_0$ is the relative code delay between the pth multipath interference and the LOS signal. The discrimination function of the code tracking loop in the GNSS receiver can be represented as [3, 4]

$$\begin{aligned} f(\varepsilon) &= R_E(\varepsilon) - R_L(\varepsilon) \\ &= \sum_{p=0}^{P} \alpha'_p \left[R(\varepsilon + \Delta\tau_p - d/2) - R(\varepsilon + \Delta\tau_p + d/2) \right] \cos\left(\varphi'_p - \hat{\varphi}'_0 \right). \end{aligned} \quad (5.15)$$

In a code tracking loop, if no multipath interference exists, the positions where the tracking errors $\varepsilon = 0$ correspond to the zeros value of the discrimination function. When the discrimination function value equals to zero, the correlation values of the early and late codes are the same, and the correlation value of the prompt code, which is located in the intermediate of the early and late codes, reaches its maximum. Consequently the code tracking loop considers that the local generated C/A code is synchronized with the received one. Therefore, adjusting the local C/A code phase by the discrimination function's zero values, we can ensure the correlation values of the early and late codes are always the same. In other words, the zeros of the discrimination function always correspond to the zero tracking errors.

When multipath interferences exist, the correlation function becomes a synthesis of the LOS signal and multipath interferences. When the discrimination value is zero, the correlation values for the early and late codes are the same, and the correlation value of the prompt code, which is located at the intermediate position between the early and late codes, deviates from the maximum. Under the condition, the prompt code is not aligned with the position where the signal code delay is zero. Consequently, distorted correlation function forces the zeros of the discrimination function to deviate from zero tracking errors, instead these zeros correspond to tracking errors introduced by multipath interferences. As a result, when $f(\varepsilon) = 0$, the corresponding ε is the tracking error introduced by a multipath interference signal. Figures 5.1 and 5.2 show the impacts on correlation functions and phase discriminator curves by a constructive multipath signal (where the multipath signal and the LOS signal have the same phase or they are partially out of phase, i.e. when $0^\circ \le |\varphi'_1 - \hat{\varphi}'_0| \le 90^\circ$, leading to superimposed signal strength enhancement) and destructive signal (when $90^\circ < |\varphi'_1 - \hat{\varphi}'_0| \le 180^\circ$, leading to superimposed signal strength reduction).

When there is only the LOS signal, the corresponding error for the phase discriminator function's zero-crossing point is $\varepsilon = 0$. A Delay Lock Loop (DLL) locks the signal by tracking the zero-crossing point of a phase discriminator function. But when multipath interference are received, due to the impacts from multipath interferences, the zero-crossing point of the phase discriminator function deviates from the LOS signal's code delay. When a receiver only receives one multipath

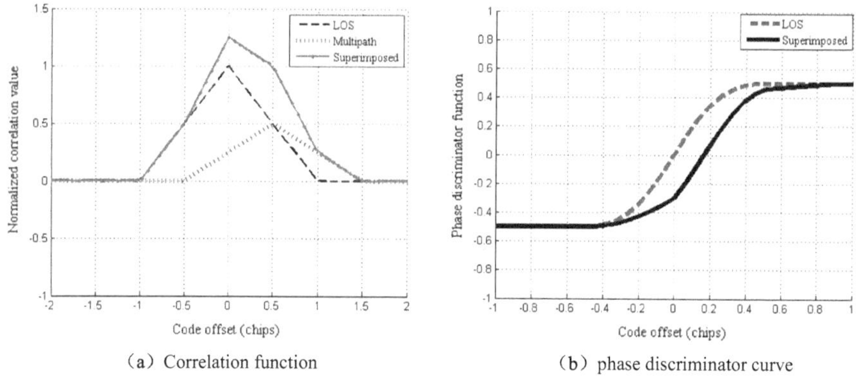

Fig. 5.1 Impacts of a constructive multipath interference signal

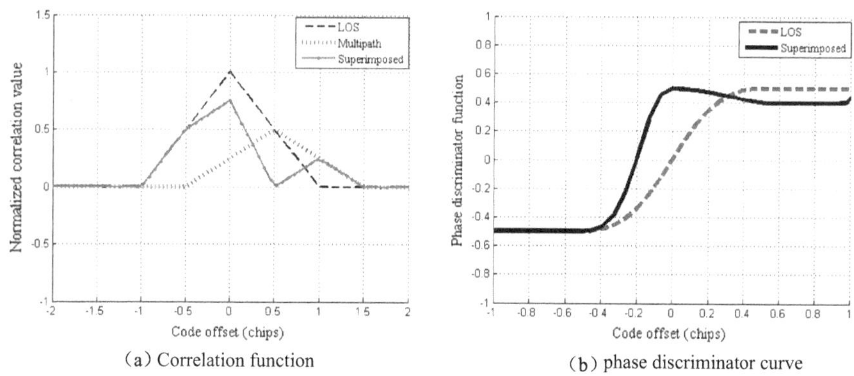

Fig. 5.2 Impacts of a destructive multipath interference signal

interference by making the discriminator function in (5.15) $f(\varepsilon) = 0$, the corresponding DLL error can be derived [3, 4] as

$$\varepsilon = \begin{cases} \dfrac{\Delta\tau_1 \cos\left(\varphi_1' - \hat{\varphi}_0'\right)}{\alpha_0'/\alpha_1' + \cos\left(\varphi_1' - \hat{\varphi}_0'\right)} & 0 < \Delta\tau_1 \leq \tau_L \\ \dfrac{d}{\alpha_0'/\alpha_1'} \cos\left(\varphi_1' - \hat{\varphi}_0'\right) & \tau_L < \Delta\tau_1 \leq \tau_H \\ \dfrac{\cos\left(\varphi_1' - \hat{\varphi}_0'\right)}{\alpha_0'/\alpha_1' - \cos\left(\varphi_1' - \hat{\varphi}_0'\right)} (d/2 + T_c - \Delta\tau_1) & \tau_H < \Delta\tau_1 \leq T_c + d/2 \\ 0 & \Delta\tau_1 > T_c + d/2 \end{cases} \tag{5.16}$$

Fig. 5.3 The error envelope curve of a traditional correlator

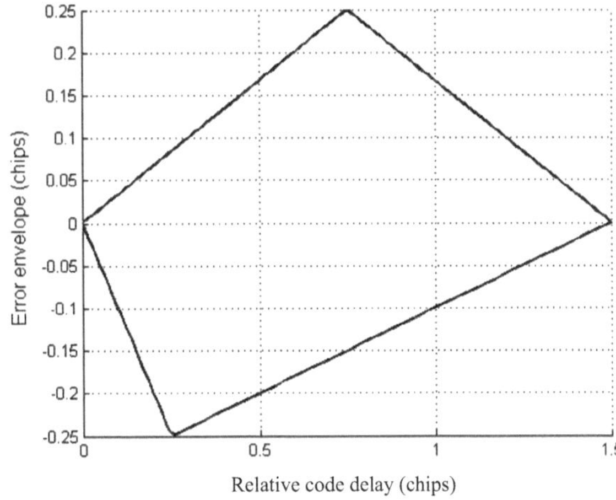

where T_c represents the chip length, then τ_L and τ_H can be represented as

$$\begin{cases} \tau_L = \dfrac{\alpha_0'/\alpha_1' + \cos\left(\varphi_1' - \hat{\varphi}_0'\right)}{\alpha_0'/\alpha_1'} d \\ \tau_H = \dfrac{d \cos\left(\varphi_1' - \hat{\varphi}_0'\right)}{2\alpha_0'/\alpha_1'} + T_c - d/2 \end{cases} \tag{5.17}$$

Assuming that only one multipath interference exists, and the Direct to Multipath Ratio (DMR) $\alpha_0'/\alpha_1' = 0.5$, in (5.16), when $\varphi_1' - \hat{\varphi}_0'$ is 0 or π (i.e. the multipath and LOS signal have the same phase or the opposite phase), by forcing the discriminant function $f(\varepsilon) = 0$, the possible maximum positive/negative multipath errors obtained by solving the formula are defined as the error envelope. Figure 5.3 shows the error envelope of the traditional correlator (i.e. correlator interval $d = 1$). When multipath interference exist, traditional correlators have greater code phase tracking errors.

5.3 Temporal Domain Multipath Interference Suppression

After analyzing the impacts on GNSS system by multipath interferences in previous sections, this section mainly discusses techniques of multipath interference suppression.

5.3.1 Narrow Correlator Technique

Narrow correlator technique is a very first multipath interference suppression technique being introduced. Its application on GPS receivers was introduced by Van Dierendonck et al. [12] from NovAtel Company. The main launching point of the narrow correlator technique is that the impacts from multipath interferences are the weakest towards the peak of a pseudo code correlation function curve. Consequently, "narrow correlator" usually means that the spacing between correlators is smaller than one chip. The arrangement is different from what is for the traditional correlator, which has $d = 1$ chip between early/late code correlators. Usually narrow correlators set the spacing between correlators as $d = 0.1$ chips [11, 12]. The multipath error envelope for narrow correlators with various spacing is shown in Fig. 5.4.

Figure 5.4 shows that multipath error envelopes become smaller as the correlators spacing becomes smaller. Compared with the traditional correlator technique, the narrow correlator technique has significantly multipath suppression effect. The maximum multipath error is proportional to the spacing d between correlators. Consequently, a narrow correlator with $d = 0.1$ can reduce the maximum multipath error to 10% [11, 12]. Another benefit of the narrow correlator technique is that when the multipath interference's code delay is longer than $1 + d/2$ chips, there is no impact on signal tracking. For traditional correlators, signal tracking is not impacted only when the multipath interference's code delay is longer than 1.5 chips.

The principle of narrow correlators' multipath interference suppression can also be explained from the perspective of discriminant functions. Figure 5.5 shows the impacts on the incoherent phase discriminator curve by correlator spacing. It can be seen that the linear operating range becomes smaller as the correlator spacing becomes smaller. And this means that a correlator spacing reduction decreases a code tracking loop's tolerance on noise and user dynamics. But, we need to point

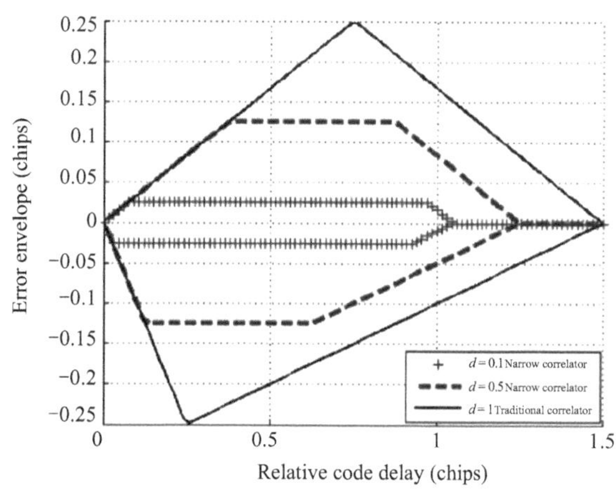

Fig. 5.4 Error envelopes for narrow correlators with various spacing

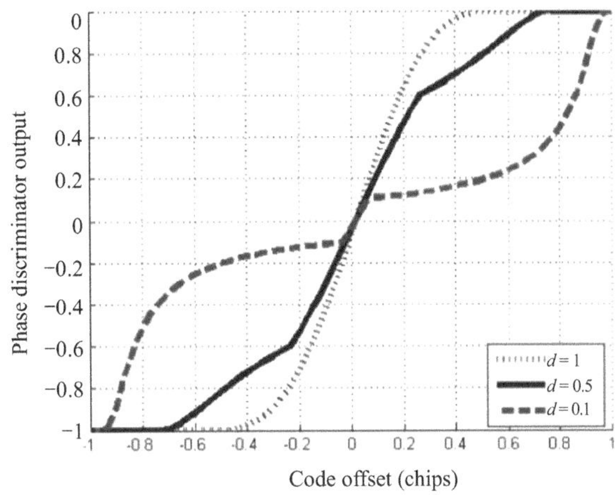

Fig. 5.5 Impacts on non-coherent phase discriminator curves by different correlator spacing

out that even though the narrow correlator technique can greatly suppress code tracking errors caused by multipath interference and improve measurement precision on code phase, it has no multipath suppression effect for the carrier tracking loop [11, 12].

In actual receivers, the front end filter has limited bandwidth so it can filter out sidelobes of signal power spectrum. The Fourier transform of an autocorrelation function is the power spectrum, so the bandwidth of the front end filter has impacts on the autocorrelation function of the signal. A narrower bandwidth leads to a flatter correlation peak for a C/A code autocorrelation function curve, as shown in Fig. 5.6. In this flat region, when a correlation value changes not sufficiently obvious, the phase discriminator output does not change, leading to a reduction on phase discrimination sensitivity, and a larger phase discrimination error. That has

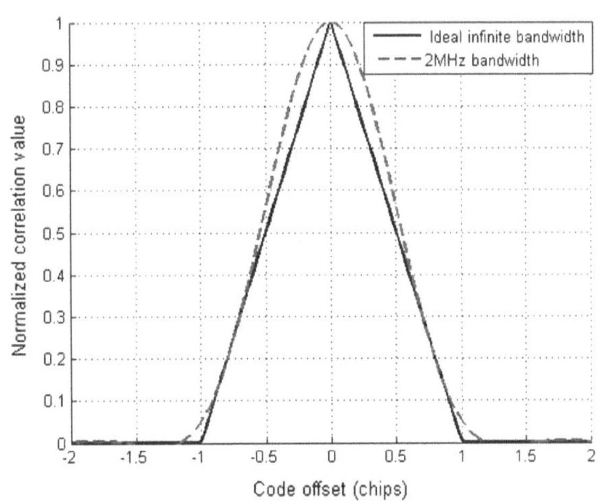

Fig. 5.6 Impacts on C/A code autocorrelation function by receiver bandwidth

serious impacts on the signal tracking sensitivity of the code tracking loop. Consequently, the narrow correlator technique requires a larger front end bandwidth.

As described before, to accurately track the spreading code, it is required to increase receiver bandwidth and increase the sampling frequency of a receiver's A/D converter. But, wider bandwidth could bring in more noise, making it more susceptible to RF interferences. That imposes higher requirements on the performance of a receiver's RF front end filter. For RF front end, filters with a flat passband, a steep transition zone and good filtering performance need to be selected, to ensure the spreading code autocorrelation function being formed has good shape. In general, narrow correlators, designed on the basis of the code tracking loop used by present-day receivers, can dramatically improve multipath suppression performance by adjusting correlator spacing, and have good compatibility with present-day receivers. Consequently narrow correlators have been widely applied.

5.3.2 *Multipath Estimate Delay-Lock-Loop (MEDLL)*

MEDLL is a multipath suppression technique founded on statistics theories [13]. The main principle is to obtain multiple correlation function samples using multiple correlators. By performing iterative computations based on the maximum likelihood criterion, it can achieve smaller code phase tracking errors, compared to what the narrow correlator technique can obtain. For the rest of this section, we introduce the basic principle of MEDLL.

In reality, the signal in (5.10) also includes a noise component $\tilde{e}(t)$ (which can usually be assumed as a white Gaussian noise), so the received signal containing noise can be represented as

$$x(t) = \sum_{p=0}^{P} \alpha'_p c(t - \tau_p) \cos\left(\varphi'_p - \hat{\varphi}'_0\right) + \tilde{e}(t) \tag{5.18}$$

By making $\varphi_p = \varphi'_p - \hat{\varphi}'_0$, and performing Hilbert transformation on the data, the corresponding complex data is

$$x(t) = \sum_{p=0}^{P} \alpha'_p c(t - \tau_p) \exp(\mathrm{j}\varphi_p) + e(t) \tag{5.19}$$

where $e(t)$ is the result of performing Hilbert transformation on $\tilde{e}(t)$. By further sampling the data in (5.19), we can obtain

$$x(n) = \sum_{p=0}^{P} \alpha'_p c(n - \tau_p) \exp(j\varphi_p) + e(n), \quad n = 1, 2, \ldots, N \tag{5.20}$$

where N denotes the total number of sample points.

Based on the maximum likelihood estimation theory, and under the condition of an additive Gaussian white noise background, to apply the maximum likelihood criterion is equivalent to solve the following nonlinear least-squares problem

$$L_1\left(\left\{\alpha'_p, \tau_p, \varphi_p\right\}_{p=0}^{P}\right) = \sum_{n=1}^{N} [x(n) - \xi(n)]^2 \tag{5.21}$$

where

$$\xi(n) \triangleq \sum_{p=0}^{P} \xi_p(n) = \sum_{p=0}^{P} \alpha'_p c(n - \tau_p) \exp(j\varphi_p) \tag{5.22}$$

Since the cost function $L_1\left(\left\{\alpha'_p, \tau_p, \varphi_p\right\}_{p=0}^{P}\right)$ contains unknown parameters for $(P+1)$ signals, so it cannot be solved directly. For this reason, we can first assume that other than the unknown parameters for the pth signal, parameters for all the other signals going through different paths $\left\{\hat{\alpha}'_q, \hat{\omega}_q, \hat{\varphi}_q\right\}_{q=0, q\neq p}^{P}$ are known or have already been estimated. By defining

$$x_p(n) = x(n) - \sum_{\substack{q=0 \\ q\neq p}}^{P} \hat{\alpha}'_q c(n - \hat{\tau}_p) \exp(j\hat{\varphi}_p) \tag{5.23}$$

and substituting the above formula into (5.21), the following cost function can be obtained

$$L_2\left(\alpha'_p, \tau_p, \varphi_p\right) = \sum_{n=1}^{N} \left[x_p(n) - \xi_p(n)\right]^2 \tag{5.24}$$

The pth signal's parameters can be estimated by calculating the partial derivative of $L_2\left(\alpha'_p, \tau_p, \varphi_p\right)$, and make the corresponding partial derivative as 0, i.e. let $\frac{\partial L_2\left(\alpha'_p, \tau_p, \varphi_p\right)}{\partial \alpha'_p} = 0$, $\frac{\partial L_2\left(\alpha'_p, \tau_p, \varphi_p\right)}{\partial \tau_p} = 0$, $\frac{\partial L_2\left(\alpha'_p, \tau_p, \varphi_p\right)}{\partial \varphi_p} = 0$, whereby

$$\hat{\tau}_p = \arg\max_{\tau}\left\{ \mathrm{Re}\left\{ \left[R_x(\tau) - \sum_{\substack{k=0 \\ k \neq p}}^{P} \alpha'_k R(\tau - \tau_k) \exp(\mathrm{j}\varphi_k) \right] \right\} \exp(-\mathrm{j}\varphi_p) \right\} \tag{5.25}$$

$$\hat{\alpha} = \mathrm{Re}\left\{ \left[R_x(\tau_p) - \sum_{\substack{k=0 \\ k \neq p}}^{P} \alpha'_k R(\tau_p - \tau_k) \exp(\mathrm{j}\varphi_k) \right] \exp(-\mathrm{j}\varphi_p) \right\} \tag{5.26}$$

$$\hat{\varphi}_p = \mathrm{angle}\left[R_x(\tau_p) - \sum_{\substack{k=0 \\ k \neq p}}^{P} \alpha'_k R(\tau_p - \tau_k) \exp(\mathrm{j}\varphi_k) \right] \tag{5.27}$$

where $R(\tau)$ is the C/A code's autocorrelation function, with the maximum normalized amplitude is 1, and the relative code delay is 0, so it can be used as the autocorrelation function for reference (also known as the reference correlation function). $R_x(\tau)$ is the actual autocorrelation by actual measurement, i.e. the calculated correlation between the received signal and the local reference signal.

MEDLL needs to obtain estimation values for signals arrived from different paths through iterations. When there are only the LOS signal and one multipath interference, the iterative procedure can be described as below

(1) Assuming the received data only contains the LOS signal, i.e. $x(n) = \xi_0(n)$, then its estimated correlation is $\hat{R}_0(\tau) = R_x(\tau)$. By sending $\hat{R}_0(\tau)$ into the parameter estimation module, a set of parameters $(\hat{\alpha}'_0, \hat{\tau}_0, \hat{\varphi}_0)$ can be estimated. Based on the parameter set, and the reference correlation function, the correlation derivation can be further configured as $\hat{R}_0(\tau) = \hat{\alpha}'_0 R(\tau - \hat{\tau}_0) \exp(\mathrm{j}\hat{\varphi}_0)$.
(2) By subtracting the reconstructed correlation function $\hat{R}_0(\tau)$ from the actual measured correlation function $R_x(\tau)$, the difference can be treated as the correlation function of the second path's signal $\xi_1(n)$, i.e. $\hat{R}_1(\tau) = R_x(\tau) - \hat{R}_0(\tau)$. By substituting it into the parameter estimation module, a set of estimated parameters $(\hat{\alpha}'_1, \hat{\tau}_1, \hat{\varphi}_1)$ for the second path's signal $\xi_1(n)$ can be obtained. Similarly, based on this set of estimated results, the correlation function of $\xi_1(n)$, can be reconstructed from $\hat{R}_1(\tau) = \hat{\alpha}'_1 R(\tau - \hat{\tau}_1) \exp(\mathrm{j}\hat{\varphi}_1)$.
(3) By subtracting the reconstructed correlation function $\hat{R}_1(\tau)$ from the actual measured correlation function $R_x(\tau)$, the difference obtained can be treated as the correlation function of $\xi_0(n)$, i.e. $\hat{R}_0(\tau) = R_x(\tau) - \hat{R}_1(\tau)$. By substituting the correlation function into the parameter estimation module, a set of parameters $(\hat{\alpha}'_0, \hat{\tau}_0, \hat{\varphi}_0)$ can be estimated. Using the set of estimated results, the correlation function $\hat{R}_0(\tau)$ for $\xi_0(n)$ can be reconstructed.

(4) Repeat steps (2) and (3) to complete iteration optimization of the signals on the two paths till the iteration condition cannot be met, then jump out of the loop.

The above iterative procedure can be directly extended to scenarios where multiple multipath interferences exist. But under normal circumstances, assuming the conditions of longer propagation code delay and larger number of reflections, the subsequent overall multipath interferences become weaker. In addition, a larger number of paths being tracked increase the computation loads. As a result, we usually only consider the LOS signal and one multipath interference.

Noticing that for each iteration (corresponding to estimating the parameters on one multipath interference signal), mutual couplings among unknown parameters in (5.25)–(5.27) are hard to be solved. Consequently, the following processing is performed.

To obtain $\left(\hat{\alpha}'_p, \hat{\tau}_p, \hat{\varphi}_p\right)$, at first the sampling point $R_p(\tau_{\max})$ corresponding to the maximum of the correlation function $R_p(\tau)$ is found. For a complex domain model, the maximum of the function $\mathrm{Re}^2\left[R_p(\tau)\right] + \mathrm{Im}^2\left[R_p(\tau)\right]$ should be found, the τ corresponding to the maximum value is $\tau_{\max}$ (e.g. which is the roughly estimated code delay). The process can be expressed as below

$$\tau_{\max} = \arg\max_{\tau}\left\{\left[\mathrm{Re}\left(R_p(\tau)\right)\right]^2 + \left[\mathrm{Im}\left(R_p(\tau)\right)\right]^2\right\} \tag{5.28}$$

Then we can estimate the phase $\hat{\varphi}_p$ based on the formula below

$$\hat{\varphi}_p = \mathrm{atan}\left(\frac{\mathrm{Im}\left[R_p(\tau_{\max})\right]}{\mathrm{Re}\left[R_p(\tau_{\max})\right]}\right), \quad \text{when } \mathrm{Re}\left[R_p(\tau_{\max})\right] > 0 \tag{5.29}$$

$$\hat{\varphi}_p = \mathrm{atan}\left(\frac{\mathrm{Im}\left[R_p(\tau_{\max})\right]}{\mathrm{Re}\left[R_p(\tau_{\max})\right]}\right) + \pi, \quad \text{when} \quad \mathrm{Re}\left[R_p(\tau_{\max})\right] < 0 \tag{5.30}$$

$$\hat{\varphi}_p = \frac{\pi}{2}, \quad \text{when} \quad \mathrm{Re}\left[R_p(\tau_{\max})\right] = 0 \quad \text{and} \quad \mathrm{Im}\left[R_p(\tau_{\max})\right] > 0 \tag{5.31}$$

$$\hat{\varphi}_p = -\frac{\pi}{2}, \quad \text{when} \quad \mathrm{Re}\left[R_p(\tau_{\max})\right] = 0 \quad \text{and} \quad \mathrm{Im}\left[R_p(\tau_{\max})\right] < 0 \tag{5.32}$$

After obtaining the phase $\hat{\varphi}_p$, a phase rotation is performed on the correlation function $R_p(\tau)$ to obtain

$$\bar{R}_p(\tau) = \mathrm{Re}\left[R_p(\tau)\right]\cos\left(\hat{\varphi}_p\right) + \mathrm{Im}\left[R_p(\tau)\right]\sin\left(\hat{\varphi}_p\right) \tag{5.33}$$

At this point, the correlation function is a real-value correlation function that can be further used to estimate code delay and amplitude.

Under most circumstances, the code delay of the LOS path is not equal to an integer multiple of the correlator spacing. The actual obtained complex sampling points might not include the maximum points of the actual correlation function. To obtain more accurate estimation on the code delay, interpolation can be performed on the actual measured correlation function, and then obtain a discriminant function $f(\tau_x)$ based on the computation. Furthermore the estimated code delay can be obtained by a lookup table on pre-stored $f'(\tau_x')$ values. The details can be expressed below:

As shown in Fig. 5.7, if A and B are the top two maximum sampling points for $R(\tau')$ (located near the actual peak value), the corresponding code delays can be represented, respectively, as τ_a' and τ_b'. Given that the correlator spacing is $\Delta\tau$, i.e. $\tau_b' - \tau_a' = \Delta\tau$, and the code delay corresponding to the correlation function's peak value is denoted as τ_0', it can be observed that $\tau_c' = (\tau_a' + \tau_b')/2$, $\tau_0' = \tau_c' + \tau_x'$. Consequently the needed code delay can be derived by obtaining τ_x'.

For that purpose, firstly $f'(\tau_x')$ can be obtained using the reference correlation function $R(\tau)$

$$f'(\tau_x') = \frac{R(\tau_0' - \tau_x' - \frac{1}{2}\Delta\tau) - R(\tau_0' - \tau_x' + \frac{1}{2}\Delta\tau)}{|R(\tau_0' - \tau_x' - \frac{1}{2}\Delta\tau)| + |R(\tau_0' - \tau_x' + \frac{1}{2}\Delta\tau)|} \tag{5.34}$$

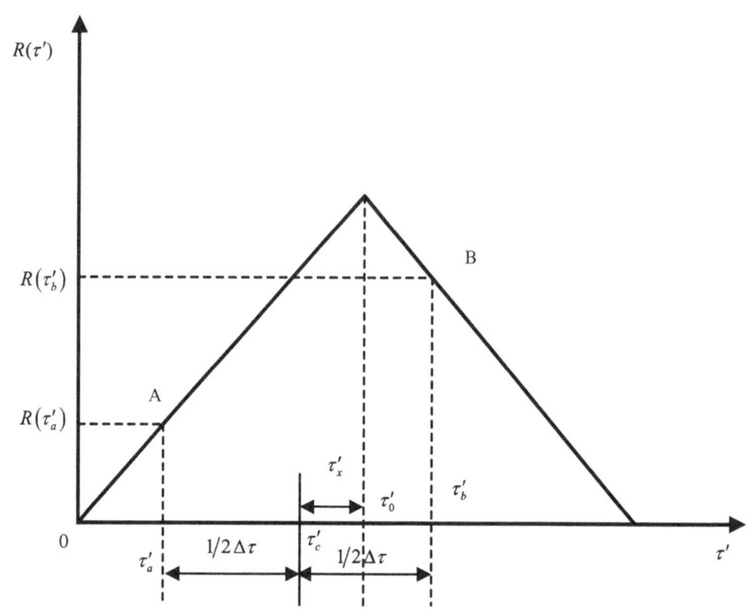

Fig. 5.7 Diagram of correlator sampling point positions

As shown in Fig. 5.7, $\tau'_a = \tau'_0 - \tau'_x - \frac{\Delta\tau}{2}$; $\tau'_b = \tau'_0 - \tau'_x + \frac{\Delta\tau}{2}$; for a given τ'_0, the range of τ'_x is $\tau'_0 - \frac{\Delta\tau}{2} \le \tau'_x \le \tau'_0 + \frac{\Delta\tau}{2}$, and furthermore the τ'_a, τ'_b and the corresponding $f'(\tau'_x)$ can be determined. And we pre-store the $f'(\tau'_x)$ as the reference discriminant function, ready to be used for the lookup table.

The discriminant function obtained based on the actual correlation function after the phase rotation $\bar{R}_p(\tau)$ is

$$f(\tau_x) = \frac{\bar{R}_p(\tau_a) - \bar{R}_p(\tau_b)}{|\bar{R}_p(\tau_a)| + |\bar{R}_p(\tau_b)|} \tag{5.35}$$

where τ_a and τ_b correspond to the τ'_a and τ'_b in $f'(\tau'_x)$. If the corresponding code delay for actual measured correlation function's peak value is denoted as τ_0, the code delay is the one to be estimated. By comparing $f(\tau_x)$ to every $f'(\tau'_x)$ in (5.34), when $\mathrm{abs}[f(\tau_x) - f'(\tau'_x)]$ reaches the minimum value, approximately it can be regarded that $\tau_x = \tau'_x$, and $\hat{\tau}_0 = \tau_c + \tau_x$ can be estimated.

In the process of phase estimation, we have already found the sampling point $\tau_{\max}$ corresponding to the maximum value. Consequently the amplitude can be estimated using $\hat{\alpha}'_p = \bar{R}_p(\tau_{\max})$. This is a method that can be applied when the correlator spacing is small, and the subsequent error is tolerable. The amplitude can also be determined using (5.26).

It can be seen that the solution process of MEDLL is to perform nonlinear curve fitting, by identifying a set of reference correlation functions, and making their amplitude, phase and code delay as the best fit for the correlation function of the input signal [13]. Essentially, a traditional receiver does the same thing, but only on one signal (the LOS signal). When multipath interferences exist, MEDLL can improve the performance of curve fitting by adding the number of estimated signals, thereby it can accurately estimate the multipath interference and then separate them from the superimposed signal. But since the MEDLL needs multiple correlators, it is more complex to implement compared to traditional receivers.

5.3.3 *Multipath Interference Suppression Based on WRELAX*

The WRELAX algorithm based on nonlinear least-squares criterion is an effective time delay estimation algorithm. It converts a multi-dimensional problem into a series of one-dimensional optimizations, significantly reducing the computation load [20–33]. The WRELAX is also a method based on the decoupled parameter estimation theory. We can expand the application of the WRELAX algorithm to GNSS multipath interference suppression. The main idea is that after acquiring the GNSS signal, the WRELAX algorithm can be used to estimate GNSS signal's code delay, and then the estimated code delay information can be used to suppress multipath interferences.

By combing the phases $\exp(j\varphi_p)$ and α'_p in (5.20) into a new unknown variable, denoted as α_p, (5.20) can be represented as

$$x(n) = \sum_{p=0}^{P} \alpha_p c(n - \tau_p) + e(n) \tag{5.36}$$

The parameters to be estimated for the above formula are $\{\alpha_p, \tau_p\}_{p=0}^{P}$. To facilitate further analyses and estimate the code delay parameters, a discrete Fourier transform is performed on the above formula and we can have

$$X(k) = C(k) \sum_{p=0}^{P} \alpha_p e^{j\omega_p k} + E(k), \quad -N/2 \le k \le N/2 - 1 \tag{5.37}$$

where $X(k)$, $C(k)$ and $E(k)$ are the discrete Fourier transformations of $x(n)$, $c(n)$ and $e(n)$ respectively; $\omega_p = -2\pi\tau_p f_s/N$; f_s denotes the sampling frequency. Now, by converting the code delay estimate to an angular frequency estimate, and using the relationship $\tau_p = -\omega_p N/2\pi f_s$, the code delay can be obtained.

The estimated value $\{\hat{\alpha}_p, \hat{\omega}_p\}_{p=0}^{P}$ can be obtained by minimizing the nonlinear least squares cost function

$$Q_1\left(\{\alpha_p, \omega_p\}_{p=0}^{P}\right) = \sum_{k=-N/2}^{N/2-1} \left| X(k) - C(k) \sum_{p=0}^{P} \alpha_p e^{j\omega_p k} \right|^2 \tag{5.38}$$

To achieve that, at first we let

$$\mathbf{C} = \text{diag}[C(-N/2), \quad C(-N/2+1), \quad \ldots, \quad C(N/2-1)] \tag{5.39}$$

$$\mathbf{X} = [X(-N/2), \quad X(-N/2+1), \quad \ldots, \quad X(N/2-1)]^{\mathrm{T}} \tag{5.40}$$

$$\mathbf{a}(\omega_p) = \left[e^{j\omega_p(-N/2)}, \quad e^{j\omega_p(-N/2+1)}, \quad \ldots, \quad e^{j\omega_p(N/2-1)}\right]^{\mathrm{T}} \tag{5.41}$$

Then the minimization described in formula (5.38) is equivalent to minimize the cost function below

$$Q_1\left(\{\alpha_p, \omega_p\}_{p=0}^{P}\right) = \left\| \mathbf{X} - \sum_{p=0}^{P} \alpha_p \mathbf{C}\mathbf{a}(\omega_p) \right\|^2 \tag{5.42}$$

Assuming that $\left\{\hat{\alpha}_q, \hat{\omega}_q\right\}_{q=0,q\neq p}^{P}$ is known or has already been estimated, we can define

$$\mathbf{X}_p = \mathbf{X} - \sum_{\substack{q=0 \\ q\neq p}}^{P} \hat{\alpha}_q\left[\mathbf{Ca}\left(\hat{\omega}_p\right)\right] \tag{5.43}$$

By substituting the above formula into (5.42), we can obtain

$$Q_2\left(\alpha_p, \omega_p\right) = \left\|\mathbf{X}_p - \alpha_p\mathbf{Ca}\left(\omega_p\right)\right\|^2 \tag{5.44}$$

By minimizing the above cost function, we can obtain

$$\hat{\omega}_p = \arg\max_{\omega_p}\left|\mathbf{a}^{\mathrm{H}}\left(\omega_p\right)\left(\mathbf{C}^*\mathbf{X}_p\right)\right|^2 \tag{5.45}$$

$$\hat{\alpha}_p = \left.\frac{\mathbf{a}^{\mathrm{H}}\left(\omega_p\right)\left(\mathbf{C}^*\mathbf{X}_p\right)}{\|\mathbf{C}\|_F^2}\right|_{\omega_p=\hat{\omega}_p} \tag{5.46}$$

where $\|\cdot\|_F^2$ represents the Frobenius norm. It can be seen from (5.45) that, $\hat{\omega}_p$ can be obtained by finding the position corresponding to the maximum of the periodogram $\left|\mathbf{a}^{\mathrm{H}}\left(\omega_p\right)\left(\mathbf{C}^*\mathbf{X}_p\right)\right|^2$. And this can be achieved by using a one-dimension zero-padding FFT operation. $\hat{\alpha}_p$ is the complex amplitude of $\frac{\mathbf{a}^{\mathrm{H}}\left(\omega_p\right)\left(\mathbf{C}^*\mathbf{X}_p\right)}{\|\mathbf{C}\|_F^2}$.

For the scenario with the LOS signal and one multipath interference, the estimation of $\left\{\hat{\alpha}_p, \hat{\omega}_p\right\}_{p=0}^{1}$ can be implemented using the following steps:

(1) Firstly assuming only the LOS signal exists, by using $\mathbf{X}$ and based on (5.45) and (5.46), $\{\hat{\alpha}_0, \hat{\omega}_0\}$ can be estimated.
(2) Using the calculated $\{\hat{\alpha}_0, \hat{\omega}_0\}$, calculate $\mathbf{X}_1$ based on (5.43); then using $\mathbf{X}_1$ to estimate $\{\hat{\alpha}_1, \hat{\omega}_1\}$ based on (5.45) and (5.46).

Subsequently we can recalculate $\mathbf{X}_0$ using $\{\hat{\alpha}_1, \hat{\omega}_1\}$ and (5.43). Then we re-estimate $\{\hat{\alpha}_0, \hat{\omega}_0\}$ using $\mathbf{X}_0$ and based on (5.45) and (5.46).

We can repeat the above process till it converges. Then the final $\left\{\hat{\alpha}_p, \hat{\omega}_p\right\}_{p=0}^{1}$ can be obtained.

The above procedure can be extended to the scenarios where multiple multipath interference signals exist [21–36]. The flow chart of the complete algorithm is shown in Fig. 5.8.

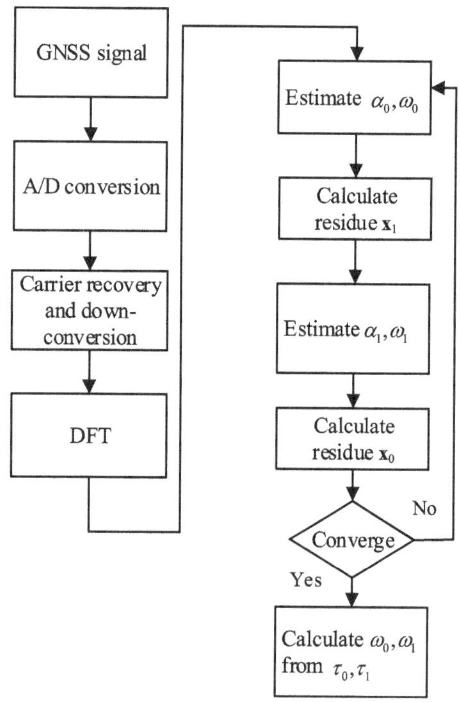

Fig. 5.8 WRELAX flow chart

Fundamentally, the multipath interference suppression based on WRELAX is a parameter estimation approach based on the nonlinear least squares criterion, while the MEDLL is an estimation method based on the maximum likelihood theory. Under the condition of Gaussian white noise background, the maximum likelihood estimation is equivalent to the nonlinear least squares criterion. Consequently MEDLL, in theory, is equivalent to the multipath interference suppression based on WRELAX. To acquire signal parameters for different paths, both MEDLL and WRELAX need iterations. But the two methods have very different approaches to solve multi-dimensional nonlinear problems for the iterations. MEDLL uses the curve fitting method to estimate parameters, so it is not precise; while WRELAX uses the zero-padding FFT to achieve accurate parameter estimation. In addition, MEDLL performs temporal domain curve fitting, and needs interpolation to enhance the accuracy of the parameter estimation, and finally obtains the estimated parameter through a lookup table. On the other hand, WRELAX obtains the accurate parameter estimation using the one-dimensional zero-padding FFT. Consequently, for the scenarios where we can assume the numbers of iterations for the two algorithms are equal to each other, the computation load needed for WRELAX is lighter.

In addition, all the above multipath suppression algorithms based on parameter estimation assume accurate carrier recovery. The carrier frequency of MEDLL can usually be obtained using a carrier tracking loop. But WRELAX does not need a carrier tracking loop, it can be obtained using a joint parameter estimation method that is much easier to implement, and the related details can be found in reference [28, 30].

5.3.4 Simulation Results

The LOS signal and one multipath interference are simulated. The amplitude attenuations of the multipath signal relative to the LOS signal DMR are chosen as 0.1, 0.3, 0.5, 0.7, and the relative delays $\tau_1 - \tau_0$ are 0–1.5 chips. Under the condition of noise-free and infinite bandwidth, and the correlator spacing $d = 1, 0.5, 0.1$ chips, the multipath error envelopes obtained by MEDLL and WRELAX algorithms are shown in Fig. 5.9.

It can be seen in Fig. 5.9 that, the traditional correlator has a larger code delay tracking error, and the narrow correlator can reduce code tracking errors by reducing the traditional correlator's correlation spacing. As the correlator spacing

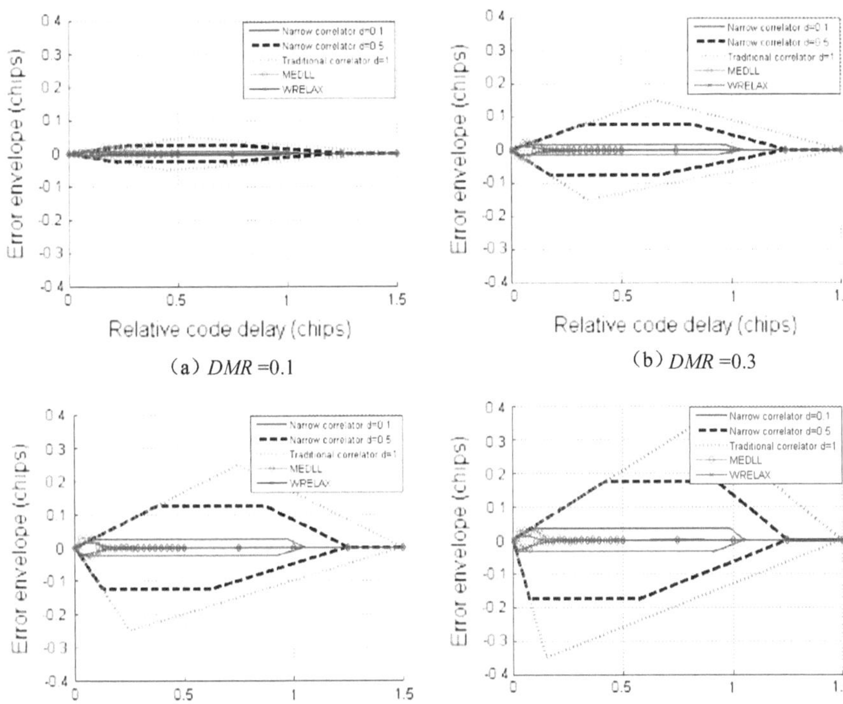

(a) *DMR* =0.1 (b) *DMR* =0.3 (c) *DMR* =0.5 (d) *DMR* =0.7

Fig. 5.9 Error envelopes for various methods under different *DMRs*

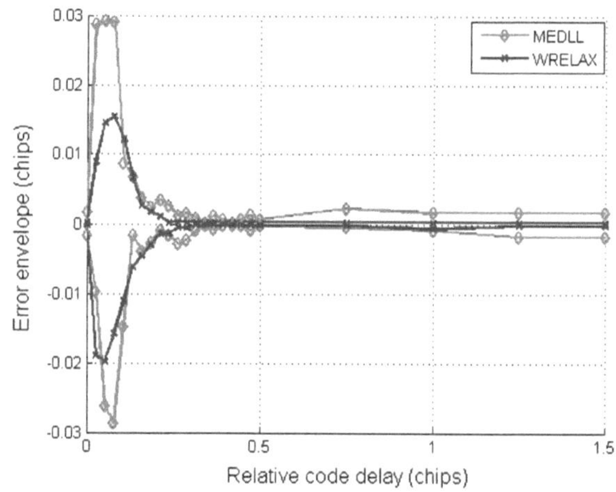

Fig. 5.10 Enlarged view on the error envelope curve when *DMR* = 0.5

becomes smaller, the tracking error can be further reduced. The performances of the MEDLL and WRELAX algorithms, compared with the narrow correlator, can further reduce the code delay tracking error.

To further compare the multipath suppression performances of the MEDLL and the WRELAX algorithms. Figure 5.10 shows the enlarged view on the error envelopes for the two methods based on parameter estimation when *DMR* = 0.5. It can be seen that, compared with the MEDLL, the multipath suppression performance of the WRELAX shows slight improvement.

5.4 Spatial Domain Multipath Interference Suppression

Since the LOS signal and the multipath interferences arrive from different directions, the spatial domain multipath interference suppression method taking advantage of this feature is introduced in this section.

5.4.1 Multipath Interference Suppression Based on Antenna Design

For the multipath interference suppression method based on antenna design, a choke ring is used to improve antenna gain pattern, and manufacture special antenna design and arrangement, as shown in Fig. 5.11. In addition, by selecting an antenna that can take advantage of the signal polarization characteristics can also eliminate multipath interferences. If a propagated GNSS signal is a right-hand polarized signal, its polarization characteristics can change after being reflected. For example, after a mirror reflection, it is converted to a left-hand polarized signal.

Fig. 5.11 A GNSS antenna with a choke ring [5]

We can also use digital filers, wideband antennas, multipath suppression plates that can absorb RF frequencies, and enhanced choke ring antennas (including advanced dual frequency choke ring) to reduce the impacts of multipath interferences. The enhanced choke ring uses a spiral arm. The newly developed choke ring antenna has the advantages of a sharper radiation roll-off model design (it can reduce the susceptibility of the multipath). There is no displaced phase center between two carriers, so there is no need for any antenna pointing provisions (i.e. required direction pointing towards due north due to symmetry requirement) and a flat structural design.

At a lower elevation angle or negative elevation angle, multipath errors happen very often. Under these conditions, the antenna multipath suppression plate can also reduce interferences on GNSS signal. But in reality, the impacts from multipath interferences cannot be completely eliminated by solely relying on the antenna design [37]. Consequently, we study more effective spatial domain multipath interference suppression techniques.

5.4.2 Multipath Interference Suppression Based on Low Sidelobe Weighting

If a GNSS receiver's antenna is an M-ary uniform linear array, and a multipath interference is composed of P reflecting paths, then the received signal can be represented as

$$\mathbf{x}(t) = \sum_{p=0}^{P} \alpha_p \mathbf{a}(\theta_p) s_p(t) + \mathbf{e}(t) \tag{5.47}$$

where $s_p(t)$ is the signal of the pth path, $p = 0, 1, \ldots, P$. For the same satellite, the spreading codes for signals from different paths are the same, but code delays are different, i.e. $s_p(t) = c(t - \tau_p)$; $\mathbf{e}(t)$ is the thermal noise vector of the receiver; $\mathbf{a}(\theta_p)$ is the corresponding array steering vector, and it can be represented as

$$\mathbf{a}(\theta_p) = \left[1, \quad \mathrm{e}^{-\mathrm{j}\pi \sin\theta_p}, \quad \ldots, \quad \mathrm{e}^{-\mathrm{j}\pi(M-1)\sin\theta_p}\right]^{\mathrm{T}} \tag{5.48}$$

where θ_p is the DOA of the pth signal.

Since strong correlations exist between the LOS signal and the multipath interference, the traditional adaptive beam forming algorithm fails. Some beamforming methods that can suppress the coherent interference usually require that the interference directions are known (especially DOAs of multipath interference signals in this case). Since multipath environments are usually very complex and GNSS signals are weak, the information on the DOAs of multipath interference signals is difficult to obtain. Consequently, the traditional adaptive beamforming technique has difficulty to suppress multipath interference.

For some GNSS systems, the DOAs of the GNSS signals can be estimated by using the inertial navigation aiding or the attitude determination technique to derive the receiver's location information, and then combining the derived receiver's location information with the ephemeris. Using this information, the low sidelobe beamforming with known LOS signal directions can be used here. The low sidelobe weighting is to perform a dot product of the original array steering vector and a low sidelobe taper weight. After the tapering, the LOS signal's steering vector becomes

$$\mathbf{b}(\theta_0) = \mathbf{w}_{taper} \odot \mathbf{a}(\theta_0) \tag{5.49}$$

where $\odot$ is the Hadamard product, and $\mathbf{w}_{taper}$ is the taper coefficient vector. A frequently used low sidelobe taper weight is the Dolph-Chebyshev weight, and $\mathbf{a}(\theta_0)$ is the weighted vector of the regular beamforming, and it is also the steering vector of the LOS signal. In this case, the array antenna pattern's main lobe points to the direction of the LOS signal, thereby significantly enhancing the LOS signal. In addition, the low level of the sidelobes can suppress the multipath interference. In this case, even though the multipath interference cannot be eliminated completely, its signal level is much lower than that of the LOS signal, so its impact on the signal is weak.

Figure 5.12 shows that the results of performing conventional beamforming and low sidelobe beamforming on the received signal using 10-ary uniform linear array when the DOA of the LOS is 20°, and its signal-to-noise ratio is $SNR = -20\,\mathrm{dB}$, and the DOA of the multipath interference signal is −20°, and its interference-to-noise ratio $INR = -20\,\mathrm{dB}$. It can be seen that, by using the

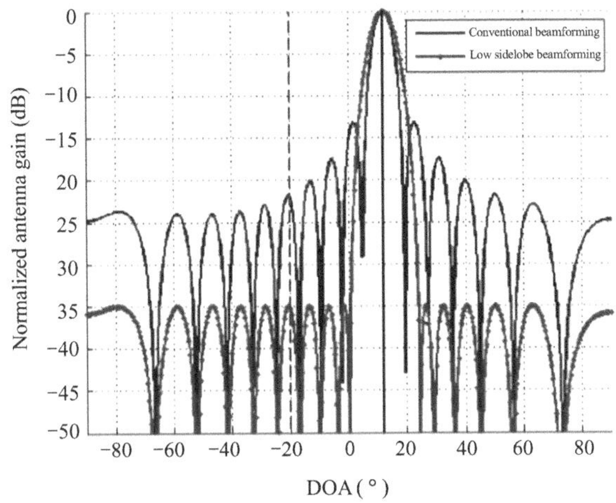

Fig. 5.12 Beam patterns before and after the low sidelobe weighting

conventional beamforming, it is very difficult to form a null towards the direction of the multipath interference. But since the low sidelobe beamforming is used, the low sidelobe beamforming after Chebyshev weighting can significantly reduce the multipath signal level, achieving the objective of multipath interference suppression.

The multipath interference suppression based on low sidelobe weighting is easy to implement, and has good performance. But the method is founded on the premise that the direction of the arrived GNSS signal is known, so it is only suitable for specific application scenarios. When the GNSS signal is unknown, we must consider other methods.

5.4.3 Multipath Interference Suppression Based on Spatial Domain Decoupled Parameter Estimation Theory

In this section, we take advantage of GNSS spreading code's periodic repetitive characteristics, and combine it with the RELAX technology from the decoupled parameter estimation theory to estimate the directions of the LOS signal and the multipath interference signals, then perform beamforming based on the estimated directions, thereby suppressing the multipath interferences. To solve the problem of weak GNSS signal, we apply the approach of using the C/A code's periodic repetitive characteristics to improve the SNR, as described in Sect. 2.4.3. Thereby the accuracy of direction estimation can be improved. Finally, the Linearly Constrained Minimum Variance (LCMV) beamforming based on the estimated direction can form a gain towards the direction of the satellite, while a null can be generated along the multipath interference's direction. Thereby the objective of multipath suppress can be achieved. A detailed flow chart is shown in Fig. 5.13. For the rest of the chapter we provide detailed introductions of the algorithm.

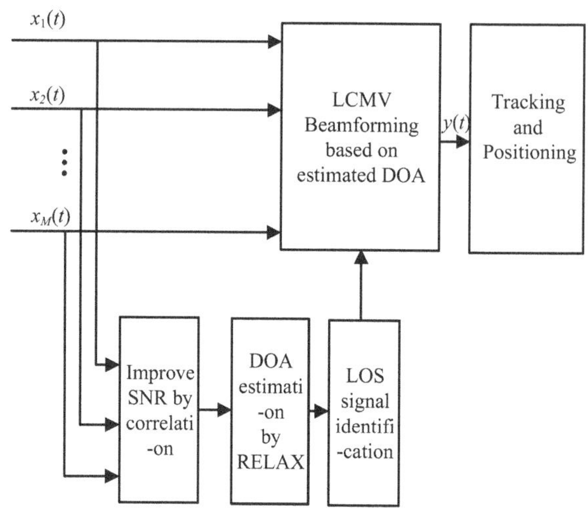

Fig. 5.13 Flow chart for the principle of the algorithm

1. DOA Estimation Using the Periodic Repetitive Characteristics of Spreading Codes

Here, we can reference the GNSS DOA estimation principle (described in Sect. 2.4.3) that uses the spreading code's periodic repetitive characteristics to estimate the DOA of the LOS signal and the multipath interference.

For a uniform linear array, the position of the first array element is used as the reference point, and the signal for the array element is denoted as $x_1(t)$, by giving $x_1(t)$ a code delay with a duration lasting one C/A code period T, we have

$$\tilde{x}(t) = x_1(t-T) = \sum_{p=0}^{P} \mathbf{a}_p(1) s_p(t-T)\, \mathrm{e}^{-\mathrm{j}2\pi f_{dp}(t-T)} + e(t-T) \tag{5.50}$$

where $\mathbf{a}_p(1)$ is the first element of the LOS signal and the multipath interference's steering vector. Since the period of the C/A code is 1 ms and the period of a navigation message symbol is 20 ms, within a symbol the C/A code repeats 20 times, showing the characteristics of periodic repetitive characteristics. Consequently, within a D code period, $s_p(t-T) = s_p(t)$, then (5.50) can be represented as

$$\tilde{x}(t) = \sum_{p=0}^{P} \mathbf{a}_p(1) s_p(t)\, \mathrm{e}^{-\mathrm{j}2\pi f_d(t-T)} + e(t-T) \tag{5.51}$$

The cross-correlation between $\tilde{x}(t)$ and $\mathbf{x}(t)$ can be represented as below

$$\begin{aligned}\mathbf{r} &= E\{\mathbf{x}(t)\tilde{x}^*(t)\} \\ &= \sum_{p=0}^{P}\left(\sigma_p^2\mathbf{a}_p^*(1)\mathrm{e}^{-\mathrm{j}2\pi f_d T} + \sum_{\substack{i=0\\ i\neq p}}^{P}\rho_{pi}\mathbf{a}_i^*(1)\right)\mathrm{e}^{-\mathrm{j}2\pi f_d T}\mathbf{a}_p \\ &\triangleq \sum_{p=0}^{P}\beta_p\mathbf{a}_p\end{aligned} \tag{5.52}$$

where $\beta_p \triangleq \sigma_p^2\mathbf{a}_p^*(1)\mathrm{e}^{-\mathrm{j}2\pi f_d T} + \sum\limits_{i=0,i\neq p}^{P}\rho_{pi}\mathbf{a}_i^*(1)$; σ_p^2 represents the power of the pth signal; ρ_{pi} represents the cross-correlation between the pth and ith signals. It can be seen from (5.52) that the directions of the LOS signal and the multipath interference signals can be estimated by minimizing the formula below

$$g_1\left(\{\theta_p,\beta_p\}_{p=0}^{P}\right) = \min_{\theta_p,\beta_p}\left\|\mathbf{r} - \sum_{p=0}^{P}\beta_p\mathbf{a}(\theta_p)\right\|_2^2 \tag{5.53}$$

Usually for actual applications, there is no means to obtain the theoretical value of $\mathbf{r}$, but it can be substituted by using the estimated value of the cross-correlation vector between $\tilde{x}(t)$ and $\mathbf{x}(t)$: $\hat{\mathbf{r}} = \frac{1}{N}\sum_{n=1}^{N}\tilde{x}^*(n)\mathbf{x}(n)$.

Let $\boldsymbol{\beta} = [\beta_0,\beta_1,\ldots,\beta_P]^{\mathrm{T}}$, (5.53) can be expanded into the following format

$$\begin{aligned}g_1\left(\{\theta_p,\beta_p\}_{p=0}^{P}\right) &= (\hat{\mathbf{r}} - \mathbf{A}\boldsymbol{\beta})^{\mathrm{H}}(\hat{\mathbf{r}} - \mathbf{A}\boldsymbol{\beta}) \\ &= \left[\boldsymbol{\beta} - (\mathbf{A}^{\mathrm{H}}\mathbf{A})^{-1}\mathbf{A}^{\mathrm{H}}\hat{\mathbf{r}}\right]^{\mathrm{H}}(\mathbf{A}^{\mathrm{H}}\mathbf{A})\left[\boldsymbol{\beta} - (\mathbf{A}^{\mathrm{H}}\mathbf{A})^{-1}\mathbf{A}^{\mathrm{H}}\hat{\mathbf{r}}\right] + \hat{\mathbf{r}}^{\mathrm{H}}\hat{\mathbf{r}} - \hat{\mathbf{r}}^{\mathrm{H}}\mathbf{A}(\mathbf{A}^{\mathrm{H}}\mathbf{A})^{-1}\mathbf{A}^{\mathrm{H}}\hat{\mathbf{r}}\end{aligned} \tag{5.54}$$

where $\mathbf{A} = [\mathbf{a}(\theta_0),\mathbf{a}(\theta_1),\ldots,\mathbf{a}(\theta_P)]$. To minimize (5.54) is equivalent to making the first term equals to 0 and maximizing the last term, i.e. $\boldsymbol{\theta}$ can be obtained using the cost function of the formula below

$$\hat{\boldsymbol{\theta}} = \arg\max_{\boldsymbol{\theta}}\left[\hat{\mathbf{r}}^{\mathrm{H}}\mathbf{A}(\mathbf{A}^{\mathrm{H}}\mathbf{A})^{-1}\mathbf{A}^{\mathrm{H}}\hat{\mathbf{r}}\right] \tag{5.55}$$

Accordingly, there is

$$\hat{\boldsymbol{\beta}} = (\mathbf{A}^{\mathrm{H}}\mathbf{A})^{-1}\mathbf{A}^{\mathrm{H}}\hat{\mathbf{r}}\Big|_{\boldsymbol{\theta}=\hat{\boldsymbol{\theta}}} \tag{5.56}$$

When there are multiple signals in the data (in this context, multiple signals refer to the direct LOS signal and multipath interferences), the cost function in (5.56) has a complex multi-peak shape, so to directly solve it faces greater difficulty. Consequently we apply the RELAX technique based on the decoupled parameter estimation theory, to implement parameter estimation [33–36, 38].

Again, here we only consider the case that only the LOS signal and one multipath interference exist, i.e. $P = 1$.

Let

$$\hat{\mathbf{f}}_k = \hat{\mathbf{f}} - \sum_{i=0,i\neq k}^{P} \hat{\beta}_i \mathbf{a}\left(\hat{\theta}_i\right) \tag{5.57}$$

Here we can assume $\left\{\hat{\theta}_i, \hat{\beta}_i\right\}_{i=0,i\neq k}^{P}$ is known or has already been estimated, we can define

$$g(\theta_k, \beta_k) = \|\hat{\mathbf{f}}_k - \beta_k \mathbf{a}(\theta_k)\|^2 \tag{5.58}$$

Then we can obtain the minimized cost function g_2 relative to θ_k and β_k

$$\begin{aligned}\hat{\theta}_k &= \arg\min_{\theta_k} \left\| \left[\mathbf{I} - \frac{\mathbf{a}(\theta_k)\mathbf{a}^{\mathrm{H}}(\theta_k)}{N} \right] \hat{\mathbf{f}}_k \right\|^2 \\ &= \arg\max_{\theta_k} \left| \mathbf{a}^{\mathrm{H}}(\theta_k) \hat{\mathbf{f}}_k \right|^2 \end{aligned} \tag{5.59}$$

$$\hat{\beta}_k = \left. \frac{\mathbf{a}^{\mathrm{H}}\left(\hat{\theta}_k\right)\hat{\mathbf{f}}_k}{N} \right|_{\theta_k=\hat{\theta}_k} \tag{5.60}$$

Consequently, $\hat{\theta}_k$ can be obtained using the main peak value of the periodogram $\frac{\left|\mathbf{a}^{\mathrm{H}}(\theta_k)\hat{\mathbf{f}}_k\right|^2}{N}$ that can be obtained by zero-padding the data sequence $\hat{\mathbf{f}}_k$ and then perform a FFT (the purpose of zero-padding is to improve the precision of $\hat{\theta}_k$). $\hat{\beta}_k$ can be obtained by computing $\frac{\mathbf{a}^{\mathrm{H}}(\theta_k)\hat{\mathbf{f}}_k}{N}$.

After all the preparations described above, the detailed steps of implementing the DOA estimation using the RELAX algorithm are described below

(1) Assuming $P = 0$, $\hat{\theta}_0$ and $\hat{\beta}_0$ can be obtained using (5.59) and (5.60) from $\hat{\mathbf{f}}$.
(2) Assuming $P = 1$, using (5.57) and the obtained $\hat{\theta}_0$ and $\hat{\beta}_0$ from step (1), $\hat{\theta}_1$ and $\hat{\beta}_1$ can be estimated from $\hat{\mathbf{f}}_1$ based on (5.59) and (5.60).

$\hat{\mathbf{f}}_0$ can be recalculated using (5.57) and $\hat{\theta}_1$, $\hat{\beta}_1$, and then $\hat{\theta}_0$ and $\hat{\beta}_0$ can be obtained from $\hat{\mathbf{f}}_0$ based on (5.59) and (5.60). The above two subroutines can be repeated until they convergence.

The algorithm can be expanded to the scenarios where any number of multipath interferences exist [37–41].

2. Multipath Interference Identification

A multipath interference has the same spreading code as the LOS signal, therefore the multipath interference signal cannot be identified from the spreading code. But generally speaking, a comparison between the multipath interference signal and the LOS signal reveals the following characteristics

(1) The multipath interference usually has a longer code delay than that of the LOS signal.
(2) The multipath interference is usually weaker than the LOS signal.
(3) The multipath interference is usually formed by ground reflections. While the satellite is far away from the ground, the elevation angle of the LOS signal is higher than the elevation angle of the multipath interference.

To summarize the above, the arrival time, amplitude, and DOA of a signal can be used to differentiate a multipath interference signal from the LOS signal.

3. Multipath Interference Suppression Using LCMV Beamforming

On the basis of successful estimations on the LOS signal and multipath interferences, it is desirable to design the array antenna's response so that a high gain can be formed along the direction of the LOS signal while nulls can be formed along the directions of the multipath interference. This is why LCMV beamformers are used to calculate the adaptive weights of the antenna. The antenna adaptive weight $\mathbf{w}$ satisfying the above requirements can be depicted as the solution for the following constrained problem

$$\begin{cases} \min \ \mathbf{w}^{\mathrm{H}}\mathbf{R}_x\mathbf{w} \\ \text{s.t.} \ \ \mathbf{w}^{\mathrm{H}}\mathbf{C} = \mathbf{f} \end{cases} \tag{5.61}$$

where $\mathbf{C} = \left[\mathbf{a}\left(\hat{\theta}_0\right), \mathbf{a}\left(\hat{\theta}_1\right), \ldots, \mathbf{a}\left(\hat{\theta}_P\right)\right]$; $\mathbf{f} = [1, 0, \ldots, 0]^T$; $\hat{\theta}_0$ represents the estimated DOA of the LOS signal; $\hat{\theta}_1, \ldots, \hat{\theta}_P$ represent estimated DOAs for the multipath interference signals; $\mathbf{R}_x$ is the covariance matrix of the received data by the antenna $\mathbf{x}(t)$, and in practical applications it is usually substituted using its estimated value $\hat{\mathbf{R}}_x$.

By solving (5.61), we can obtain

$$\mathbf{w} = \hat{\mathbf{R}}_x^{-1}\mathbf{C}\left(\mathbf{C}^{\mathrm{H}}\hat{\mathbf{R}}_x^{-1}\mathbf{C}\right)^{-1}\mathbf{f} \tag{5.62}$$

By using the above weight vector on the received signal of the antenna $\mathbf{x}(t)$, the signal after the interference suppression can be obtained by

$$y(t) = \mathbf{w}^{\mathrm{H}}\mathbf{x}(t) \tag{5.63}$$

Lastly, the signal after the multipath interference suppression $y(t)$ is sent into the tracking and positioning module, to complete the process of GNSS tracking and position.

5.4.4 Simulation Results

The received signals used in the experiments are composed of the LOS signal, multipath interference, and noise. The IF frequency is 4.309 MHz, and the sampling frequency is 5.714 MHz. To verify the performance of DOA estimation based on RELAX as given in this section, the antenna array under consideration is a 10-ary uniform linear array with a half-wavelength array element spacing. The received data by the antenna is composed of a LOS signal sent by a satellite and one multipath interference signal from the satellite. The LOS signal incidents on the array along the 0° direction, and the multipath interference signal incidents on the array along the −20° direction. The corresponding LOS signal phase $\varphi_0 = 0$, and the phase of the multipath interference signal $\varphi_1 = 0.9\pi$, and the difference of code delays between the two signal $\tau_1 - \tau_0 = 0.09T_c$, and $DMR = 2$dB. We compare the performances among the RELAX algorithm, the commonly used MUSIC algorithm, and the MUSIC algorithm after de-coherence. The results are shown in Fig. 5.14. Among them, subfigures (a) and (b) show the DOA estimation performance for the LOS signal and the multipath interference as SNR changes.

It can be seen from the experiment results in Fig. 5.14, the estimated DOA results directly using the MUSIC algorithm have larger errors. Using the MUSIC algorithm after de-coherence can significantly improve the performance of DOA estimation (here we use the spatial smoothing technique for de-coherence, and the number of overlapping sub-arrays is selected as 2) [39, 40]. The estimation performance of the RELAX algorithm is even better than the MUSIC algorithm after de-coherence.

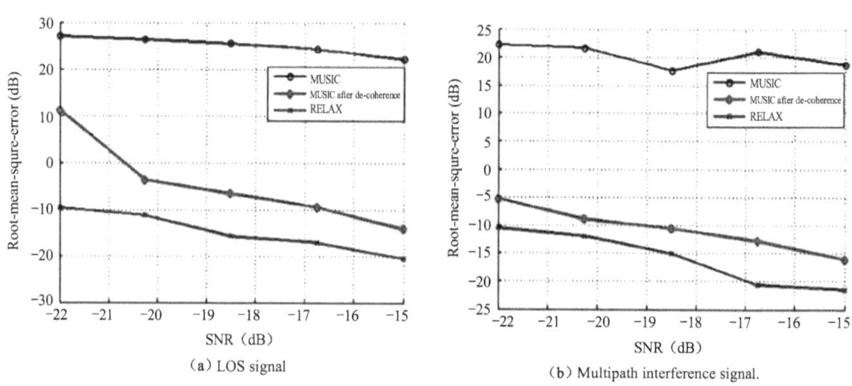

Fig. 5.14 The curve of estimated DOA as SNR changes

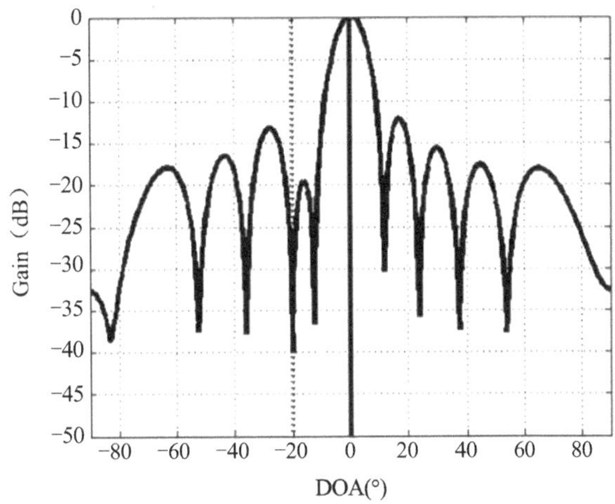

Fig. 5.15 LCMV beam pattern

Then the DOA estimation results of the RELAX algorithm are used as premise, to perform a follow-up experiment. The LOS signal SNR are selected to be $SNR = -20\,\text{dB}$, $DMR = 2\,\text{dB}$,while the rest of the parameters are the same as those used in the simulations above.

Figure 5.15 shows the results of performing LCMV beamforming using the estimated DOA. It can be seen that after estimating the DOA of the multipath interference and the LOS signal, the LCMV can form a gain in the direction of the LOS signal and form a null at the direction of the multipath interference. Figure 5.16 shows the correlation functions before and after multipath interference suppression. It can be seen that, due to impacts from multipath interferences, the correlation functions are distorted. But after using the algorithm described in this

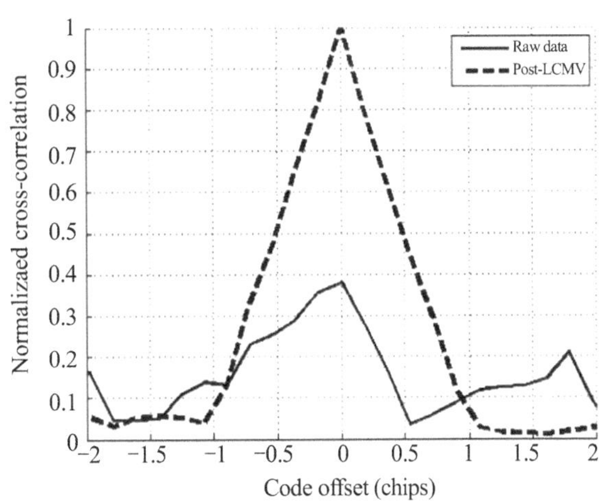

Fig. 5.16 Correlation functions before and after the multipath interference suppression

Fig. 5.17 Error envelopes before and after multipath suppression

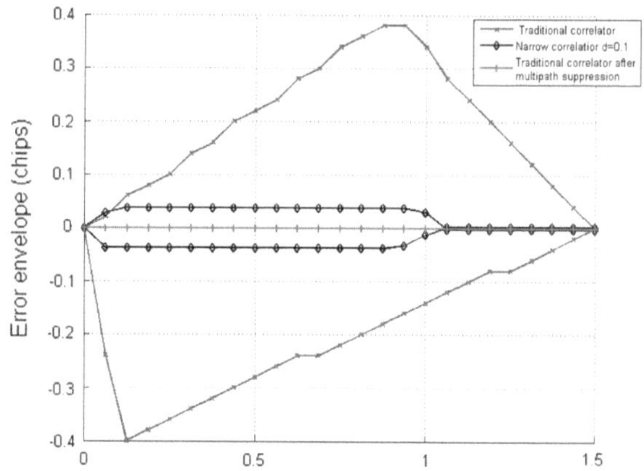

section for multipath interference suppression, the correlation function shows the shape of a standard symmetric equilateral triangle. This result further verifies the effectiveness of the algorithm. In addition, since the phase difference between the multipath and the LOS signal is 0.9π, leading to the phenomenon of amplitude cancellation, the results show that the amplitude of the correlation function for the raw data is much smaller than that for the "post-LCMV".

Next, the results on the code delay tracking errors of the GNSS receiver loop for simulating the LOS signal and one multipath interference are listed. The parameters are selected so that $DMR = 2\,\mathrm{dB}$, and the relative code delay $\tau_1 - \tau_0$ is 0–1.5 chips, and the phase difference is set to $\Delta\varphi = 0°$ and $\Delta\varphi = 180°$ respectively. Under the condition of noiseless and infinite bandwidth, Fig. 5.17 shows the code delay tracking error envelope curve on the data without multipath interference suppression using traditional correlator and narrow correlator, and the code delay tracking error envelope curve of the traditional correlator on the data after multipath interference suppression using the proposed algorithm described in this section.

In Fig. 5.17, when a multipath interference signal exists, the traditional correlation technique has a larger code phase tracking error. Even though the narrowband correlator method can greatly reduce the error, it cannot reduce the error unlimited. The proposed multipath interference suppression method based on the spatial signal decoupled parameter estimation theory suppresses the multipath interference first, and then sends the data after interference suppression to the traditional tracking loop, and then the error of code delay estimation becomes very small.

5.5 Summary

In this chapter, we discuss the impacts of multipath interferences on GNSS system. We mainly investigate temporal and spatial types of multipath interference suppression techniques in the temporal domain and spatial domain. Through theoretic analyses and simulation results, we can conclude that for temporal domain processing, the algorithm based on WRELAX, compared with the existing MEDLL technique, can lower the computation complexity while improving performance slightly. For spatial domain processing, the technique based on RELAX, by estimating the DOA of the LOS signal and multipath interference signals first and performing beamforming as a follow-up step, can achieve better multipath suppression effects.

References

1. Van Nee RDJ. Multipath effects on GPS code phase measurements. In: Proceedings of the 4th international technical meeting of the satellite division of the Institute of Navigation, Albuquerque, NM. 1991. p. 915–24.
2. Kos T, Markezic I, Pokrajcic J. Effects of multipath reception on GPS positioning performance. In: Proceedings of the ELMAR. 2010. p. 399–402.
3. Braasch MS. GPS multipath model validation. In: Proceedings of the position location and navigation symposium. IEEE; 1996. p. 672–8.
4. Kalyanarman SK, Braasch MS, Kelly JM. Code tracking architecture influence on GPS carrier multipath. IEEE Trans Aerosp Electron Syst. 2006;42(2):548–61.
5. Scire-Scappuzzo F, Makarov SN. A low-multipath wideband GPS antenna with cutoff or non-cutoff corrugated ground plane. IEEE Trans Antennas Propag. 2009;57(1):33–46.
6. Lopez AR. GPS landing system reference antenna. Antennas Propag Mag. 2010;52(1):104–13.
7. Maqsood M, Gao S, Brown T, et al. Effects of ground plane on the performance of multipath mitigating antennas for GNSS. In: Antennas & propagation conference, Loughborough. 2010. p. 241–4.
8. Daneshmand S, Jafarnia-Jahromi A, Broumandan A, et al. A GNSS structural interference mitigation technique using antenna array processing. In: Proceedings of the sensor array and multichannel signal processing workshop. 2014. p. 109–12.
9. Cebrian JM, Picanyol J, Seco-Granados G. Adibeam: design and experimental validation of a robust beamforming platform for Galileo reference ground stations. In: Proceedings of the satellite navigation technologies and European workshop on GNSS signals and signal processing. 2010. p. 1–8.
10. Maqsood M, Gao S, Brown TWC, et al. A compact multipath mitigating ground plane for multiband GNSS antennas. Antennas Propag. 2013;61(5):2775–82.
11. Cannon ME, Lachapelle G, Qiu W, et al. Performance analysis of a narrow correlator spacing receiver for precise static GPS positioning. In: Proceedings of the position location and navigation symposium. 1994. p. 355–60.
12. Van Dierendonck AJ, Fenton P, Ford T. Theory and performance of narrow correlator spacing in a GPS receiver. Navig: J Inst Navig. 1992;39(3):265–84.
13. Van Nee RDJ. The multipath estimating delay lock loop. In: 1992 ISSTA 92 IEEE second international symposium on proceedings of the spread spectrum techniques and applications. 1992. p. 39–42.

14. Chen K. Studies on GNSS multipath estimation theory and methods. PhD dissertation. Zhengzhou: PLA Information Engineering University; 2012.
15. Sahmoudi M, Amin MG. Optimal robust beamforming for interference and multipath mitigation in GNSS arrays. In: IEEE international conference on proceedings of the acoustics, speech and signal processing. 2007. vol 3, p. 693–6.
16. Moelker DJ. Multiple antennas for advanced GNSS multipath mitigation and multipath direction finding. In: Proceedings of the 10th international technical meeting of the satellite division of (ION GPS 1997). Kansas City, MO: The Institute of Navigation; 1997. p. 541–50.
17. Haimovich A, Bar-Ness Y. Adaptive antenna arrays using eigenvector methods. In: Proceedings of the antennas and propagation society international symposium. 1988. p. 980–3.
18. Buckley KM. Broad-band beamforming and the generalized sidelobe canceller. IEEE Trans Acoust Speech Signal Process. 1986;34(5):1322–3.
19. Ray JK, Cannon ME, Fenton P. GPS code and carrier multipath mitigation using a multi antenna system. IEEE Trans Aerosp Electron Syst. 2001;37(1):183–95.
20. Li J, Wu RB. An efficient algorithm for time delay estimation. IEEE Trans Signal Process. 1998;46(8):2231–5.
21. Wu RB, Li J. Time-delay estimation via optimizing highly oscillatory cost functions. IEEE J Ocean Eng. 1998;23(3):235–44.
22. Wu R, Li J. Time delay estimation with multiple looks in colored Gaussian noise. IEEE Trans Aerosp Electron Syst. 1999;35(4):1354–61.
23. Wu R, Li J, Liu Z. Super resolution time delay estimation via MODE-WRELAX. Aerosp Electron Syst. 1999;35(1):294–307.
24. Li X, Wu R, Sheplak M, et al. Multi-frequency CW-based time-delay estimation for proximity ultrasonic sensors. IEE Proc Radar Sonar Navig. 2002;149(2):53–9.
25. Wu R, Li X, Li J. Continuous pavement profiling with ground-penetrating radar. IEE Proc Radar Sonar Navig. 2002;149(3):183–93.
26. Li X, Wu R, Rasmi S, et al. An acoustic proximity ranging system for monitoring the cavity thickness. Ultrason Ferroelectr Freq Control. 2003;50(7):898–910.
27. Li X, Wu R, Rasmi S, et al. Acoustic proximity ranging in the presence of secondary echoes. Instrum Meas. 2003;52(5):1593–605.
28. Li J. Studies on some key techniques of GNSS anti-jamming. PhD dissertation. Tianjin: Tianjin University; 2013.
29. Jia Q, Wu R, Wang W, et al. Multipath interference mitigation in GNSS via WRELAX. GPS Solut. 2017;21(2):487–98.
30. Li J, Wu R, Wang W, et al. GPS fine time delay estimation based on signal separation estimation theory. In: 2012 11th international conference on signal processing proceedings. 2012. p. 236–40.
31. Li J, Wu R, Wang W, et al. A novel GPS signal acquisition algorithm. Adv Inf Sci Serv Sci. 2012;4(17):597–604.
32. Wu R, Jia Q, Wang W. Efficient FFT-based algorithms for multipath interference mitigation in GNSS. Invited Chapter in "Fourier transform—signal processing and physical sciences". INTECH Open Access Publisher; 2015.
33. Li J, Stoica P. Efficient mixed-spectrum estimation with applications to target feature extraction. IEEE Trans Signal Process. 1996;44(2):281–95.
34. Wu R, Li J. Adaptive ground bounce removal for landmine detection with ground penetrating radar. Electron Lett. 2001;37(20):1250–2.
35. Wu R, Wang W, Jia Q. FFT-based efficient algorithms for time delay estimation. Invited Chapter in "Fourier transform/book 1". INTECH Open Access Publisher; 2011.
36. Stoica P, Li H, Li J. Amplitude estimation of sinusoidal signals: survey, new results, and an application. IEEE Trans Signal Process. 2000;48(2):338–52.
37. Hofmann-Wellenhof B, Lichtenegger H. Global navigation satellite system. 2008.
38. Li J, Stoica P. Angle and waveform estimation via RELAX. IEEE Trans Aerosp Electron Syst. 1997;33(3):1077–87.

39. Pillai SU, Kwon BH. Forward/backward spatial smoothing techniques for coherent signal identification. Acoust Speech Signal Process. 1989;37(1):8–15.
40. Haber F, Zoltowski M. Spatial spectrum estimation in a coherent signal environment using an array in motion. IEEE Trans Antennas Propag. 1986;34(3):301–10.
41. Lu D, Wu RB, Liu HT. Global positioning system anti-jamming algorithm based on period repetitive clean. IET Radar Sonar Navig. 2013;7(2):164–9.

Chapter 6
Pulse Interference Suppression Techniques

6.1 Introduction

Intentional interferences mainly include multipath interference and pulse interference. We studied multipath interference in the last chapter. In this chapter we study the suppression of the other type of intentional interference—the pulse interference.

To better meet the demands for various applications, the GPS system has added several new frequency bands during its modernization process, e.g. GPS L5 signal. Similar to Galileo E5 band and Beidou B2 band, the main objective of the added band is to provide better services related to the Safety of Life Service (SoL). All these signals work on the Aeronautical Radio Navigation Service (ARNS) frequency band (962–1213 MHz). However, some other systems already work within this band [1] such as the civil Distance Measuring Equipment (DME), military Tactical Air Navigation (TACAN), Secondary Surveillance Radar (SSR), Traffic Collision and Avoidance System (TCAS), and Automatic Dependent Surveillance-Broadcast (ADS-B). As a result, these systems could generate certain degrees of interferences on the GNSS signal located within the same band. Among these systems, DME/TACAN equipment is the main factor impacting GPS L5, Galileo E5, and Beidou B2 signals [2]. The overlapping working frequency bands between the DME system and the GPS L5, Galileo E5 and Beidou B2 signals are illustrated in Fig. 6.1. As a result, high-power pulses from DME transmitters can produce pulse interference on the GNSS signals located at the above frequencies. Similar to jamming on continuous wave, serious high power pulse interference can impact the GNSS receivers' capabilities of acquisition, tracking and positioning. But since the pulse interference has the characteristics of high power, short duration, wide bandwidth and randomness, it is not as stable as regular interference signals, thereby the interference suppression algorithms relying on the stability of the signal are not effective.

Conventional pulse interference suppression algorithms include temporal domain pulse blanking [3–6], frequency domain notch filtering [7–11] and

R. Wu et al., *Adaptive Interference Mitigation in GNSS*, Navigation: Science and Technology, https://doi.org/10.1007/978-981-10-5571-3_6

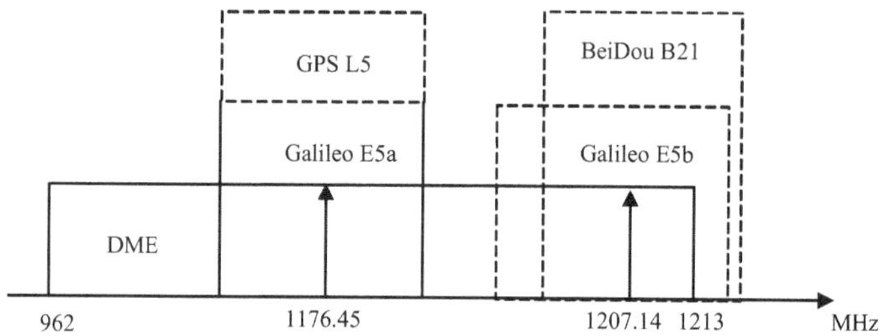

Fig. 6.1 Overlapping frequency bands between GNSS signals and the DME system

temporal-frequency domain hybrid filtering [7–12]. For interference pulses with low density, these methods are simple and effective. But, as the pulse density increases, the GNSS signal losses caused by these conventional pulse interference suppression algorithms increase accordingly, and thereby the GNSS performance degrades. In this case, the pulse interference suppression algorithms based on parameter estimation [13] and wavelet packet transformation [14–19] can be used to preserve many more GNSS signals, so better pulse interference suppression performance can be achieved.

In this chapter, DME pulse interference is used as an example to study the pulse interference suppression problem. We mainly study temporal domain pulse blanking algorithm, frequency domain notch filtering algorithm, temporal-frequency domain hybrid filtering algorithm, parameterized algorithm, and wavelet packet transformation based algorithm. The related principles can also be expanded to other pulse interference suppression methods.

6.2 Conventional Pulse Interference Suppression

DME signal has a pulse pair structure [20], a baseband DME pulse pair signal can be represented as

$$s(t) = \mathrm{e}^{-\frac{\alpha}{2}t^2} + \mathrm{e}^{-\frac{\alpha}{2}(t-\Delta t)^2} \tag{6.1}$$

where $\alpha = 4.5 \times 10^{11}\ \mathrm{s}^{-2}$ can ensure that the half-amplitude bandwidths for the two Gaussian pulses are both 3.5 μs; Δt is the internal interval between the pulse pair. For the X wave channel overlapping with the GNSS signal frequency band, $\Delta t = 12$ μs. The mathematical model for the DME signal after modulation can be represented as

$$s_m(t) = \left(\mathrm{e}^{-\frac{\alpha}{2}t^2} + \mathrm{e}^{-\frac{\alpha}{2}(t-\Delta t)^2}\right)\mathrm{e}^{\mathrm{j}(\omega t + \theta)} \tag{6.2}$$

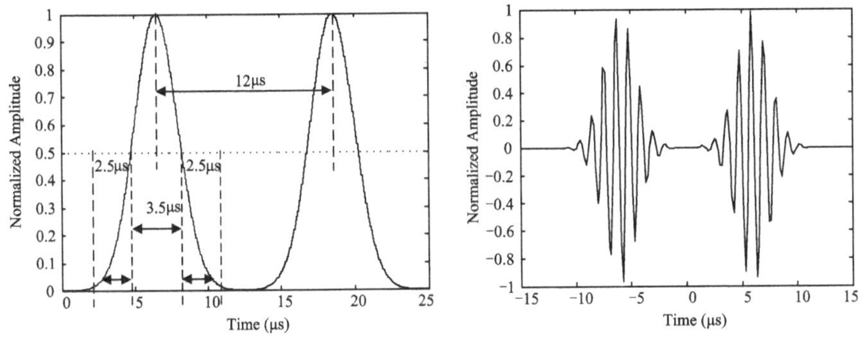

(a) DME baseline signal waveform
(b) DME signal temporal domain waveform after modulation

Fig. 6.2 DME pulse pair signal waveform

where ω and θ are carrier wave frequency and phase. Figure 6.2 shows the baseline DME signal waveform and the DME signal waveform after modulation.

By combining the Fig. 6.2a and (6.1), within the DME waveform envelop, the time duration for it to drop from the half-amplitude bandwidth to the point where the amplitude is 0.017 of the amplitude peak is 2.5 μs. We can approximately calculate the time duration for the DME pulse pair as

$$2.5 + \frac{3.5}{2} + 12 + \frac{3.5}{2} + 2.5 = 20.5\ \mu\text{s} \tag{6.3}$$

The tails of Gaussian pulses approach zero asymptotically, and parts of the tails are below the noise level. When performing DME pulse interference detection, to ensure that a complete pulse pair is contained inside a detection window, the detection window's length can be selected as 25 μs.

In the temporal domain, a DME signal is a Gaussian pulse pair with a certain trailing tail, and in the frequency domain it is a signal associated with a certain bandwidth as 99% of the energy concentrates within 0.8 MHz [21]. Its power spectrum density is shown in Fig. 6.3.

The mathematical model of the received signal under the impacts of Q *DME* transmitters can be represented as

$$x(t) = s_{GNSS}(t) + \sum_{q=1}^{Q}\sum_{u=1}^{U_q} \sqrt{P_q}\, s(t - t_{q,u})\, e^{j(\omega_{q,u}t + \theta_{q,u})} + e(t) \tag{6.4}$$

where $s_{GNSS}(t)$ is the received GNSS signal; $s(t)$ is the baseband DME signal; U_q is the total number of pulse pairs received within the observation time duration transmitted by the qth DME transmitter; P_q is the received power from the qth DME transmitter; $t_{q,u}$ is the arrival time of the uth pulse pair generated by the qth DME

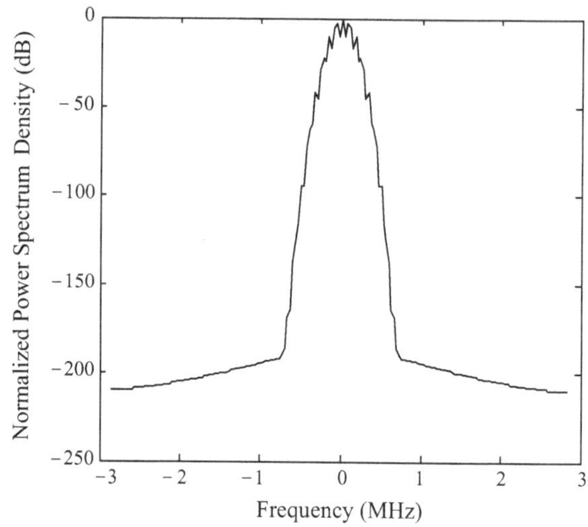

Fig. 6.3 Power spectrum of the DME signal

transmitter; $\omega_{q,u}$ is the corresponding carrier frequency; $\theta_{q,u}$ is the corresponding carrier phase; $e(t)$ is the receiver noise, and we can assume that it is a Gaussian white noise.

6.2.1 *Temporal Domain Pulse Blanking Method*

The most frequently used pulse interference suppression algorithm is the temporal domain pulse blanking method. The approach of the temporal domain pulse blanking method is to perform amplitude detection at every sample point for the received GNSS signal. If the amplitude surpasses a preset threshold (usually related to the noise power) then a zero-setting process can be performed, consequently suppressing the interferences higher than the preset threshold. The processing method for the temporal domain pulse blanking can also be used to compute the pulse interference's duty cycle (the ratio between the number of samples higher than the threshold and the total number of observed samples). A higher duty cycle means a larger number of DME interference pulses within a time unit, or a higher density of DME interference pulse. The processing on the interference signal using the temporal domain pulse blanking method is shown in Fig. 6.4, where the Fig. 6.4a is the temporal domain waveform of the received signal under the impact of interference. The Fig. 6.4b is the signal waveform after the interference suppression using the temporal domain pulse blanking method, and it can be seen that the signals larger than the threshold have all been set to zero.

The temporal domain pulse blanking method is easy to implement and has high efficiency. Under the condition of a small duty cycle, it can also have good interference suppression performance. But when the duty cycle is higher, this method

Fig. 6.4 Temporal domain pulse blanking method for DME interference suppression

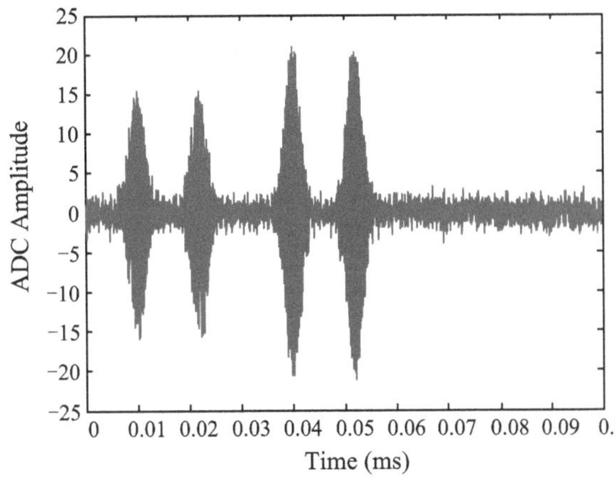

(a) Temporal domain signal before the interference suppression.

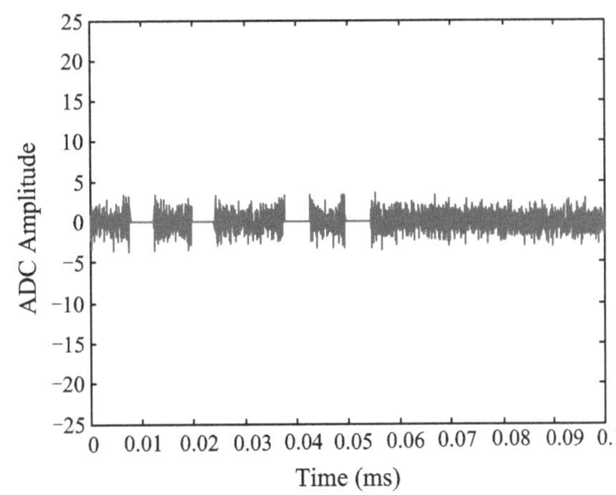

(b) Temporal domain signal after the interference suppression.

can lead to a significant loss of authentic signals when it performs zero-setting on the interference, consequently degrading the acquisition performance on authentic GNSS signals by the receiver.

The selection of a proper threshold is one of the crucial factors for the performance of the temporal domain pulse blanking method. The selection is based on the noise level. Since authentic GNSS signals are buried in the noise, both noise and the GNSS signals are regarded as "noise signals", and the Constant False Alarm Rate (CFAR) criterion can be used to calculate the threshold. Since the noise level fluctuates all the time, when the noise amplitude surpasses the detection threshold, the detection system believes that a target (or an interference signal) has been found. This type of error is defined as a false alarm, and the probability of its occurrence is known as CFAR.

Given the real and imaginary parts of a complex Gaussian noise $e(t)$ are $e_r(t)$ and $e_i(t)$, the probability density functions of $e_r(t)$ and $e_i(t)$ both follow a Gaussian distribution with zero mean and σ^2 variance, then we have $e(t) = e_r(t) + je_i(t)$, consequently the envelop of the complex noise follows a Raleigh distribution. Firstly, we introduce some features of the Raleigh distribution: Given the probability density function of a random variable w

$$f(w) = \begin{cases} \frac{w}{a^2} \mathrm{e}^{-\frac{w^2}{2a^2}}, & w \geq 0 \\ 0, & w < 0 \end{cases} \tag{6.5}$$

where a $(a > 0)$ is a constant, then w follows a Raleigh distribution with parameter a, and its expectation is

$$E[w] = a\sqrt{\frac{\pi}{2}} \tag{6.6}$$

If the CFAR larger than w_0 is P_{fa}, then

$$P_{fa} = \int_{w_0}^{\infty} \frac{w}{a^2} \exp\left(-\frac{w^2}{2a^2}\right) \mathrm{d}w = \exp\left(-\frac{w_0^2}{2a^2}\right) \tag{6.7}$$

The threshold w_0 calculated based on (6.7) is

$$w_0 = a\sqrt{-2\ln(P_{fa})} \tag{6.8}$$

Based on (6.8), for the Raleigh distribution, given a certain CFAR, the threshold w_0 is related to the parameter a. If a can be solved, the threshold can also be obtained. In the example, the envelop of complex noise $|e(t)|$ follows a Raleigh distribution, and the corresponding parameter a is σ. As a result, for a complex Gaussian noise with zero mean, the temporal domain pulse blanking threshold Th_0 corresponding to the CFAR P_{fa} is

$$Th_0 = \sigma\sqrt{-2\ln(P_{fa})} \tag{6.9}$$

6.2.2 *Frequency Domain Notch Filtering Algorithm*

Since the DME pulse interference is manifested as a narrowband interference with a fixed bandwidth in the frequency domain, the suppression can be performed on the DME interference, and the frequency domain notch filtering algorithm is a method based on such an approach. There are two types of implementation approaches for the frequency domain notch filtering algorithm: the approach based on notch filter and the approach based on FFT. The filtering approach based on notch filter uses a pre-designed notch filter (the interference frequency is the stop band center frequency, and the stop band bandwidth is the same as the interference bandwidth) on the received signal. The filtering approach based on FFT filtering converts the received signals to the frequency domain using FFT, and in the frequency domain it suppresses the frequency components higher than a threshold, then converts the results back the temporal domain using IFFT to achieve interference suppression. During the airplane's flight process, since different DME transmitters have different carrier frequencies, it is very hard to construct the notch filter beforehand. In addition, since the number of transmitters associated with DME interferences may be more than one, a notch filter with multiple stop bands needs to be built, which is harder to implement. Therefore, the frequency domain notch filtering approach based on FFT is usually used to suppress the DME interference. The frequency domain notch filtering algorithm can remove the GNSS located at the same frequencies when it filters out the interferences. By using the same data as in Sect. 6.2.1, Fig. 6.5 illustrates the frequency domain notch filtering algorithm used for interference suppression.

Figure 6.5a is the signal spectrum for the signal plotted in Fig. 6.4a under the influence of interference, and Fig. 6.5b is the signal spectrum after a threshold is set in the frequency domain to filter out the interference frequency; Fig. 6.5c is the temporal domain waveform of the signal after the interference suppression. Next, we discuss the threshold selection for the frequency domain notch filtering algorithm based on FFT, since the selection of threshold is related to the noise model, we can still assume that the noise is a complex Gaussian white noise.

Given that the complex noise $e(t)$ has N points altogether, and both the real part $e_r(t)$ and imaginary part $e_i(t)$ follow Gaussian distributions with zero mean and variance σ^2. Then the $E(k)$ derived based on $e(t)$ using DFT transform still follows a complex Gaussian distribution. The $E(k)$'s envelop $|E(k)|$ follows a Raleigh distribution, and the mean of $|E(k)|$ is

$$E[|E(k)|] = \sqrt{\frac{N\pi}{2}}\sigma \tag{6.10}$$

Let (6.10) equals (6.6), in this case we can derive the parameter corresponding to the Raleigh distribution $a = \sigma\sqrt{N}$. Based on (6.8), the threshold Th_1 corresponding to the CFAR P_{fa} after transforming the complex Gaussian noise onto the frequency domain is

$$Th_1 = \sigma\sqrt{-2N\ \ln(P_{fa})} \tag{6.11}$$

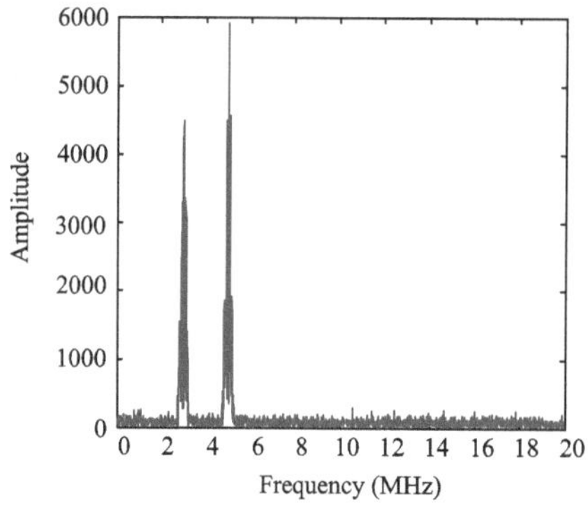

(a) Signal spectrum before interference suppression.

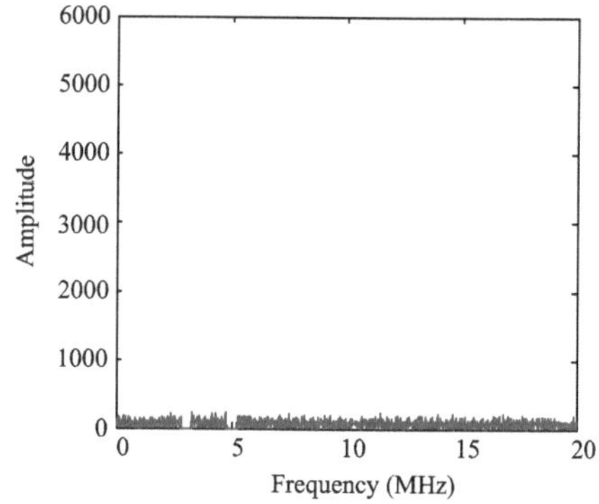

(b) Signal spectrum after interference suppression.

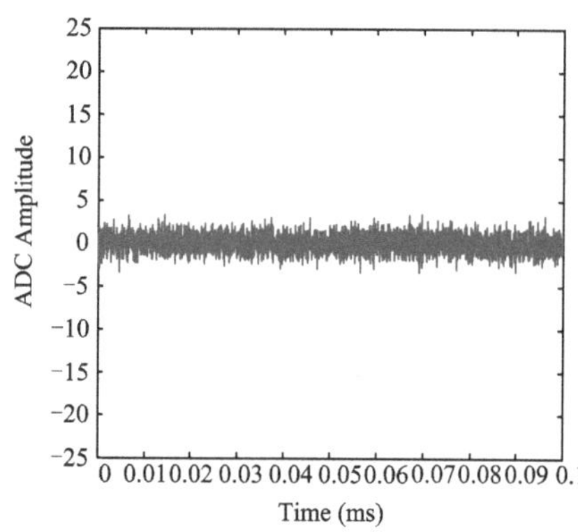

(c) Temporal domain waveform after interference suppression.

Fig. 6.5 Frequency domain notch filtering algorithm used to suppress DME interference

6.2.3 Temporal-Frequency Hybrid Filtering Algorithm

Temporal-frequency hybrid filtering method algorithm is a combination of temporal domain pulse blanking method and the frequency domain filtering. When the signal is received by the receiver, a detection window with a certain time width is used to detect the DME interference. If an interference is detected, then the data under the influence of interference within the window can be transformed onto the frequency domain where the frequency domain filtering algorithm can be used for interference suppression. Otherwise, if no interference detected, no processing is needed. The temporal-frequency hybrid filtering algorithm can preserve more authentic signals compared to the temporal domain pulse blanking algorithm and the frequency domain filtering algorithm. But the temporal-frequency hybrid filtering algorithm still inevitably filters out valid GNSS signals sharing the same frequencies with the

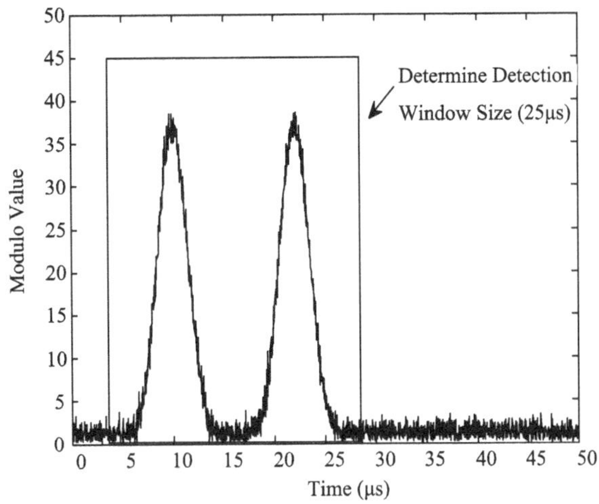

Fig. 6.6 DME pulse pair detection

interference. Figure 6.6 shows how a sliding window is used to detect the DME pulse interference.

Figure 6.7 illustrates the interference suppression using the temporal-frequency hybrid filtering algorithm. Figure 6.7a illustrates how to detect interference using a sliding window in the temporal domain, where the scopes defined by the two rectangular areas are where the interferences are detected, i.e. the two interference pulse pairs detected in the example. Figure 6.7b, c are the frequency spectrums for DME signals corresponding to the two detection windows. The frequency spectrums after a frequency domain threshold is set to filter out DME interferences are shown in Fig. 6.7d, e. Please pay attention to the difference on the ordinate scales for these two subfigures as opposed to those for the Fig. 6.7b, c. Using the frequency domain data after filtering out the interference, the temporal domain data can be obtained using the IFFT transform. Then the data inside the detection windows can be filled back, to obtain the signal's temporal domain waveform after the temporal-frequency domain hybrid filtering, as shown in the Fig. 6.7f.

In terms of the threshold selection for the temporal-frequency domain hybrid filtering algorithm, we can reference (6.11) in which the threshold selection is determined for the frequency domain notch filtering algorithm. The only difference is that the number of samples for the temporal-frequency domain hybrid filtering algorithm is related to the corresponding number of samples in the interference detection window.

One drawback of implementing the temporal-frequency domain hybrid filtering algorithm is that when aliasing exists in the pulse pair, it is very difficult to have accurate positioning on the DME pulses because within a detection window there might be more than one DME pulse pair, and existing DME pulse pairs might not be complete. An incomplete DME pulse pair, after transforming to the frequency domain, can lead to a frequency spectrum leak by the DME pulse pair. As results, the width the frequency spectrum main lobe increases, and more authentic signals

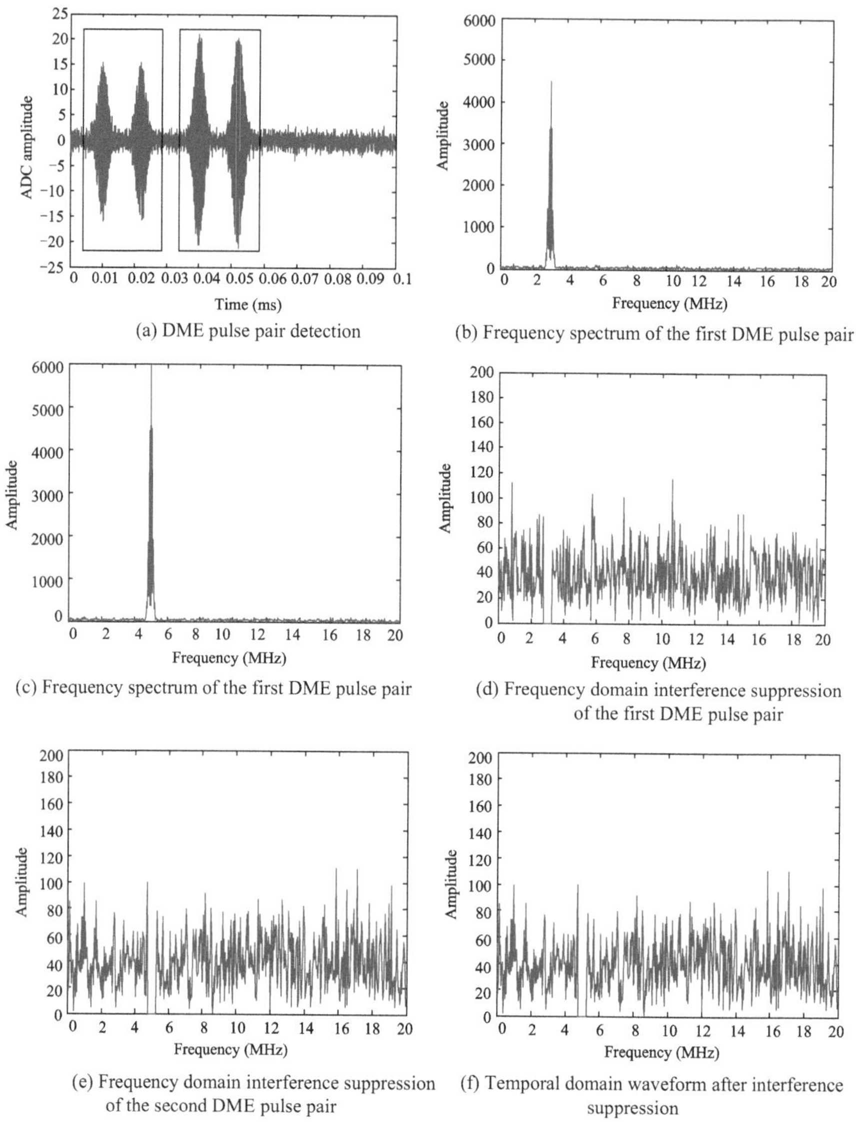

(a) DME pulse pair detection

(b) Frequency spectrum of the first DME pulse pair

(c) Frequency spectrum of the first DME pulse pair

(d) Frequency domain interference suppression of the first DME pulse pair

(e) Frequency domain interference suppression of the second DME pulse pair

(f) Temporal domain waveform after interference suppression

Fig. 6.7 Interference suppression effect of the temporal-frequency domain hybrid filtering algorithm

are removed due to the interference suppression in the frequency domain, and the interference suppression effect is not good, as illustrated in Fig. 6.8. Figure 6.8a shows that DME pulse associated with aliasing is detected in the detection window region; Fig. 6.8b shows the frequency spectrum after the signal inside the detection window is transformed into the frequency domain. Two peaks can be seen, showing

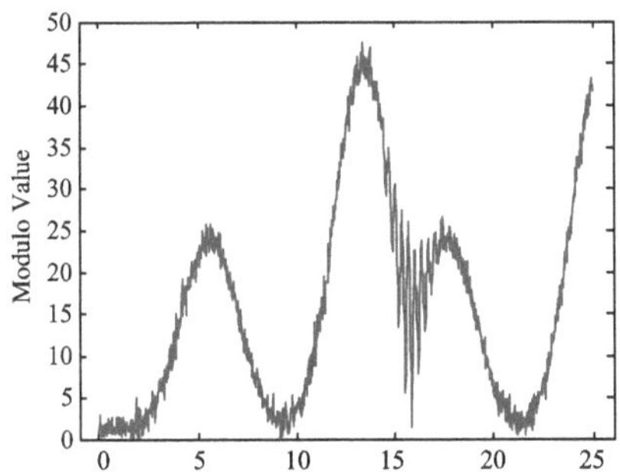

(a) Interference signal with aliasing in the detection window

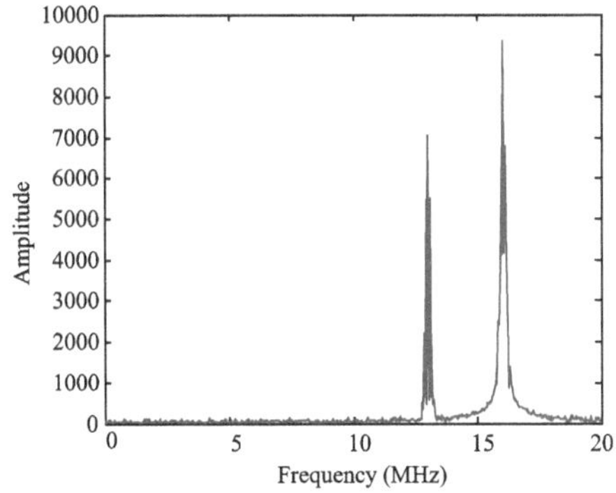

(b) Interference signal frequency spectrum

Fig. 6.8 Interference signal's temporal domain and frequency domain waveforms for pulse pair aliasing

that two DME pulse signals are included in the detection window, and since one DME pulse pair is incomplete in the detection window, after FFT transformation, we have frequency band broadening.

6.3 Pulse Interference Suppression Based on Parameterized Algorithm

The above mentioned algorithms for conventional DME interference suppressions are all non-parametric methods, which have not taken advantage of the a priori information on the DME signal waveform. If we can estimate the frequency, complex amplitude, and time delay of the DME signal, we can reconstruct the DME signal and use it for interference suppression. The methods based on parameterized algorithm can preserve more authentic GNSS signals compared with the non-parametric methods, and have better interference suppression performance. The basic principle is that if the DME interference has no aliasing in the temporal domain, a DME pulse pair is used as the detection object. In the temporal domain, sliding windows are used for interference detection to confirm the window zone in which the interference is detected as the region of interest. Then in the region of interest we estimate the frequency, complex amplitude, and time delay on the DME interference. Then the estimated parameters are used to reconstruct the DME signals in the region of interest. Finally the reconstructed DME signal is subtracted to obtain the signal after interference suppression. Below we introduce the method in detail.

The width of the interference detection window is set as the duration of a DME pulse pair. We assume that there is no aliasing between DME pulses, so it is easy to detect and confirm the signal range of a single complete DME pulse pair. Figure 6.6 shows the detection method. The detection window region in which a single complete DME interference is detected in the received signal is defined as the region of interest. Then the signal in the region of interest can be represented using the mathematical model below:

$$y(t) = \tilde{\beta} s\,(t-\tau)\mathrm{e}^{\mathrm{j}\omega(t-\tau)} + e(t) \tag{6.12}$$

where $\tilde{\beta}$ is the complex amplitude; $s(t)$ is the baseband DME signal as expressed in (6.1); τ is the time delay of the DME signal; ω is the center frequency of the received DME signal; $e(t)$ is the complex noise and GNSS signals buried inside the noise, which usually is assumed to be a Gaussian white noise with zero mean. The expression of the $s(t-\tau)$ is

$$s(t-\tau) = \mathrm{e}^{-\frac{\alpha}{2}(t-\tau)^2} + \mathrm{e}^{-\frac{\alpha}{2}(t-\Delta t-\tau)^2} \tag{6.13}$$

where $\alpha = 4.5 \times 10^{11}\ \mathrm{s}^{-2}$; $\Delta t = 12\ \mu\mathrm{s}$.

Using a sampling interval T_s, the continuous signal in (6.13) can be sampled to obtain the following formula

$$y(n) = \tilde{\beta} s(n_s - \tau)\mathrm{e}^{\mathrm{j}\omega(n_s-\tau)} + e(n), \quad n = 0, 1, \ldots, N-1 \tag{6.14}$$

First, we estimate the frequency ω, which is a problem of estimating the frequency of unknown signal frequency. The Nonlinear Least-Squares (NLS) method can be used for this estimation on frequency. By simplifying (6.14), $\mathrm{e}^{-\mathrm{j}\omega\tau}$ and $\tilde{\beta}$ can be combined as β, then the following minimization problem can be defined [22]

$$\min_{\beta,\omega} \sum_{n=0}^{N-1} \left| y(n_s) - \beta s(n_s - \tau)\mathrm{e}^{\mathrm{j}n\omega} \right|^2 \tag{6.15}$$

To write the above formula in matrix format, let $\mathbf{y} = [y(0), y(1), \ldots, y(N-1)]^{\mathrm{T}}, \mathbf{z} = [s(-\tau), s(1-\tau), \ldots, s((N-1)-\tau)]^{\mathrm{T}}$, $\mathbf{D} = \mathrm{diag}\left\{1, \mathrm{e}^{\mathrm{j}\omega}, \ldots, \mathrm{e}^{\mathrm{j}\omega(N-1)}\right\}$, then (6.15) can be written as

$$\min_{\beta,\omega} \|\mathbf{y} - \beta\mathbf{D}\mathbf{z}\|^2 \tag{6.16}$$

First we fix ω, to solve the above formula, the estimated β can be obtained as

$$\hat{\beta} = \mathbf{z}^{\mathrm{T}}\mathbf{D}^{\mathrm{H}}\mathbf{y}\left(\mathbf{z}^{\mathrm{T}}\mathbf{z}\right)^{-1} \tag{6.17}$$

Since the unknown frequency ω is contained in $\mathbf{D}$, another solution is needed. By substituting (6.17) into (6.16), the following minimization problem can be obtained

$$\min_{\omega}\left\|\left(\mathbf{I}-\mathbf{D}\mathbf{z}\mathbf{z}^{\mathrm{T}}\mathbf{D}^{\mathrm{H}}\left(\mathbf{z}^{T}\mathbf{z}\right)^{-1}\right)\mathbf{y}\right\|^{2} \tag{6.18}$$

It is equivalent to the optimization of the formula below

$$\mathbf{y}^{\mathrm{H}}\mathbf{D}\mathbf{z}\mathbf{z}^{\mathrm{T}}\mathbf{D}^{\mathrm{H}}\mathbf{y}/\mathbf{z}^{\mathrm{T}}\mathbf{z}=\left|\mathbf{z}^{\mathrm{T}}\mathbf{m}\right|^{2}/\mathbf{z}^{\mathrm{T}}\mathbf{z} \tag{6.19}$$

where $\mathbf{m}=\mathbf{D}^{\mathrm{H}}\mathbf{y}$ is a function of unknown frequency ω. Since $\mathbf{m}$ is a complex vector, it can be represented as

$$\mathbf{m}=\mathbf{m}_r+\mathrm{i}\mathbf{m}_i \tag{6.20}$$

where $\mathbf{m}_r$ denotes the real part, and $\mathbf{m}_i$ denotes the imaginary part. Thereby

$$\begin{aligned}\left|\mathbf{z}^{\mathrm{T}}\mathbf{m}\right|^{2}&=\left(\mathbf{z}^{\mathrm{T}}\mathbf{m}_r\right)^{2}+\left(\mathbf{z}^{\mathrm{T}}\mathbf{m}_i\right)^{2}\\&=\mathbf{z}^{\mathrm{T}}\left[\mathbf{m}_r\quad\mathbf{m}_i\right]\begin{bmatrix}\mathbf{m}_r^{\mathrm{T}}\\\mathbf{m}_i^{\mathrm{T}}\end{bmatrix}\mathbf{z}\\&=\mathbf{z}^{\mathrm{T}}\mathbf{M}\mathbf{M}^{T}\mathbf{z}\end{aligned} \tag{6.21}$$

where $\mathbf{M}=\left[\mathbf{m}_r,\mathbf{m}_i\right]$. Notice that $\mathbf{M}^{\mathrm{T}}\mathbf{M}$ is a 2×2 real matrix, and the eigenvalues can be represented as

$$\mathbf{M}^{\mathrm{T}}\mathbf{M}=\mathbf{U}\begin{bmatrix}\lambda_1&0\\0&\lambda_2\end{bmatrix}\mathbf{U}^{\mathrm{T}}=\mathbf{U}\Lambda\mathbf{U}^{\mathrm{T}} \tag{6.22}$$

where $\mathbf{U}=\left[\mathbf{u}_1,\mathbf{u}_2\right]$ is an orthogonal matrix composed of two eigenvectors, and we can assume that the eigenvalues satisfy the condition that $\lambda_1>\lambda_2$. The following formula can be computed

$$\begin{aligned}\left(\mathbf{M}\mathbf{M}^{\mathrm{T}}\right)\underbrace{\mathbf{M}\left(\mathbf{M}^{\mathrm{T}}\mathbf{M}\right)^{-1/2}\mathbf{U}}_{\mathbf{V}}&=\mathbf{M}\left(\mathbf{M}^{\mathrm{T}}\mathbf{M}\right)^{1/2}\mathbf{U}\\&=\mathbf{M}\left(\mathbf{M}^{\mathrm{T}}\mathbf{M}\right)^{-1/2}\mathbf{U}\underbrace{\mathbf{U}^{\mathrm{T}}\left(\mathbf{M}^{\mathrm{T}}\mathbf{M}\right)\mathbf{U}}_{\Lambda}\\&=\mathbf{V}\Lambda\end{aligned} \tag{6.23}$$

That means that for the eigenvalues of the $\mathbf{M}\mathbf{M}^{T}$, $N-2$ of them are equal to zero, and the other two non-zero eigenvalues are λ_1 and λ_2 respectively. The corresponding eigenvectors for the eigenvalue λ_1 are

$$\mathbf{v}_1 = \mathbf{M}\left(\mathbf{M}^{\mathrm{T}}\mathbf{M}\right)^{-1/2}\mathbf{u}_1 \tag{6.24}$$

Thereby, the solutions of $\mathbf{z}$ and ω are

$$\hat{\mathbf{z}} = \mathbf{v}_1 \tag{6.25}$$

$$\hat{\omega} = \arg\max_{\omega} \lambda_1(\omega) \tag{6.26}$$

Next we derive the detailed expression for deriving $\lambda_1(\omega)$. Since

$$\begin{aligned} &\begin{vmatrix} \mathbf{m}_r^{\mathrm{T}}\mathbf{m}_r - \lambda & \mathbf{m}_r^{\mathrm{T}}\mathbf{m}_i \\ \mathbf{m}_r^{\mathrm{T}}\mathbf{m}_i & \mathbf{m}_i^{\mathrm{T}}\mathbf{m}_i - \lambda \end{vmatrix} \\ &= \lambda^2 - \lambda\left(\mathbf{m}_r^{\mathrm{T}}\mathbf{m}_r - \mathbf{m}_i^{\mathrm{T}}\mathbf{m}_i\right) + \left(\mathbf{m}_r^{\mathrm{T}}\mathbf{m}_r\right)\left(\mathbf{m}_i^{\mathrm{T}}\mathbf{m}_i\right) - \left(\mathbf{m}_r^{\mathrm{T}}\mathbf{m}_i\right)^2 \end{aligned} \tag{6.27}$$

We can obtain

$$2\lambda_1 = \left(\mathbf{m}_r^{\mathrm{T}}\mathbf{m}_r + \mathbf{m}_i^{\mathrm{T}}\mathbf{m}_i\right) + \left[\left(\mathbf{m}_r^{\mathrm{T}}\mathbf{m}_r - \mathbf{m}_i^{\mathrm{T}}\mathbf{m}_i\right)^2 + 4\left(\mathbf{m}_r^{\mathrm{T}}\mathbf{m}_i\right)^2\right]^{1/2} \tag{6.28}$$

Notice

$$\mathbf{m}_r^{\mathrm{T}}\mathbf{m}_r + \mathbf{m}_i^{\mathrm{T}}\mathbf{m}_i = \mathbf{m}^{\mathrm{H}}\mathbf{m} \tag{6.29}$$

$$\left(\mathbf{m}_r^{\mathrm{T}}\mathbf{m}_r - \mathbf{m}_i^{\mathrm{T}}\mathbf{m}_i\right)^2 + 4\left(\mathbf{m}_r^{\mathrm{T}}\mathbf{m}_i\right)^2 = \left|\mathbf{m}^{\mathrm{T}}\mathbf{m}\right|^2 \tag{6.30}$$

As a result, ω's estimator $\hat{\omega}$ can be obtained by maximizing the following formula

$$\mathbf{m}^{\mathrm{H}}\mathbf{m} + \left|\mathbf{m}^{\mathrm{T}}\mathbf{m}\right| \tag{6.31}$$

Also since $\mathbf{m}^{\mathrm{H}}\mathbf{m}$ is not related to ω, we can obtain ω's solution $\hat{\omega}$ as

$$\hat{\omega} = \arg\max_{\omega}\left|\sum_{n=0}^{N-1} y^2(n)\mathrm{e}^{-\mathrm{j}2\omega n}\right|^2 \tag{6.32}$$

The formula for the solution above is similar to the formula of discrete Fourier transformation. As a result the solution of $\hat{\omega}$ in the above formula can be obtained using FFT, and additionally $\hat{\omega}$ is half of the frequency corresponding to the peak of the frequency spectrum obtained by performing FFT on $y^2(n)$.

Once we have the estimate on frequency, we can estimate the complex amplitude β and time delay τ using the estimated frequency. The estimation on these two parameters can also be solved using the NLS method in the frequency domain.

Given $x(n) = s(n)e^{j\omega n}$, we have $x(n-\tau) = s(n_s - \tau)e^{j\omega(n-\tau)}$. Then (6.14) can be rewritten as

$$y(n) = \tilde{\beta}x(n-\tau) + e(n), \quad n = 0, 1, \ldots, N-1 \tag{6.33}$$

By converting the signal in the (6.33) onto the frequency domain using Fourier transform, we have the following formula [23]

$$Y(k) = \tilde{\beta}X(k)e^{j\omega_d k} + E(k) \tag{6.34}$$

where $Y(k)$, $X(k)$, and $E(k)$ $(k = -\mathrm{N}/2, -\mathrm{N}/2+1, \ldots, \mathrm{N}/2-1)$ represent the discrete Fourier transform for $y(n)$, $x(n)$ and $e(n)$; and $\omega_d = -2\pi\tau/N$. The estimation on time delay is included in ω_d. Below we solve the problem related to β and ω_d, which can be regarded as the least squares problem below

$$\min_{\tilde{\beta},\omega_d} \sum_{k=-N/2}^{N/2-1} \left|Y(k) - \tilde{\beta}X(k)e^{j\omega_d k}\right|^2 \tag{6.35}$$

If we let
$\mathbf{Y} = [Y(-N/2), Y(-N/2+1), \ldots, Y(N/2-1)]^{\mathrm{T}}$,
$\mathbf{X} = \mathrm{diag}\{X(-N/2), X(-N/2+1), \ldots, X(N/2-1)\}$,
$\mathbf{a}(\omega_d) = \left[e^{j\omega_d(-N/2)}, e^{j\omega_d(-N/2+1)}, \ldots, e^{j\omega_d(N/2-1)}\right]^{\mathrm{T}}$, then (6.35) can be written as

$$\min_{\tilde{\beta},\omega_d} \left\|\mathbf{Y} - \tilde{\beta}\mathbf{X}\mathbf{a}(\omega_d)\right\|^2 \tag{6.36}$$

Based on the formula above, the solutions on $\tilde{\beta}$ and ω_d can be given by the two formulas below

$$\hat{\omega}_d = \arg\max_{\omega_d} \left|\mathbf{a}^{\mathrm{H}}(\omega_d)(\mathbf{X}^*\mathbf{Y})\right|^2 \tag{6.37}$$

$$\hat{\tilde{\beta}} = \left.\frac{\mathbf{a}^{\mathrm{H}}(\omega_d)(\mathbf{X}^*\mathbf{Y})}{\|\mathbf{X}\|^2}\right|_{\omega_d=\hat{\omega}_d} \tag{6.38}$$

Based on (6.37), we can obtain the time delay τ's estimate $\hat{\tau}$

$$\hat{\tau} = -\frac{\hat{\omega}_d N}{2\pi f_s} \tag{6.39}$$

Based on (6.32), (6.38) and (6.39), we can obtain the DME interference signal's frequency, complex amplitude and time delay $\hat{\omega}_d$, $\hat{\tilde{\beta}}$ and $\hat{\tau}$ in the region of interest; then the reconstructed DME signal $\hat{y}_{DME}(n)$ can be obtained by using these three parameters

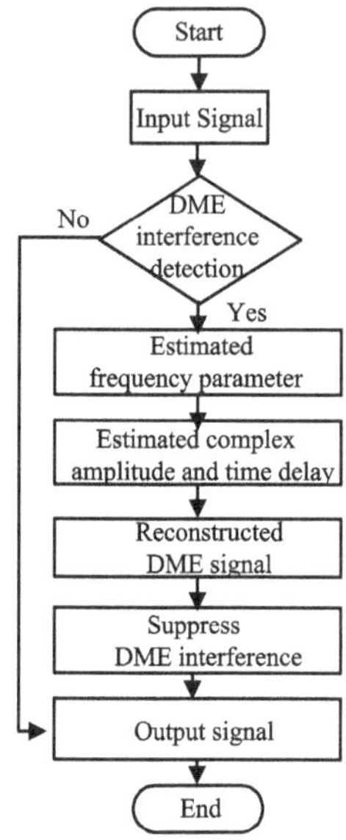

Fig. 6.9 Flow chart for DME interference suppression based on parameterized algorithm

$$\hat{y}_{DME}(n) = \hat{\hat{\beta}} s(n - \hat{\tau}) e^{j\hat{\omega}_I(n-\hat{\tau})} \tag{6.40}$$

Finally in the temporal domain, the reconstructed DME signal can be subtracted to obtain the signal after the interference suppression $\hat{y}(n)$

$$\hat{y}(n) = y(n) - \hat{y}_{DME}(n) \tag{6.41}$$

Figure 6.9 shows the flow chart for DME interference suppression based on parameterized algorithm.

6.4 Pulse Interference Suppression Based on Wavelet Packet Transformation

In the two previous sections, we discussed several different pulse interference suppression algorithms. In Sect. 6.2, we have studied temporal domain pulse blanking, frequency domain filtering and temporal-frequency hybrid filtering algorithms. When the DME interference pulse density is higher, these algorithms can bring larger losses on the authentic GNSS signals, thereby negatively impact the performance of pulse interference suppression. The pulse interference suppression algorithm based on parameterized algorithm in Sect. 6.3 takes advantage of the known a priori information on the DME pulse waveform, so it reduces the GNSS signal loss while suppressing the interference. But, when there is overlapping between DME pulses, or the DME pulse's waveform is incomplete in the received signal, the performance of the pulse interference suppression algorithm based on parameterized algorithm degrades. In this section we implement DME interference suppression from the perspective of wavelet packet transformation. First the wavelet theory is introduced, where we introduce the basic principle of DME pulse interference suppression based on wavelet packet transformation, then we discuss the parameter selection, including the selections on wavelet packet coefficient domain threshold, wavelet packet decomposition series and wavelet function type.

6.4.1 *Wavelet Theory*

1. Continuous Wavelet Transformation

Wavelet transformation is done using localized analyses based on time and frequency. It performs multi-scale decomposition on the signals using scaling and shifting operations. In the low frequency part, there is higher frequency resolution and lower time resolution; in the high frequency part, there is higher time resolution and lower frequency resolution. In other words, the wavelet transformation provides a temporal-frequency window that is fully scalable, so it can automatically adapt to the requirements on non-stationary signal analyses.

The continuous wavelet transformation of a function $f(t)$ can be defined as [24]

$$\gamma(s,\tau) = \int_{-\infty}^{\infty} f(t)\psi_{s,\tau}^{*}(t)\mathrm{d}t \tag{6.42}$$

where $(\cdot)^*$ denotes the complex conjugate; $\psi_{s,\tau}(t)$ is the wavelet's basis function, also known as wavelet. It is generated from a single mother wavelet function $\psi(t)$ by scaling and shifting, as shown in the formula below

$$\psi_{s,\tau}(t) = \frac{1}{\sqrt{s}}\psi\left(\frac{t-\tau}{s}\right) \tag{6.43}$$

Here $\psi_{s,\tau}(t)$ relies on parameter s and τ, where s is the scale factor and τ is the shift factor. $1/\sqrt{s}$ is used to implement energy normalization among different scales. The function of the scale factor s is to scale the mother wavelet function $\psi(t)$, stretching when $s > 1$ and shrinking when $s < 1$. The larger s is, the wider $\psi\left(\frac{t}{s}\right)$ becomes, which means that a general observation is made on the signal. When s is smaller, a detailed observation is made on the signal. Equation (6.42) shows how a function $f(t)$ is decomposed into a series of wavelet basis functions. Figure 6.10 shows a continuous wavelet transformation operation.

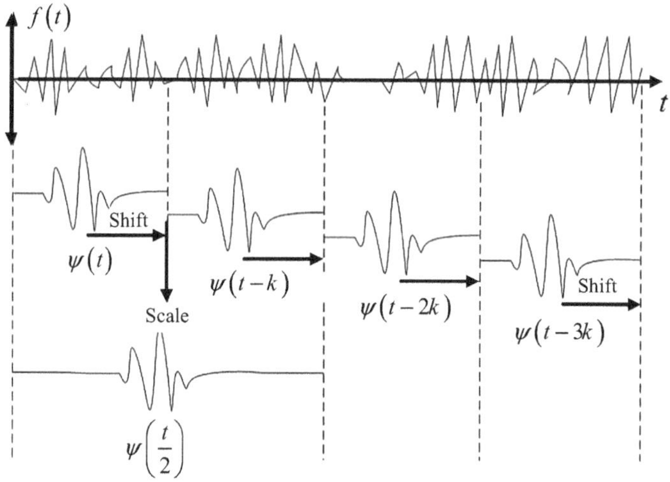

Fig. 6.10 Continuous wavelet transformation operations

Given that the Fourier transform function of the mother wavelet function is $\Psi(\omega)$, then the mother wavelet function $\psi(t)$ needs to satisfy the following conditions [24]

(1) Square integrable.
(2) $\int_{-\infty}^{\infty}\psi(t)\,\mathrm{d}t = 0$ [i.e. $\psi(0) = 0$].
(3) Permitting condition: $C_\psi = \int_{+\infty}^{-\infty}\frac{|\Psi(\omega)|^2}{\omega}\mathrm{d}\omega < \infty$.

The inverse transformation of a continuous wavelet is

$$f(t) = \iint \psi_{s,\tau}^{*}(t)\gamma(s,\tau)\,\mathrm{d}s\mathrm{d}\tau \tag{6.44}$$

In (6.42)–(6.44), no wavelet basis function has been specified. That is the difference between wavelet transformation, Fourier transformation, and other transformations. Wavelet transformation theory only studies the general properties of wavelet and wavelet transformation. In actual applications, proper mother wavelet function can be selected or a detailed mother wavelet function can be designed for different application scenarios.

2. Discrete Wavelet Transformation

In (6.42), the computation of the wavelet transformation can be realized by continuously shifting a scalable function on the signal being analyzed and calculating the correlation between the two. Wavelet maps a one-dimension signal to a two-dimensional time-scale joint function, but this expression is highly redundant. For most applications we want to use as few components as possible to describe a signal, therefore discrete wavelet is introduced. A discrete wavelet cannot continuously scale and shift. It can only scale and shift with discrete steps.

Discrete wavelet can be represented as

$$\psi_{j,k}(t) = \frac{1}{\sqrt{s_0^j}}\psi\left(\frac{t - k\tau_0 s_0^j}{s_0^j}\right) \tag{6.45}$$

where j and k are integers, $s_0 > 1$ is a fixed scaling step; the shifting factor τ_0 determines the scaling step. The effect of discretizing the wavelet is equivalent to sampling the time-scale space at discrete intervals. Usually we select $s_0 = 2$, $\tau_0 = 1$, i.e. defining $\psi_{j,k}(t)$ as

$$\psi_{j,k}(t) = \frac{1}{\sqrt{2^j}}\psi\left(\frac{t}{2^j} - k\right) \tag{6.46}$$

Therefore a function $f(t)$'s binary discrete wavelet transform is defined as

$$\gamma(j,k) = \int_{-\infty}^{\infty} f(t)\psi_{j,k}^*(t)\,\mathrm{d}t \tag{6.47}$$

i.e., the discrete wavelet transform performs discrete processing on the scale factor s and shift factor τ for a continuous waveform transformation's wavelet function.

Even if discrete wavelet function is used, we still need an infinite number of scaling and shifting to calculate the wavelet transformation, so we need to find a method to convert this infinite number to a limited number for practical implementations. Since a wavelet has a bandpass type of frequency spectrum. Fourier transform theory tells us that the scaling on the temporal domain is equivalent to the compression and shifting towards a low frequency band in the frequency domain. Given a function $f(t)$'s Fourier transformation is $F(\omega)$, then the Fourier transformation of the function $f\left(\frac{t}{a}\right)$ is

$$F\left[f\left(\frac{t}{a}\right)\right] = |a|F(a\omega) \tag{6.48}$$

Equation (6.48) illustrates the scaling on the function with a factor $a = 2$ on the temporal domain. It compresses the function's frequency spectrum by a factor of 2 and shifts it towards the low frequency. Given wavelet functions $\psi_{1,k}(t)$, $\psi_{2,k}(t)$, $\psi_{3,k}(t)$ and $\psi_{4,k}(t)$ have frequency spectrum functions of Ψ_1, Ψ_2, Ψ_3 and Ψ_4, then the frequency and bandwidth for Ψ_2 are only half of the frequency and bandwidth for Ψ_1. And the frequency and bandwidth for Ψ_3 are half of the frequency and bandwidth for Ψ_2, i.e. the ratios of center frequencies and the bandwidths (also known as quality factor) do not change. As a result, a scalable wavelet frequency

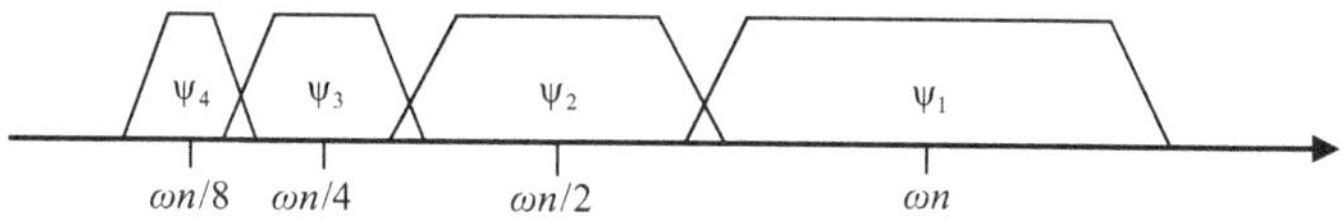

Fig. 6.11 Overlapping wavelet frequency spectrums obtained by scaling the wavelet in the temporal domain

spectrum can be used to cover a signal's limited spectrum. The scaled wavelet frequency spectrum should overlap with each other, as shown in Fig. 6.11.

But, if we only use the wavelet function for segmentation, every time the frequency spectrum is divided and only a half remains, the wavelet frequency spectrum cannot cover the whole frequency down to the zero frequency. But the number of functions after the segmentations is still infinite. After the segmentation reaches a certain frequency, the segmentation stops, and all the remaining low frequency parts need to be represented using a function with a low pass frequency spectrum. This function is the scale function $\varphi(t)$ which can be represented using the corresponding wavelet

$$\varphi(t) = \sum_{j,k} \gamma(j,k)\psi_{j,k}(t) \tag{6.49}$$

Scale functions $\varphi_{j,k}(t)$ for different scales can be defined as

$$\varphi_{j,k}(t) = \frac{1}{\sqrt{2^j}}\varphi\left(\frac{t}{2^j} - k\right) \tag{6.50}$$

Figure 6.12 illustrates the scale functions. The scale functions cover all the areas that are not covered by the wavelet function frequency spectrums until the scale j, consequently the number of wavelets reduces from infinite to finite.

By adding a wavelet function to the original scale function, we can derive a new scale function with a frequency spectrum which doubles that of the original scale function. It is equivalent to say that the original scale function can be expressed by the new scale function. Since the scale function has a low pass type of frequency spectrum, based on the new scale function, we can derive the original scale function using a low pass filtering. Then the equations of two-scale related to $\varphi(t)$ can be expressed as

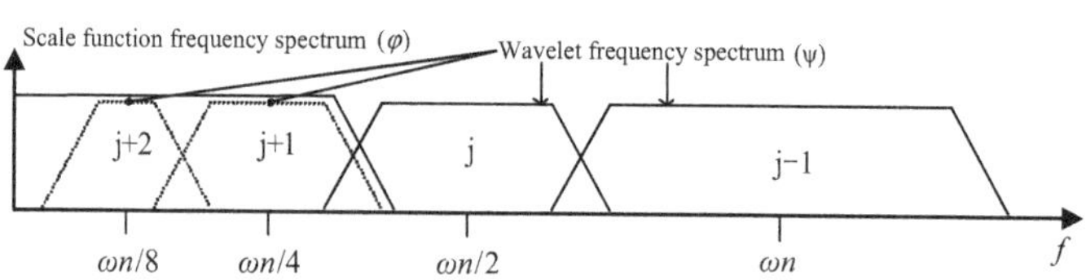

Fig. 6.12 Introduction of scale functions

$$\varphi_{j,0}(t) = \sum_{k} h_{j-1}(k)\varphi_{j-1,k}(t) \tag{6.51}$$

where $h_{j-1}(k)$ can be regarded as the low pass filter's $(j-1)$ th stage. The above formula shows that a scale function at one level can be expressed using the scale function with its previous level. For the same reason, since the wavelet has a band-pass type of frequency spectrum, and the new scale function contains the added wavelet function, then by using the new scale function, we can derive the added wavelet function using a high-pass filter. As a result, the wavelet function can be expressed as

$$\psi_{j,0}(t) = \sum_{k} g_{j-1}(k)\varphi_{j-1,k}(t) \tag{6.52}$$

where $g_{j-1}(k)$ is the $(j-1)$th high-pass filter. The above formula is also the two-scale relation between the wavelet function and the scale function. Next, we study the relations among low-pass filters with various scales and various high-pass filters with various scales. To make it easy for derivation, for the derivation process below we assume that the scale function and wavelet function are all real numbers. Based on (6.51), we can have the derivation below

$$\begin{aligned} h_{j-1}(k) &= \langle \varphi_{j,0}(t), \varphi_{j-1,k}(t) \rangle = \int \varphi_{j,0}(t)\varphi_{j-1,k}(t)\mathrm{d}t \\ &= \int \frac{1}{\sqrt{2^j}}\varphi\left(\frac{t}{2^j}\right)\frac{1}{\sqrt{2^{j-1}}}\varphi\left(\frac{t}{2^{j-1}} - k\right)\mathrm{d}t \\ &= \int \frac{1}{\sqrt{2}}\varphi\left(\frac{t'}{2}\right)\varphi(t'-k)\mathrm{d}t' \quad \left[\frac{t}{2^{j-1}} = t'\right] \\ &= \langle \varphi_{1,0}(t), \varphi_{0,k}(t) \rangle = h_0(k) \end{aligned} \tag{6.53}$$

The above formula shows that the low-pass filter corresponding to any scale is equivalent to $h_0(k)$. Similarly we can derive that

$$g_{j-1}(k) = g_0(k) \tag{6.54}$$

i.e., the high-pass filter equivalent to any scale is equivalent to $g_0(k)$. As a result we can uniformly represent the low-pass filters and high-pass filters with any scales as $h(k)$ and $g(k)$. Then by substituting (6.46) and (6.50) into (6.51) and (6.52) and after simplifying, the following two-scale equation can be derived

$$\varphi\left(\frac{t}{2^j}\right) = \sqrt{2}\sum_{k} h(k)\varphi\left(\frac{t}{2^{j-1}} - k\right) \tag{6.55}$$

$$\psi\left(\frac{t}{2^j}\right) = \sqrt{2}\sum_{k} g(k)\varphi\left(\frac{t}{2^{j-1}} - k\right) \tag{6.56}$$

Thereby, for function $f(t)$, either we can express it as a scale function with the scale $j-1$

$$f(t) = \sum_m \lambda_{j-1}(m)\varphi_{j-1,m}(t) \tag{6.57}$$

Or we can jointly express it using scale functions up to the scale j and the wavelet function

$$f(t) = \sum_k \lambda_j(k)\varphi_{j,k}(t) + \sum_k \gamma_j(k)\psi_{j,k}(t) \tag{6.58}$$

where $\lambda_{j-1}(m)$, $\lambda_j(k)$ is known as the coarse coefficient or low frequency coefficient; $\gamma_j(k)$ is known as the detail coefficient or high-frequency coefficient. We can also regard $\sum_k \lambda_j(k)\varphi_{j,k}(t)$ as an coarse approximation of signal $f(t)$, and regard $\sum_k \gamma_j(k)\psi_{j,k}(t)$ as a detail approximation of the signal $f(t)$. If we let (6.57) equals (6.58), we can have the following equation

$$\sum_k \gamma_j(k)\psi_{j,k}(t) = \sum_m \lambda_{j-1}(m)\varphi_{j-1,m}(t) - \sum_k \lambda_j(k)\varphi_{j,k}(t) \tag{6.59}$$

The above equation shows that the detail approximation is the difference between the two-scale approximations, i.e. reflecting the detailed difference between the two approximations.

3. Multi-resolution Analysis

If the scale function $\varphi_{j,k}(t)$ and the wavelet function $\psi_{j,k}(t)$ are orthogonal, i.e. the inner product $\langle \varphi_{j,k}(t), \psi_{j,k}(t) \rangle = 0$; then the coefficients λ_j and γ_j can be obtained using the inner products between $f(t)$ and the scale function, and between the $f(t)$ and the wavelet function

$$\lambda_j(k) = \langle f(t), \varphi_{j,k}(t) \rangle \tag{6.60}$$

$$\gamma_j(k) = \langle f(t), \psi_{j,k}(t) \rangle \tag{6.61}$$

Next we solve for $\lambda_j(k)$ and $\gamma_j(k)$. $\lambda_j(k)$ is solved first, and then by substituting (6.58) into (6.61), we have

$$\begin{aligned} \lambda_j(k) &= \left\langle \sum_m \lambda_{j-1}(m)\varphi_{j-1,m}(t), \varphi_{j,k}(t) \right\rangle \\ &= \sum_m \langle \varphi_{j-1,m}(t), \varphi_{j,k}(t) \rangle \lambda_{j-1}(m) \end{aligned} \tag{6.62}$$

By solving for $\langle \varphi_{j-1,m}(t), \varphi_{j,k}(t) \rangle$ independently, we have

$$\begin{aligned}
\langle \varphi_{j-1,m}(t), \varphi_{j,k}(t) \rangle &= \int \varphi_{j-1,m}(t) \varphi_{j,k}(t) \mathrm{d}t \\
&= \int \frac{1}{\sqrt{2^{j-1}}} \varphi\left(\frac{t}{2^{j-1}} - m\right) \frac{1}{\sqrt{2^j}} \varphi\left(\frac{t}{2^j} - k\right) \mathrm{d}t \\
&= \int \frac{1}{\sqrt{2^{j-1}}} \varphi\left[\frac{t'}{2^{j-1}} - (m-2k)\right] \frac{1}{\sqrt{2^j}} \varphi\left(\frac{t'}{2^j}\right) \mathrm{d}t' \qquad \left[令 \frac{t}{2^j} - k = \frac{t'}{2^j}\right] \\
&= \int \frac{1}{\sqrt{2^j}} \varphi\left(\frac{t'}{2^j}\right) \frac{1}{\sqrt{2^{j-1}}} \varphi\left[\frac{t'}{2^{j-1}} - (m-2k)\right] \mathrm{d}t' \\
&= \langle \varphi_{j,0}(t), \varphi_{j-1,m-2k}(t) \rangle = h(m-2k)
\end{aligned} \tag{6.63}$$

Then we have

$$\lambda_j(k) = \sum_m h(m-2k) \lambda_{j-1}(m) \tag{6.64}$$

Next we solve for $\gamma_j(k)$ by substituting (6.58) into (6.61), we have

$$\begin{aligned}
\gamma_j(k) &= \left\langle \sum_k \lambda_j(k) \varphi_{j,k}(k) + \sum_k \gamma_j(k) \psi_{j,k}(t), \psi_{j,k}(t) \right\rangle \\
&= \left\langle \sum_k \gamma_j(k) \psi_{j,k}(t), \psi_{j,k}(k) \right\rangle \\
&= \left\langle \sum_m \lambda_{j-1}(m) \varphi_{j-1,m}(t) - \sum_k \lambda_j(k) \varphi_{j,k}(t), \psi_{j,k}(t) \right\rangle \\
&= \left\langle \sum_m \lambda_{j-1}(m) \varphi_{j-1,m}(t), \psi_{j,k}(t) \right\rangle \\
&= \sum_m \langle \varphi_{j-1,m}(t), \psi_{j,k}(t) \rangle \lambda_{j-1}(m)
\end{aligned} \tag{6.65}$$

Now we solve $\langle \varphi_{j-1,m}(t), \psi_{j,k}(t) \rangle$ independently, which is similar to the solution process in (6.63), and the result is

$$\langle \varphi_{j-1,m}(t), \psi_{j,k}(t) \rangle = g\,(m-2k) \tag{6.66}$$

By substituting (6.66) into (6.65), we can derive the solution of $\gamma_j(k)$ as

$$\gamma_j(k) = \sum_m g(m - 2k)\lambda_{j-1}(m) \tag{6.67}$$

From (6.64) and (6.67), the scale function coefficient $\lambda_j(k)$ and wavelet function coefficient $\gamma_j(k)$ on a certain scale can be obtained by applying the low-pass filter $h(k)$ and the high-pass filter $g(k)$ on the scale function coefficient $\lambda_{j-1}(k)$ at the previous scale level. Equations (6.64) and (6.67) can jointly form the stage of an iterative digital filter, as the result the wavelet transform can be implemented using an iterative digital filter. After every iteration, the number of samples for the next stage reduces by half, turning a discrete wavelet decomposition problem into a simple inner product operation.

If we regard the sampled signal $f(k)$ as the scale function coefficient λ_{j-1} at the next scale stage, we can decompose the signal using the known decomposition low-pass filter $h(k)$ and decomposition high-pass filter $g(k)$. Given that the total bandwidth occupied by the original signal is $(0 \sim f/2)_s$, we can define the space V_0, where f_s is the sampling frequency. After one-stage decomposition, V_0 can be divided into two subspaces of low frequency V_1 $(0 \sim f_s/4)$ and high-frequency W_1 $(f_s/4 \sim f_s/2)$. After the second-stage decomposition, the low-frequency part V_1 can be further divided into two subspaces of low frequency V_2 $(0 \sim f_s/8)$ and high frequency W_2 $(f_s/8 \sim f_s/4)$, as shown in Fig. 6.13. By iteratively decomposing the signal into different frequency bands, multi-resolution analysis can be realized. The wavelet reconstruction on the decomposed signals is an inverse process of the wavelet decomposition. After performing a 1:2 interpolation on the next stage's low frequency coefficients $\lambda_j(k)$ and high frequency coefficients $\gamma_j(k)$, the summation of the outputs derived by the reconstructed low-pass filter $\overline{h}(k)$ and the reconstructed high-pass filer $\overline{g}(k)$ is the low-frequency coefficients $\lambda_{j-1}(k)$ of the previous stage. Figure 6.13 illustrates two-stage wavelet decomposition (wavelet transform). Figure 6.14 illustrates a two-stage wavelet reconstruction (wavelet inverse transform).

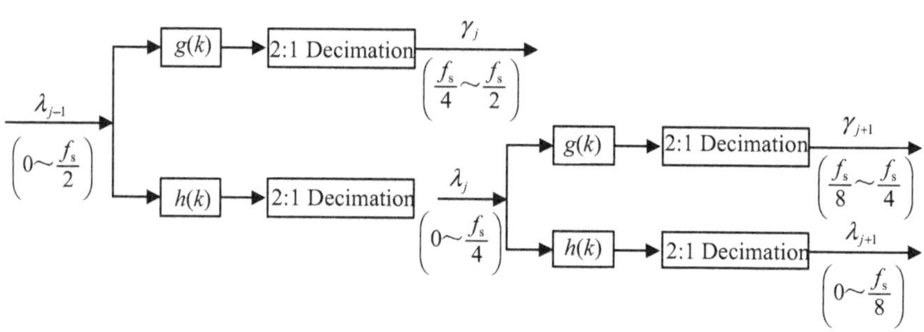

Fig. 6.13 Two-stage wavelet decomposition

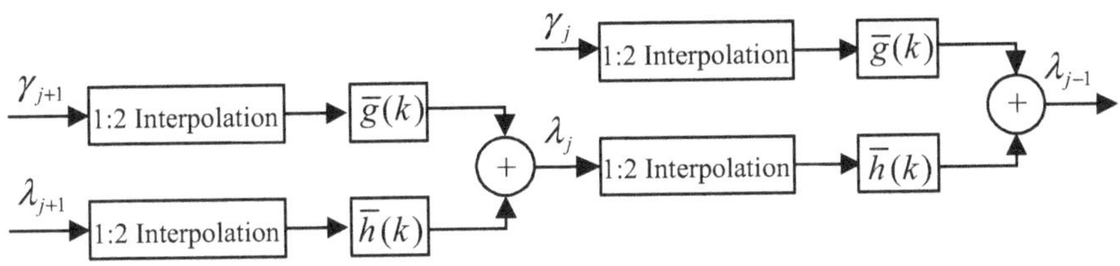

Fig. 6.14 Two-stage wavelet reconstructions

4. Wavelet Packet Transformation

For multi-resolution analysis based on wavelet transformation introduced in the previous section, every decomposition step only decomposes the low-frequency sub-band, and does not decompose the high-frequency sub-band, leading to worse frequency resolution in the high frequency band. In practice, the space corresponding to the low frequency part of the wavelet decomposition is represented as V_i, and the high frequency part of the wavelet decomposition is represented as W_i. The high and low frequencies here are in terms of the frequency band of the previous stage. Wavelet transformation only performs binary sub-division on various V_i spaces, and the same operation can be performed on various W_i spaces as well. Thereby, the wavelet packet transformation is introduced, i.e. the wavelet transformations are performed simultaneously on the low frequency space V_i and high frequency space W_i, consequently more refined signal decomposition can be achieved compared with the wavelet decomposition, to improve the temporal-frequency resolution.

Given that every space has its corresponding integer displacement orthonormal basis, which can be represented using $w_n^{(2^j)}(t), n = 0 \sim 2^j - 1$, or can be abbreviated as $w_n^{(j)}$; the superscript (2^j) denotes the scale; j is the scale level; and the subscript n is the space serial number for that level. For example, for j = 2, there are 4 sets of orthonormal bases $w_0^{(4)}(t), \ldots, w_3^{(4)}$ (only using the base wavelet with zero displacement as the example); The set of orthonormal bases can form a wavelet packet. Given that the wavelet packet takes the forms of the set of all subspaces at jth stage and the corresponding filters, then the basic wavelets for various spaces $w_n^{(j)}(t)$ have the following definitions [24]: the even components are

$$w_{2n}^{(j)}(t) = \sqrt{2}\sum_k h(k) w_n^{(j)}(2t - k) \tag{6.68}$$

And the odd components are

$$w_{2n+1}^{(j)}(t) = \sqrt{2}\sum_k g(k) w_n^{(j)}(2t - k) \tag{6.69}$$

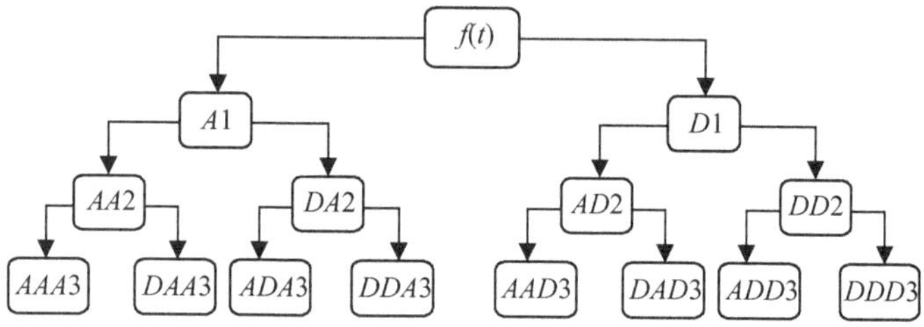

Fig. 6.15 Three-stage wavelet packet decomposition tree

where $j, k \in Z$ (Z is an integer set); $n \in N$ (N is a non-negative integer set); k is the time or position parameter, $h(k)$ and $g(k)$ are a set of conjugate mirror filters. When $n = 0$, the relation in (6.68) corresponds to the two-scale equation related to the scale function $\varphi(t)$ in (6.55); $w_0^{(j)}(t)$ is the scale function; the relation in (6.69) corresponds to the two-scale equation related to the scale function $\varphi(t)$ and the wavelet function $\psi(t)$ in (6.56); $w_1^{(j)}(t)$ is the wavelet function. And various other $w_n^{(j)}(t)$ at the same stage can be derived using (6.68) when the space serial number is an even number, or (6.69) when the space serial number is an odd number.

The process of using a filter set to implement wavelet transformation can be analyzed below: After the analyzed signals are processed by the decomposition low-pass filter and the decomposition high-pass filter, the signal frequency band is evenly divided into two frequency bands (low-frequency and high-frequency). The low-frequency signal and high-frequency signal, after the down-sampling, are decomposed for the next round using low-pass and high-pass filters so their low-frequency and high-frequency bands are divided again. This process repeats until the whole frequency band is divided into an even number of frequency bands. 2^N subband coefficients can be obtained by performing N-stage wavelet packet transform. Figure 6.15 shows a three-stage wavelet packet decomposition tree for a signal $f(t)$ where A denotes the low-frequency coefficients when the signal goes through a low-frequency filter, and D denotes the high-frequency coefficients when the signal goes through a high-frequency filter. The serial number at the end denotes the number of stages for the wavelet packet transform.

6.4.2 Suppression Method Based on Wavelet Packet Transformation

1. Algorithm Principle

Based on the introduction in the previous section, the wavelet transformation, essentially, is a method of measuring the degree of similarity on the waveforms. The more similar a signal is to a wavelet, the bigger the coefficient of the wavelet packet becomes. To perform DME interference suppression for GNSS, the GNSS signal is buried in the noise, and the similarity between the wavelet and the noise is smaller than the similarity between the wavelet and the DME signal. Therefore we can convert the received signal under the impact of DME interference onto the wavelet packet coefficient domain using wavelet packet transform. The coefficients with larger absolute values characterize the DME interference signals, and the coefficients with smaller absolute values characterize the noise and authentic GNSS signals. By setting a threshold in the wavelet packet domain, we can detect and suppress the interferences by performing zero-setting on coefficients with larger absolute values. Finally we can perform wavelet packet inverse transformation on the processed wavelet packet coefficients to obtain time-domain signal after interference suppression. The DME interference suppression algorithm based on wavelet packet transformation mainly has the following three steps:

(1) Wavelet packet transformation: select a proper wavelet function and wavelet packet decomposition level, to perform L-level wavelet packet transformation on the received signal under the impact of DME interference.
(2) Detect and suppress the interference in the wavelet packet domain. In the wavelet packet coefficient domain, a proper threshold is selected, to zero-set all the coefficients larger than the threshold, thereby achieving the objective of suppressing the DME interference.
(3) Wavelet packet inverse transformation: using the L-level wavelet packet coefficient obtained after the process in step (2), wavelet packet inverse transformation is used to obtain the temporal domain waveform of the signal after interference suppression.

It can be seen from the above algorithm steps that the selections on wavelet function type, wavelet packet decomposition level and wavelet packet coefficient domain threshold are the critical parameters for the wavelet packet transformation algorithm. The selections on these parameters can affect the effects of DME interference suppression using the wavelet packet transformation. Next, we discuss the selections on wavelet packet coefficient domain threshold, wavelet packet decomposition level, and wavelet function type.

2. Selection on Wavelet Packet Coefficient Domain Threshold

The overall principle of setting the threshold in the wavelet packet domain is to find a threshold that can differentiate noise and DME interference. First, we need to consider the characteristics of noise after the wavelet packet transformation. The implementation of wavelet packet transformation is based on filtering, thereby we need to consider the characteristics of the noise after filtering. Since the GNSS signals are buried in the noise, the GNSS signal and noise can be modeled together as a zero-mean Gaussian noise ε. After filtering, the noise's probability density distribution still follows a Gaussian distribution [21]. We give the proof below.

Given ε's variance as σ^2, then its power spectral density is $S_\varepsilon(\omega) = \sigma^2$. The variance of the wavelet coefficient $W_1\varepsilon$, after the ε is transformed by 1 level of wavelet packet transformation using the filter G_0, is σ_1^2. The frequency response of the given G_0's is [24]

$$G_0(\omega) = \sum_{n=-\infty}^{\infty} g_0(n)\mathrm{e}^{-\mathrm{j}n\omega}, \quad \omega \in [-\pi, \pi] \tag{6.70}$$

where $g_0(n)$ is the unit pulse response sequence. Then the power spectral density of $W_1\varepsilon$ is

$$S_{W_1}(\omega) = |G_0(\omega)|^2 S_\varepsilon(\omega) \tag{6.71}$$

and $W_1\varepsilon$'s autocorrelation function is

$$R_{W_1}(\tau) = \frac{1}{2\pi}\int_{-\infty}^{\infty} S_{W_1}(\omega)\mathrm{e}^{\mathrm{j}\omega\tau}\mathrm{d}\omega \tag{6.72}$$

Since the mean value of a zero-mean noise after filtering is zero, then we have

$$\begin{aligned}
\sigma_1^2 = R_{W_1}(0) &= \frac{1}{2\pi}\int_{-\infty}^{\infty} |G_0(\omega)|^2 S_\varepsilon(\omega)\mathrm{d}\omega \\
&= \frac{\sigma^2}{2\pi}\int_{-\pi}^{\pi}\left[\sum_{n=-\infty}^{\infty} g_0(n)\mathrm{e}^{-\mathrm{j}n\omega}\right]\left[\sum_{-\infty}^{\infty} g_0(n)\mathrm{e}^{\mathrm{j}n\omega}\right]\mathrm{d}\omega \\
&= \frac{\sigma^2}{2\pi}\int_{-\pi}^{\pi}\left[\sum_{n=-\infty}^{\infty}(g_0(n))^2 + \sum_{n=-\infty}^{\infty}\sum_{\substack{m=-\infty \\ m\neq n}}^{\infty} g_0(n)g_0(m)\mathrm{e}^{-\mathrm{j}(n-m)\omega}\right]\mathrm{d}\omega \\
&= \frac{\sigma^2}{2\pi}\sum_{n=-\infty}^{\infty}(g_0(n))^2\int_{-\pi}^{\pi}\mathrm{d}\omega + \frac{\sigma^2}{2\pi}\int_{-\pi}^{\pi}\sum_{n=-\infty}^{\infty}\sum_{\substack{m=-\infty \\ m\neq n}}^{\infty}\mathrm{e}^{-\mathrm{j}(n-m)\omega}\mathrm{d}\omega \\
&= \sigma^2\|g_0\|^2
\end{aligned} \tag{6.73}$$

In the simulations, the filters used for wavelet packet decomposition are all filters after energy normalization, i.e. $\|g_0\|^2 = 1$; then $\sigma_1^2 = \sigma^2$. As a result, for a zero-mean Gaussian white noise, after various levels of wavelet packet decomposition, all its wavelet packet coefficients at every scale all follow zero-mean Gaussian distribution with σ^2 variance as well.

Therefore we can set a uniform threshold in the wavelet packet coefficient domain to suppress DME interference. Based on the characteristics that the noise probability density follows Gaussian distribution, the selection of threshold can be set based on the false alarm rate criterion described in Sect. 6.2.1.

3. Selection of Wavelet Packet Decomposition Level

Wavelet packet transformation divides the signal frequency band evenly using a set of decomposition filters. For every decomposition, the frequency range of the signal is divided into two halves, the frequency band resolution of the Level-L wavelet packet decomposition BW_L is defined as

$$BW_L = \frac{f_s/2}{2^L} \tag{6.74}$$

where f_s is the sampling frequency.

As the number of decomposition levels increases, we do finer division on the signal, and the effect of interference suppression is better. But for every decomposition there is a down-sampling process, and the number of points for useful information decreases. After certain levels of decomposition, further decomposition cannot bring significant performance improvement and the computation complexity increases. In addition, due to non-ideal characteristics of wavelet packet filtering, the increase of decomposition levels can lead to the problem of producing sub-band frequency spectrum aliasing. For the selection of wavelet packet decomposition level, we need to consider both the bandwidth and computation complexity of the interference signal, matching the decomposed wavelet packet frequency band resolution and the bandwidth of the interference signal being suppressed. For a single DME transmitter, the bandwidth of the DME interference signal concentrates within an 1 MHz bandwidth. Consequently for DME interference suppression based on wavelet packet transformation application, a proper decomposition level needs to be selected to match the frequency resolution of the wavelet packet to the 1 MHz bandwidth.

4. Selection of Wavelet Function Type

Based on the definition of wavelet packet transformation, as a mother function of the wavelet becomes more similar to the signal being analyzed, the coefficients after wavelet packet transformation become more localized. In another word, if fewer points with more concentrated energy in the wavelet coefficient domain are used to represent interference, interference suppression can be better achieved. But in real applications, some wavelet functions do not have explicit expression, so it is very hard to derive the similarity between the wavelet function and the function being analyzed. To perform multi-resolution analysis using quick algorithm of wavelet transform, we use the usual selection criteria on the wavelet function: compactly supported, orthogonal or bi-orthogonal [24, 25].

Frequently used wavelet functions include: Daubechies wavelet function family (db*N*, $N = 1,2,\ldots,15$, where db1 is equivalent to the Haar wavelet), Symlet wavelet function family (sym*N*, $N = 2, 3,\ldots,15$), Coifle wavelet function family (coif*N*, $N = 1, 2, 3, 4, 5$), bior bi-orthogonal wavelet function family (bior*Nr.Nd*; *Nr.Nd*

Table 6.1 Function properties of various wavelet functions

Function property	Wavelet function					
	db1/Haar	db$N$($N \neq 1$)	sym	coif	bior/rbio	dmey
Compactly supported	Yes	Yes	Yes	Yes	Yes	Yes
Orthogonal	Yes	Yes	Yes	Yes	No	Yes
Bi-orthogonal	Yes	Yes	Yes	Yes	Yes	Yes
Discrete transformation	Yes	Yes	Yes	Yes	Yes	Yes
Fast algorithm	Yes	Yes	Yes	Yes	Yes	Yes

parameters are (1, 1/3/5), (2, 2/4/6/8), (3, 1/3/5/7/9), (4, 4), (5, 5), (6, 8), rbio inverse bi-orthogonal wavelet function family (rbio*Nr.Nd*; *Nr.Nd* parameters are (1, 1/3/5), (2, 2/4/6/8), (3, 1/3/5/7/9), (4, 4), (5, 5), (6, 8)) and discrete Meyer wavelet function "dmey". Altogether 6 types of wavelet function family, and 65 types of wavelet function are used for DME interference suppression. The properties of these wavelet function family are shown in Table 6.1 [24, 25].

From Table 6.1, the characteristics of wavelet functions satisfy the requirements of compact support, orthogonality and bi-orthogonality, and they all have quick algorithms. Since the ideal wavelet function does not exist, and some wavelet function cannot be expressed using fixed mathematical expressions, we compare interference suppression effects of different types of wavelet functions based on the acquisition properties of the signals after the interference suppression, so the most proper wavelet function can be selected.

6.5 An Integrated Pulse Interference Suppression

We have introduced temporal domain pulse blanking algorithm, frequency domain filtering and temporal-frequency hybrid filtering algorithm, and all these algorithms have good interference suppression performance when the interference duty cycle is lower, but as the interference duty cycle increases, the performance of these algorithms degrade seriously. In this section, we first introduce a method to suppress DME interferences by integrating the parameterized algorithm and the wavelet packet transformation algorithm that has better performance when the DME interference duty cycle is higher. Secondly, we consider the computational complexity of the interference suppression performance and algorithm, and propose an integrated DME pulse interference suppression method. The method combines the conventional DME pulse interference suppression method and the hybrid algorithm, which can adaptively select the proper method for interference suppression towards interference scenarios with different pulse interference duty cycles.

6.5.1 The Hybrid Algorithm

Since the interval between the DME pulse pairs transmitted by the ground DME transmitters are random, even though the pulse pair signals from the same DME transmitter have no aliasing, the signals from different DME transmitters may have aliasing. When there is no DME signal aliasing, the DME interference suppression method based on parameterized algorithm can preserve the authentic GNSS signals to the greatest extent. But when there is aliasing in the DME signals, the DME interference suppression method based on parameterized algorithm is difficult to implement. Therefore we propose a hybrid DME interference suppression method using both parameterized algorithm and wavelet packet transformation algorithm. The basic principle of the method is to add an aliasing detector before the interference suppression. If there is no aliasing detected in the DME interference, a DME interference suppression method based on parameterized algorithm is used. But if there is aliasing detected in the DME interference, the DME interference suppression method based on wavelet transformation is used. The flow chart of the hybrid algorithm is shown in Fig. 6.16.

From Fig. 6.16, the implementation steps of the hybrid algorithm, by combining the parameterized algorithm and wavelet packet transformation algorithm, are as below:

(1) When the DME interference detection is performed on the input signal using a 25 μs long sliding time window, as shown in Fig. 6.6, if interferences are detected, then interference suppression process can follow. Otherwise if no interference is detected, no processing is performed and the non-interfered signal can be directly output. Figure 6.17 shows the detection process when aliasing DME pulse pair exists.
(2) Determination has to be made on whether the pulse pair within the window in which interference has been detected has aliasing. If there is no aliasing and the waveform is complete, then the pulse interference suppression based on parameterized algorithm can be used to obtain the output signal after interference suppression. Otherwise, the pulse pair data within the window is saved as the staging data for the pulse interference's pulse region with aliasing, and the detection window's position can be recorded. Among these steps, the determination on whether aliasing exists in the pulse pair within the detection window can be made based on the number of peaks in the frequency domain. When the number of peaks is 1, there is only one DME signal source, i.e. no aliasing. When the number of peaks is larger than 1, the number of DME signal sources within the detection window is more than 1, meaning aliasing exists. Then, we determine whether the pulse waveform within the detection window is complete or not based on the peak value and known pulse width within the detection window. Figure 6.18 shows two examples of detecting an incomplete pulse pair within the detection windows. By sliding the detection window, and repeating steps (1) and (2), the final staging data in the pulse interference region

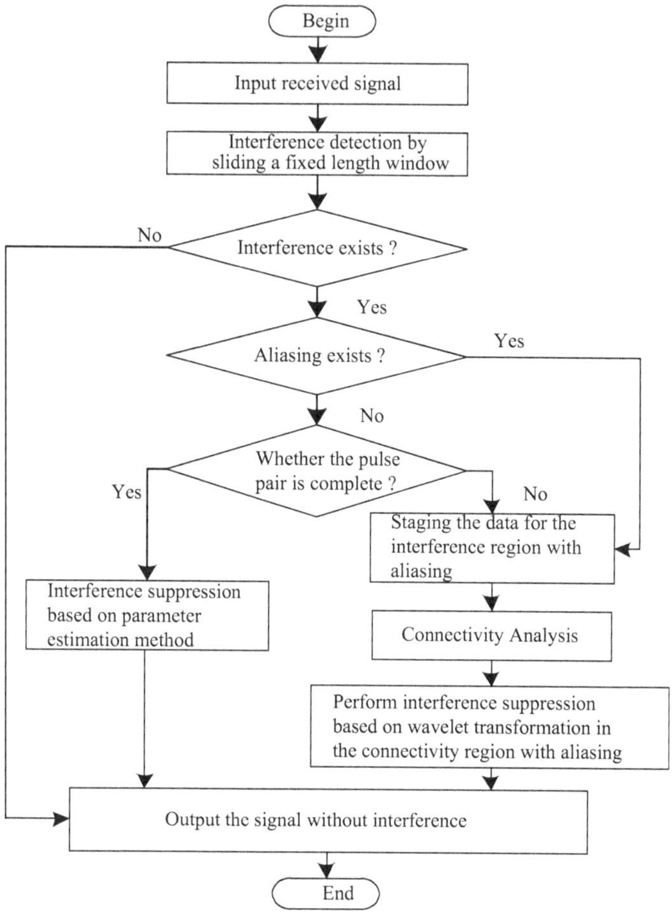

Fig. 6.16 Flow chart of the hybrid algorithm

with aliasing and the position of the corresponding detection window can be obtained at last.

(3) Connectivity analysis is performed on the staging data for pulse interference regions with aliasing. Given that the interval between two neighboring windows for the staging data is smaller than a certain distance, then the region after connecting these two windows can be used as the new pulse interference region with aliasing. Otherwise, we keep the position for the pulse interference region with aliasing.

(4) By performing DME pulse interference suppression based on wavelet transformation on the pulse interference region with aliasing after the connecting, the output signal after the interference suppression can be obtained at last.

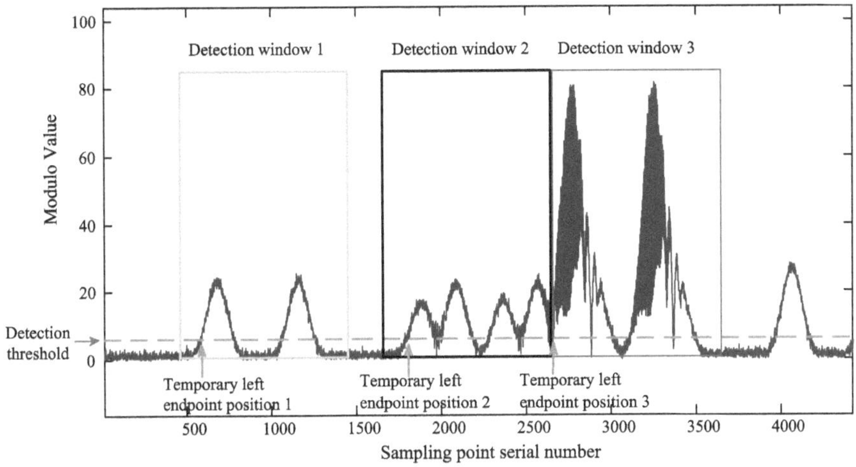

Fig. 6.17 Detection process when aliasing exists in the pulse pair

Fig. 6.18 Incomplete pulse pair detected within the detection region

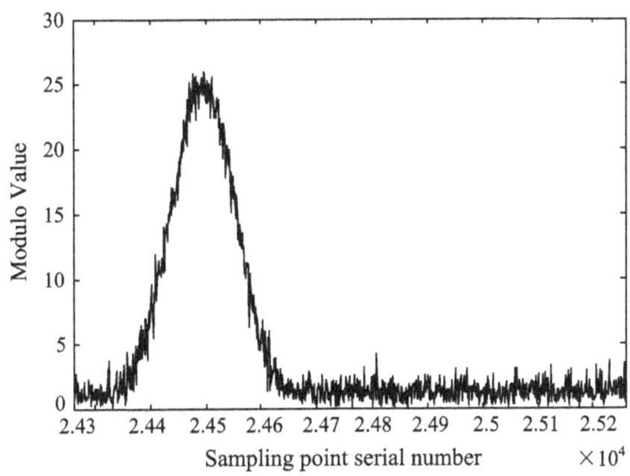

(a) Example 1

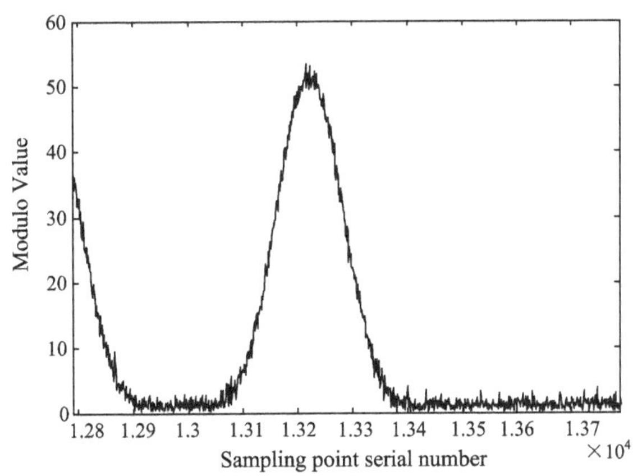

(b) Example 2

6.5.2 Integrated DME Pulse Interference Suppression Method

The method integrating the parameterized algorithm and the wavelet transformation combines the advantages of parameterized algorithm and wavelet transformation. It can have good performance when the duty cycle of the pulse interference is high. But compared with the other conventional DME interference suppression methods, it has higher complexity. It trades off between the complexity and interference suppression performance. In an actual DME interference environment, the number of interference pulses in the received signal is unknown so there is a need to estimate it. This number can be measured using the pulse interference duty cycle. By balancing computation complexity and interference suppression performance, we propose an integrated DME pulse interference suppression method in this section. This method combines the conventional interference suppression algorithm and the hybrid algorithm. It can adaptively select proper methods for interference suppression towards various pulse interference scenarios with different duty cycles. Detailed steps of the method are listed below:

First the interference duty cycle for the input signal is calculated. It can be represented using the blanking duty cycle, which is defined as the ratio between the number of samples larger than the temporal domain pulse blanking threshold and the total number of samples. Secondly, when the interference duty cycle of the input signal is smaller than the duty cycle threshold γ, the temporal-frequency hybrid filtering algorithm is used for DME pulse interference suppression. When the interference duty cycle of the input signal is larger than the duty cycle threshold γ, the hybrid algorithm is used for DME pulse interference suppression. Based on the decision on the duty cycle threshold, we can constantly switch between proper interference suppression algorithms to achieve the objective of DME interference suppression. The threshold γ can be selected based on Monte Carlo simulation, which can be determined in advance based on experience. The block diagram for implementing the algorithm is shown in Fig. 6.19.

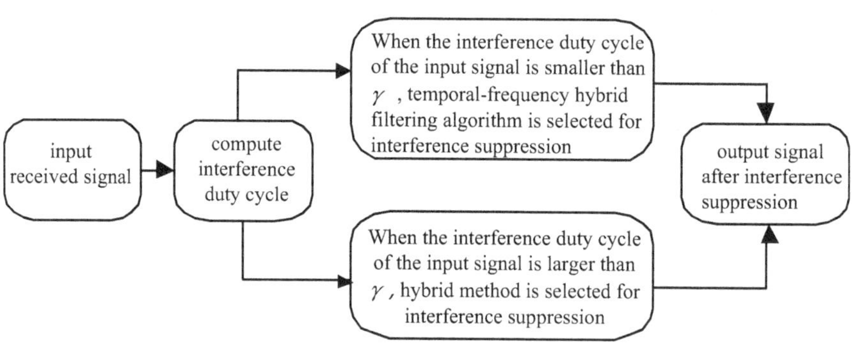

Fig. 6.19 Block diagram of integrated DME pulse interference suppression method

6.6 Simulation Results

In this section two simulation results verify the performance of the algorithms mentioned above. The first simulation verifies the performance of the DME interference suppression based on wavelet packet transformation, and the impacts on the performance by the wavelet types and decomposition levels are compared as well. The second simulation, under interference scenarios with different duty cycles, compares the performances of the temporal domain pulse blanking algorithm, the frequency domain notch filtering algorithm, the temporal domain hybrid filtering algorithm, the wavelet packet transformation algorithm, the hybrid algorithm, and the integrated algorithm.

The first simulation demonstrates the DME interference suppression algorithm based on wavelet packet, and it evaluates the interference suppression performance with different wavelet functions and wavelet packet transformation levels by setting a unified threshold under a given false alarm rate condition. To compare the interference suppression performance of implementing different wavelet packet transformation algorithms, the Correlation Peak to next Peak Ratio (CPPR) is used as the evaluation criterion for acquisition performance. A larger CPPR represents better acquisition performance, i.e. better interference suppression performance [2]. 200 Monte Carlo simulations are performed on every algorithm, and the average CPPR of the 200 trials is used to compare acquisition performance of different algorithms.

A GPS L5 GNSS signal is added into the simulation. The signal-to-noise ratio is −20 dB, and the intermediate frequency is 10 MHz. Acquisition is performed on I-branch. Its sample frequency is 40 MHz, and the observation time is 1 ms. A more realistic DME interference environment is simulated in which interferences from 18 DME transmitters are added. The repeat frequency of DME pulse pair transmitted by every transmitter is 2000 pairs/s, and the intervals between pulses changes randomly based on a Poisson distribution. The interference duty cycle is 37%. The temporal domain waveform for the signal under the impact of interference is shown in Fig. 6.20a. In the simulation, we also consider the problem of possible aliasing among signals transmitted by different DME transmitters. Figure 6.20b shows a locally amplified view with aliasing pulses.

For the simulations the wavelet packet coefficient domain threshold is set in a unified way under the condition of 0.01 false alarm rate. Figure 6.21 shows the results of comparing the performances of interference suppression for different wavelet packet decomposition levels and using different wavelet functions

In the simulations, 65 types of wavelet functions are compared for their interference suppression results. The details can be found in Sect. 6.4.2. Figure 6.21 lists performance results for 6 typical wavelet functions (db15, coif5, sym15, dmey, bior6.8 and rbio6.8) using 1–8 levels of wavelet packet decompositions. It can be seen from Fig. 6.21 that, as the decomposition levels increase, the interference suppression performance improves. But when the level of decomposition reaches a certain level, the performance improvement is not obvious anymore and it shows a

Fig. 6.20 Temporal domain waveform for received signal interfered by DME

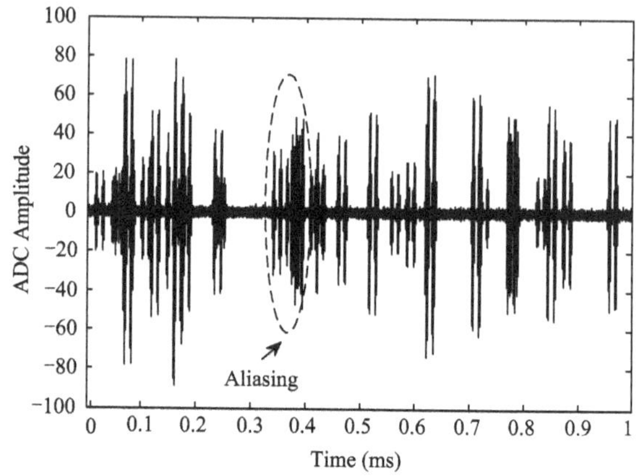

(a) Temporal domain waveform for signal before interference suppression.

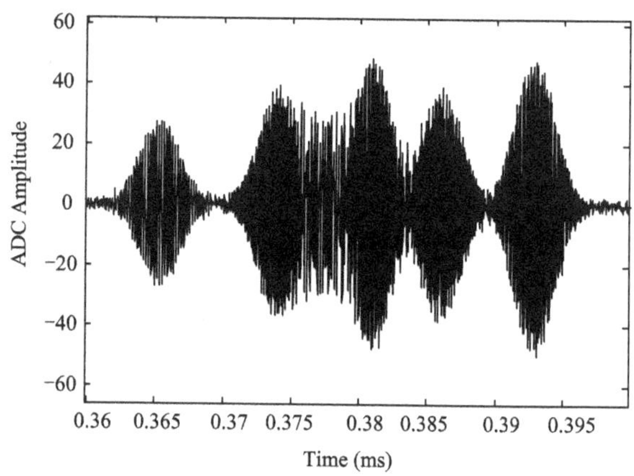

(b) Locally amplified view with aliasing pulses.

trend of degradation. From Fig. 6.21, we also know that the best performance happens by using the dmey wavelet function for a 5-level wavelet packet transform. This is because the corresponding wavelet packet frequency band resolution for the 5 level wavelet packet transformation is 625 kHz, which is equivalent to the frequency band 0.8 MHz for the DME interference. Also for these wavelet functions, the corresponding filter bank has the most number of points, so relatively speaking it has better frequency response characteristics. Figure 6.22 shows the temporal domain signal waveform after the DME interference suppression using a dmey wavelet packet for a 5-level wavelet packet transform.

A GPS L5 signal is added for the second simulation. The signal-to-noise ratio is −20 dB, the intermediate frequency is 10 MHz, the sampling frequency is 40 MHz, and the time interval of observing the signal is 1 ms. The acquisition factor CPP also serves as the criterion of evaluating the acquisition performance.

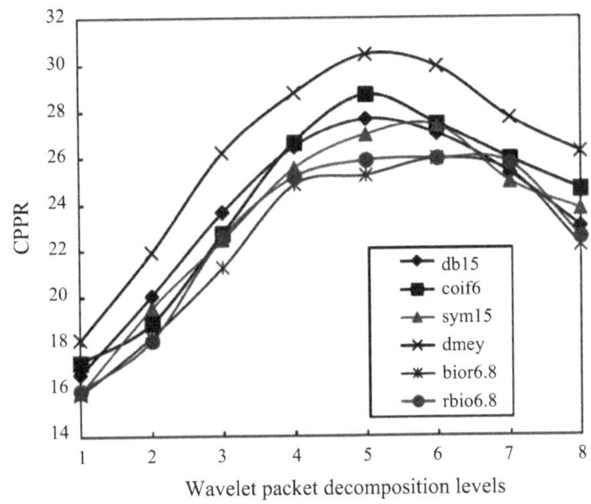

Fig. 6.21 Relations between wavelet types, decomposition levels and the interference suppression performance

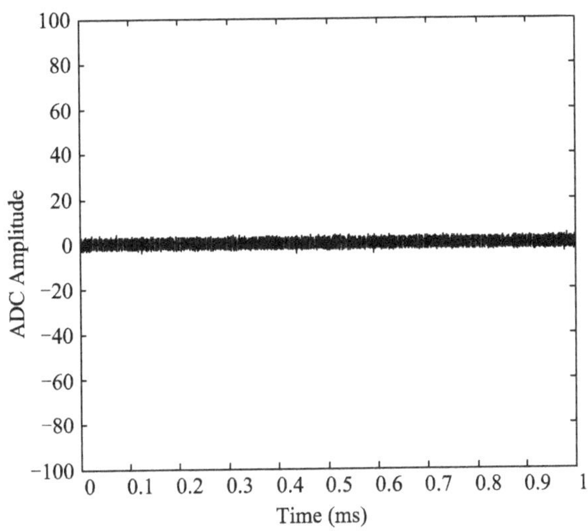

Fig. 6.22 Temporal domain wavelet after DME interference suppression using wavelet packet transformation

The interference duty cycle computed in the simulations of this chapter is the ratio between the number of samples that are higher than the temporal domain pulse blanking threshold and the total number of samples. A higher number of DME transmitters means that there is a higher number of corresponding pulse pairs DME, leading to higher interference duty cycle. For the simulation, the added number of DME transmitters changes from 1 to 18, as a result the interference duty cycle changes from 4 to 51%. The repeat frequency for the pulse pair from every DME transmitter is 2700 pairs/s, the ratio between a pulse's peak power and the noise's peak power is 20 dB, and the arrival time of the pulse pair is subject to a Poisson distribution.

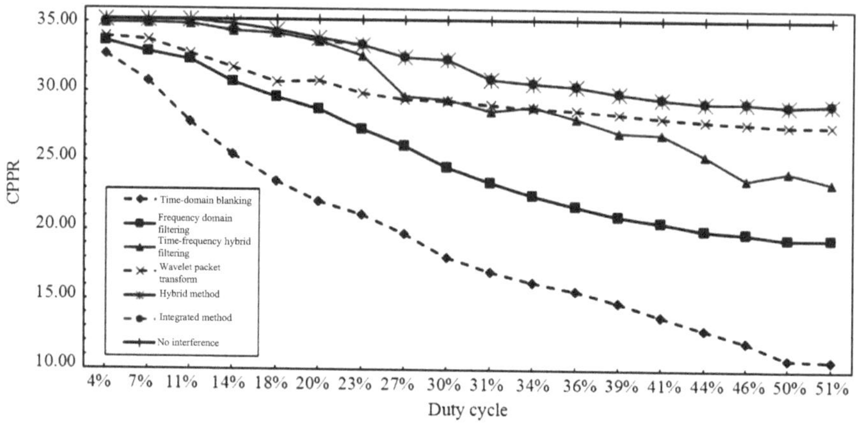

Fig. 6.23 Comparisons of interference suppression using different methods under the conditions of different duty cycles

Figure 6.23 compares the signal acquisition performance after interference suppression using temporal domain pulse blanking method, frequency domain filtering method, temporal-frequency hybrid filtering method, wavelet packet transformation method, hybrid method (combining parameterized algorithm and wavelet packet transformation method), and integrated method under interference conditions associated with different duty cycles, in which the related pulse blanking threshold setting by different methods are all obtained when the false alarm rate is set as 0.001. The Figure also shows the performance under the condition of no interference. Figure 6.23 shows that the performance of all interference suppression methods degrade as the interference duty cycle increases. When the duty cycle is smaller than 20%, the integrated method selects the temporal-frequency hybrid algorithm that has less computational complexity. If the duty cycle is bigger than 20%, the performance of the temporal-frequency hybrid filtering algorithm drops significantly. In this case, the hybrid algorithm's performance is much better than the conventional method, and consequently the integrated method selects the hybrid method.

To achieve the optimum tradeoff between interference suppression performance and computational complexity, the proposed integrated method combines the temporal-frequency hybrid filtering algorithm and hybrid algorithm. For the integrated method, the duty cycle threshold γ is set as 20%. When the duty cycle is smaller than 200%, the integrated method selects the temporal-frequency hybrid filtering method as the interference suppression method; when the duty cycle is bigger than 20%, the integrated method selects the hybrid algorithm as the interference suppression method.

6.7 Summary

In this chapter, we mainly study the pulse interference suppression algorithms for GNSS. Using the DME pulse interference as an example, we first summarize the conventional pulse interference suppression method, i.e. temporal domain pulse blanking method, frequency domain notch filtering algorithm and temporal-frequency hybrid filtering algorithm. Then we propose the pulse interference suppression algorithm based on parameterized algorithm and wavelet packet transformation. Finally, to achieve a tradeoff between performance and algorithm complexity, we propose an integrated pulse interference suppression algorithm. The algorithm adaptively selects an algorithm with good performance and low computation complexity based on different pulse interference density. It is worthy to point out that even though the DME pulse interference is used as an example in this chapter, related algorithms can also be applied to suppress other pulse interferences.

References

1. Hegarty C, Kim T, Ericson S, et al. Methodology for determining compatibility of GPS L5 with existing systems and preliminary results. In: Proceedings of the Institute of Navigation annual meeting. 1999. no 1, p. 1–10.
2. Gao GX, Heng L, Hornbostel A, et al. DME/TACAN interference mitigation for GNSS: algorithms and flight test result. GPS Solut. 2013;17(4):561–73.
3. Grabowski J, Hegarty C. Characterization of L5 receiver performance using digital pulse blanking. In: Proceedings of Institute of Navigation global positioning systems conference (ION GPS), Portland, OR. 2002. p. 1630–5.
4. Hegarty C, Van Dierendonck AJ, Bobyn D, et al. Suppression of pulse interference through blanking. In: Proceedings of IAIN world congress in association with the US ION 56th annual meeting, San Diego. 2000.
5. Bastide F. Analysis of the feasibility and interests of Galileo E5a/E5b and GPS L5 signals for use with civil aviation. In: Ecole Nationale de I'Aviation Civile (ENAC). 2004.
6. Musumeci L, Samson J, Dovis F. Performance assessment of pulse blanking mitigation in presence of multiple distance measuring equipment/tactical air navigation interference on global navigation satellite systems signals. Radar Sonar Navig IET. 2014;8(6):647–57.
7. Gao GX. DME/TACAN interference and its mitigation in L5/E5 bands. In: Proceedings of Institute of Navigation global navigation satellite systems conference (ION GNSS), Fort Worth, Texas, USA. 2007.
8. Denks H, Hornbostel A, Chopard V, et al. GNSS receiver testing by hardware simulation in an standardized pulsed and CW interference environment. In: Proceedings of European navigation conference on global navigation satellite systems (ENC GNSS), Naples. 2009.
9. Palubinskas M, Gunther C. Interference mitigation in GNSS. Technical University of Munich; 2011.
10. Raimondi M, Julien O, Macabiau C. Implementation issues and performance characterization of FDAF for a GNSS receiver. In: European navigation conference on global navigation satellite systems (ENC GNSS). 2008.

11. Gao GX, De Lorenzo D, Chen A, et al. Galileo GIOVE-A broadcast E5 codes and their application to acquisition and tracking. In: ION NTM conference, San Diego, California. 2007.
12. Cuntz M, Heckler M, Erker S, et al. Navigating in the Galileo test environment with the first GPS/Galileo multi-antenna-receiver. In: 2010 5th ESA workshop on satellite navigation technologies and European workshop on GNSS signals and signal processing (NAVITEC). IEEE; 2010.
13. Fang W, Wu RB, Wang WY, et al. DME pulse interference suppression based on NLS for GPS. In: International conference on signal processing proceedings. 2012. p. 174–8.
14. Anyaegbu E, Brodin G, Cooper J, et al. An integrated pulsed interference mitigation for GNSS receivers. J Navig. 2008;61(2):239–56.
15. Paonni M, Jang JG, Eissfeller B, et al. Innovative interference mitigation approaches: analytical analysis, implementation and validation. In: 2010 5th ESA workshop on satellite navigation technologies and European workshop on GNSS signals and signal processing (NAVITEC). IEEE; 2010. p. 1–8.
16. Musumeci L, Samson J, Dovis F. Experimental assessment of distance measuring equipment and tactical air navigation interference on GPS L5 and Galileo E5a frequency bands. In: 2012 6th ESA workshop on satellite navigation technologies and European workshop on GNSS signals and signal processing (NAVITEC). IEEE; 2012. p. 1–8.
17. Musumeci L, Dovis F. Use of the wavelet transform for interference detection and mitigation in global navigation satellite systems. Int J Navig Obs. 2014;1–14.
18. Wu R, Wang W, Li L, et al. Distance measuring equipment interference suppression based on parametric estimation and wavelet-packet transformation for global navigation satellite systems. IEEE Trans Aerosp Electron Syst. 2016;52(4):1607–17.
19. Hu T, Wang L, Wu R. Study on wideband and pulse interference mitigation techniques with RAW BD data. In: Proceedings of the 28th international technical meeting of the satellite division of the Institute of Navigation (ION GNSS+2015). 2015.
20. Zhang Q, Zheng YB, Wilson SG. Excision of distance measuring equipment interference from radio astronomy signals. Astron J. 2005;129:2933–9.
21. Yin T, Viberg M, Weber DIC. Simulator of pulsed interference environment of an airborne GNSS receiver. Chalmers University of Technology; 2007.
22. Besson O, Stoica P. Nonlinear least squares approach to frequency estimation and detection for sinusoidal signals with arbitrary envelope. Digit Signal Proc. 1999;9:45–56.
23. Li J, Wu R. An efficient algorithm for time delay estimation. IEEE Trans Signal Process. 1998;46(8):2231–5.
24. Daubechies I. Ten lectures on wavelets. In: Volume 61 of proceedings of the CBMS-NSF regional conference series in applied mathematics. Philadelphia, PA: SIAM; 1992.
25. Abry P. Ondelettes et Turbulences. Paris: Diderot Editeur; 1997.

Erratum to: Adaptive Interference Mitigation in GNSS

Erratum to:
R. Wu et al., *Adaptive Interference Mitigation in GNSS*, Navigation: Science and Technology, https://doi.org/10.1007/978-981-10-5571-3

In the original version of the book, the following belated corrections have been incorporated:

In Chapter 1, Page 13, Fig. 1.6 has been replaced with a new figure.
In Chapter 2, Page 33, the sentence "τ_l is the propagation for the *l*th satellite signal time delay;" has been modified as "τ_l is the propagation delay for the *l*th satellite signal;".

The updated online version of these Chapters can be found at
https://doi.org/10.1007/978-981-10-5571-3_1
https://doi.org/10.1007/978-981-10-5571-3_2
https://doi.org/10.1007/978-981-10-5571-3

R. Wu et al., *Adaptive Interference Mitigation in GNSS*, Navigation: Science and Technology, https://doi.org/10.1007/978-981-10-5571-3_7

FSC
www.fsc.org
MIX
Papier | Fördert
gute Waldnutzung
FSC® C083411

Kolibri 360 GmbH
Ferdinand-Jühlke-Straße 7
99095 Erfurt, Deutschland
produktsicherheit@kolibri360.de